Algorithmic Urbanism for Climate Resilience

This book offers a comprehensive exploration of how artificial intelligence (AI), ecological design, and governance reforms can be leveraged to tackle climate-induced urban heat.

The book is structured around three key themes: AI for climate adaptation, green infrastructure and design innovation, and policy and planning perspectives. It delves into how intelligent technologies like remote sensing, real-time monitoring, and machine learning are revolutionising the assessment and management of urban microclimate risks. Specific applications include enhancing urban resilience through hyperlocal weather alert systems and analysing the influence of vegetation and building distribution on urban heat for better mitigation. The book further examines how nature-based interventions, such as AI-optimised vegetation strategies and cooling infrastructure, can create thermally comfortable and ecologically robust urban environments. It highlights the growing intersection of AI and green infrastructure in climate adaptation and urban sustainability. Beyond technology and design, the author focuses on governance and policy frameworks, exploring how institutional arrangements and planning paradigms can translate digital and design solutions into equitable and effective climate action. It investigates how climate action is reflected in local urban planning, drawing insights from cities committed to global climate agreements. By integrating computational tools, bioclimatic design, and multilevel governance, the book provides a holistic framework for addressing urban heat challenges.

It is an essential resource for urban planners, climate researchers, and policymakers seeking forward-looking, inclusive, and technologically informed strategies for building heat-ready cities in a warming world.

Tan Yigitcanlar is a leading Australian scholar recognised internationally for contributions to urban studies, sustainability, technology, and planning. He is a Professor at Queensland University of Technology in Australia and serves as a Distinguished Visiting Professor at the University of Johannesburg in South Africa.

"In *Algorithmic Urbanism for Climate Resilience*, Tan Yigitcanlar brings together AI, green infrastructure, and policy in ways that speak directly to today's urban challenges. Drawing on his long-standing leadership in urban technology and sustainability, the book offers clear ideas and practical guidance for planners and policymakers working to create more equitable, heat-ready cities."
Professor Thomas Sanchez, *Texas A&M University, USA*

"This must-read, groundbreaking book integrates artificial intelligence, green infrastructure planning, and policy innovation to tackle the escalating challenge of urban heat. Spanning AI-driven climate mapping, nature-based cooling strategies, and global governance frameworks, it offers evidence-based pathways for building equitable and heat-resilient cities across diverse socio-spatial contexts."
Professor Yuan Lai, *Tsinghua University, China*

"This book provides a timely, integrated framework for urban heat resilience under the new idea of algorithmic urbanism for climate resilience. It synthesises AI-enabled heat mapping, optimisation-enhanced green infrastructure planning, and governance mechanisms that translate analytical insights into policy and practice. Its interdisciplinary approach and practice-oriented case studies enhance both methodological rigor and real-world applicability, supporting targeted, equity-centred interventions across interconnected urban socio-ecological-technological systems."
Professor Steven Jige Quan, *Seoul National University, South Korea*

"Professor Yigitcanlar's *Algorithmic Urbanism for Climate Resilience* is a groundbreaking synthesis of AI, green infrastructure, and policy innovation. This book equips planners and researchers with the tools needed to combat urban heat and provides a scalable roadmap for heat-ready cities. This book is an indispensable resource for scholars striving for urban climate resilience."
Professor Yanjie Fu, *Arizona State University, USA*

Algorithmic Urbanism for Climate Resilience

AI, Green Infrastructure, and Policy Innovation for Heat-Ready Cities

Tan Yigitcanlar

CRC Press
Taylor & Francis Group
Boca Raton London New York

CRC Press is an imprint of the
Taylor & Francis Group, an **informa** business

Designed cover image: Shutterstock

First edition published 2027
by CRC Press
2385 NW Executive Center Drive, Suite 320, Boca Raton FL 33431

and by CRC Press
4 Park Square, Milton Park, Abingdon, Oxon, OX14 4RN

CRC Press is an imprint of Taylor & Francis Group, LLC

ISBN: 978-1-041-23428-9 (hbk)
ISBN: 978-1-041-23235-3 (pbk)
ISBN: 978-1-003-73705-6 (ebk)

DOI: 10.1201/9781003737056

Typeset in Palatino
by KnowledgeWorks Global Ltd.

This book is dedicated to my daughters, Ela *and* Selin,
whose future will be shaped by a changing climate and the AI-driven cities we design today. The world you will inherit is a constant reminder that technology, no matter how advanced, must serve justice, compassion, and the well-being of generations yet to come.

Contents

Author

Tan Yigitcanlar is a leading Australian scholar internationally recognised for his contributions to urban studies, sustainability, technology, and planning. He is a Professor in the School of Architecture and Built Environment at Queensland University of Technology (QUT), Brisbane, Australia, Distinguished Visiting Professor in the Department of Urban and Regional Planning at the University of Johannesburg, Johannesburg, South Africa, and an Adjunct Professor in the Department of Urban and Regional Planning at the Sepuluh Nopember Institute of Technology, Surabaya, Indonesia, while also holding several senior leadership roles. At QUT, he leads the Urban AI Hub, and the City 4.0 Lab, and he also serves as a distinguished member of the Australian Research Council College of Experts.

With a career spanning over three decades, Tan's work has encompassed research, teaching, training, and capacity building at prestigious universities in Australia, Brazil, Finland, Indonesia, Japan, South Africa, and Turkey. His research is dedicated to addressing contemporary challenges in urban planning and development, covering a broad spectrum of economic, societal, spatial, governance, and technology-related issues. At the heart of his work lies a strong commitment to promoting smart and sustainable urbanisation.

Tan has provided research consultancy services to all levels of government, international corporations, and non-governmental organisations both in Australia and internationally. His expertise has been instrumental in helping these entities formulate forward-looking strategies, build resilience, and respond effectively to emerging disruptive challenges. He serves as the Lead Editor-in-Chief of the Elsevier Smart Cities Book Series and holds senior editorial appointments with 12 high-impact academic journals. Tan has held senior leadership roles in more than 40 major international conferences, including serving as chair for 12 of them. He has delivered more than 90 keynote and invited addresses at prominent academic and industry events.

Tan's research findings have been extensively disseminated, with over 360 articles published in high-impact journals and 36 key reference books published by esteemed international publishing houses. His work has significantly influenced urban policy, practice, and research internationally, with an h-index of over 110. He is ranked #1 in Australia and top-10 worldwide in urban and regional planning according to the 2023 Science-wide Author Databases of Standardised Citation Indicators. He was also recognised as an "Australian Research Superstar" in the Social Sciences Category by The Australian's 2020 Research Special Report and named Australia's Social Sciences "Research Field Leader" for Urban Studies and Planning in the 2024 edition of *The Australian Research Magazine*. He is acknowledged as one of the top 1% of scientists worldwide by Clarivate in 2023.

Foreword

Cities have always been experiments in how humans live with climate. For centuries, urban form—from shaded streets and courtyards to water gardens and parks—has been a search for thermal comfort under local conditions. But what happens when the climate itself is no longer a relatively stable background and becomes a volatile force to survive? Today, our experiments are running hotter, faster, and at scales that exceed what our planning institutions were built to handle.

Heat is no longer a minor design constraint; it has become a defining pressure on urban life. As the preface to this book notes, urban heat stress is a "silent but severe" threat: unlike floods or storms, it rarely produces spectacular images, yet it quietly erodes health, productivity, and liveability, and kills more people than many better-known hazards. Prolonged high temperatures exacerbate cardiovascular and respiratory conditions, reduce labour productivity, and disproportionately harm those with the fewest resources; older adults, children, low-income residents, and outdoor workers. In rapidly urbanising regions of the Global South, where informal settlements, limited cooling infrastructure, and constrained municipal capacity are common, these burdens are especially acute.

At the same time, the physical drivers of heat (impervious surfaces, building density, loss of vegetation) are tightly bound up with the ways we have planned and built our cities. Heat is not evenly distributed; some neighbourhoods, often those with fewer trees and higher densities, consistently experience higher temperatures and lower adaptive capacity. Urban heat is therefore not only a biophysical phenomenon; it is a lens on environmental justice, exposing who benefits from cooling infrastructure and who is left to cope with "silent disasters" largely on their own.

As a planning problem, heat is quintessentially "wicked." There is no simple, final solution, only better or worse configurations of risk over time under deep uncertainty (Rittel & Webber, 1973). Urban heat is shaped by land cover, behaviour, governance, and technology, and global climatic shifts that are only partially within local control. It sits at the intersection of many policy arenas: land use, transport, housing, energy, health. Responding to it requires not just new tools, but new ways of thinking about what counts as evidence, how we balance competing objectives, and how we govern systems that increasingly rely on AI and real-time data streams.

Algorithmic Urbanism for Climate Resilience: AI, Green Infrastructure, and Policy Innovation for Heat-Ready Cities takes this wicked problem as its organising focus and asks a deceptively simple question: what would it mean to take urban heat seriously as a central design and governance concern, in an era when AI and other computational tools are reshaping how we see, model, and manage cities?

The book's answer is that we must stop treating AI, green infrastructure, and governance as separate conversations. As Yigitcanlar's preface argues, the last decade has seen rapid advances in three domains: AI and remote sensing for monitoring and predicting heat; nature-based interventions such as vegetation, parks, and water-sensitive design for cooling; and new planning and policy frameworks for climate adaptation. Yet these have too often evolved in silos. AI studies focus on model accuracy but say little about implementation; green infrastructure work shows cooling potential but offers limited guidance on where and how to deploy interventions under real constraints; policy analyses examine planning instruments but rarely connect them to emerging computational capabilities.

This book proposes algorithmic urbanism for climate resilience as a way to weave those strands together: using AI and computational tools to inform and optimise ecological interventions, and embedding both in governance frameworks that can turn technical insight into equitable, durable climate action. It treats cities as socio-ecological-technological systems and heat as a stress test of whether those systems can be reoriented towards resilience and justice.

My own work has approached this terrain from the perspective of planning theory and AI governance. In a 2023 paper with colleagues, I proposed a four-phase "Pathway of Urban Planning AI" (AI-assisted, AI-augmented, AI-automated, and, in prospect, AI-autonomised planning) and used it to analyse how AI's role in plan-making is likely to evolve (Peng et al., 2023). AI-assisted planning uses AI to collect, process, and visualise data (heat maps, vulnerability indicators) under close human supervision. AI-augmented planning lets AI generate options and simulate scenarios, while planners remain firmly "in the loop," defining goals and choosing among alternatives. AI-automated planning introduces bounded automation—systems that execute certain tasks, such as triggering microclimate alerts or adjusting infrastructure operations, under "planner-on-the-loop" oversight. AI-autonomised planning, still largely speculative, imagines AI agents that can orchestrate parts of the planning process with limited human intervention.

The key argument of that work was that as AI's autonomy grows, governance must intensify: planners must move from doing every analytic task to governing the conditions under which AI is allowed to operate, while retaining sovereignty over goals, constraints, and equity criteria. AI can and should help with pattern detection, scenario exploration, and optimisation; it should not be left to redefine what we mean by "equitable," "sustainable," or "acceptable risk."

My forthcoming book, *Symbiotic Planning for Urban Futures: Empowering Collaborative Human-AI Planning* (Peng, 2026), builds on that typology by formalising a governance-first paradigm in which AI is a governed co-creator. It operates under a social contract defined by five axioms (human ends, machine means; transparency; contestability; plural values; and adaptability) and is implemented through risk-tiered planner-in/on-the-loop roles, equity guardrails, and time-bound authorisations. In this paradigm, AI extends human perception and creativity but never sets the ends of planning; it generates options and stress-test strategies, but people, through democratic processes, decide which trade-offs are acceptable.

Seen through that lens, this book is not just "about heat" but about what it means to practise planning in an age when AI is rapidly moving from peripheral tool to central actor in how we perceive, simulate, and intervene in urban systems. It offers a concrete domain, urban heat, in which to test whether a symbiotic, governed relationship between human planners and AI can actually be made to work in practice.

One of the most striking illustrations of this comes from the book's discussion of transit-oriented development (TOD) and heat. For decades, TOD has been promoted as a cornerstone of sustainable urbanism: dense, walkable, transit-rich neighbourhoods that reduce car dependence and emissions. Yet, as the Brisbane case shows, these very districts can exhibit higher levels of urban heat and faster rates of warming than more car-dependent, lower-density areas. High land use diversity, building intensity, and reduced porous surfaces (hallmarks of TOD success) can become thermal liabilities. Without conscious attention to canopy and permeable ground, the very features that make TOD attractive for mobility and carbon can intensify local heat. The lesson is not to abandon TOD, but to recognise that any one-dimensional optimisation (carbon, access, or even density) can backfire if it ignores compound risks. AI's role in this context is not to hand down an answer, but to quantify these tensions and surface them for explicit debate: how much extra heat is acceptable for a given reduction in vehicle miles travelled, and who bears that thermal cost?

Another example comes from the Brazilian climate policy analysis. Here, Yigitcanlar and colleagues examine how municipalities are incorporating climate concerns into planning instruments in the absence of strong national-level monitoring. They find important progress in mainstreaming climate language into plans and regulations, but also a familiar adaptation-over-mitigation bias and a persistent gap between formulation and implementation. These are precisely the kinds of contexts where algorithmic urbanism is most often assumed to be "too hard." Data is sparse, informality is widespread, and institutional capacity is stretched. Yet, as the chapter suggests, these constraints can catalyse innovation and forcing planners to combine remote sensing with participatory mapping, rely on "frugal AI" rather than heavy infrastructure, and design governance mechanisms that are realistic for resource-constrained institutions. If algorithmic urbanism cannot function in São Paulo or Recife, it is unlikely to be a robust paradigm for the rest of the world.

These examples help situate the book within a wider body of work. Smart city discourse has often foregrounded technological capability (sensors, platforms, optimisation) while treating equity and governance as afterthoughts. Climate adaptation planning has emphasised vulnerability and justice but not always had the analytical tools or data systems to navigate multi-objective trade-offs at the speed and scale that climate risks now demand. Environmental justice scholarship has documented unequal exposures (Bullard, 2005). But it has not yet systematically engaged with the design and governance of AI systems that could either reproduce or disrupt those patterns. What distinguishes Yigitcanlar's approach is the insistence that these domains must be addressed together: AI without governance becomes algorithmic managerialism; green infrastructure without targeting becomes an amenity for those already protected; policy without operationalised evidence becomes aspiration detached from material change.

At the same time, this book does not gloss over the tensions involved in such integration. Some of those tensions are technical, but many are fundamentally political. One concerns time: heat adaptation systems designed today will operate in climates that are still changing, using socioeconomic baselines that may quickly become obsolete. How do we design systems that are robust to such drift (technically and institutionally) without over-engineering brittle solutions? Another tension involves representation and data justice. AI systems need data, but data about whom, collected by whom, under what conditions of consent and power? As my own work and others' have emphasised, relying on data streams that over-represent some groups and systematically under-sample others (those without smartphones, those living in informal settlements, those outside official registries) risks building "data deserts" on top of already disadvantaged neighbourhoods (Kontokosta, 2018; Leszczynski, 2019). The book's emphasis on multisource integration and Global South cases is a step towards confronting this problem, but it also exposes how much work remains to make algorithmic urbanism truly inclusive.

A third tension concerns green gentrification: many of the most effective cooling interventions (trees, parks, water features) also tend to raise property values. Without explicit anti-displacement policies, the act of "cooling" a neighbourhood can become a prelude to displacing the very residents most at risk from heat. AI-optimised vegetation strategies or thermal comfort corridors must therefore be paired with land use tools (such as inclusionary zoning, community land trusts, or right-to-stay protections) if they are to serve equity rather than accelerate exclusion. A purely technical optimisation that maximises cooling per dollar without considering who remains to enjoy the benefits fails the test of climate justice, no matter how elegant the model.

Urban heat is, in many ways, an ideal test case for this emerging paradigm of symbiotic, algorithmic urbanism. Its effects are severe but often reversible; interventions (more shade, different materials, revised building forms) are tangible and widely understood; and the links between land cover, form, and temperature are increasingly well characterised. That makes it possible to build AI-based systems that can be meaningfully evaluated, both technically and ethically. If we cannot design governed, equity-conscious AI systems for a challenge as measurable as heat, where victims are identifiable and interventions exist, our prospects are dim for applying such systems to more fraught domains such as policing, housing allocation, or climate migration.

The book also points towards a research agenda. Many of its claims can be turned into testable propositions. Do AI-assisted heat mapping and microclimate alert systems, when coupled with explicit equity floors, measurably reduce heat-related morbidity and mortality compared with conventional adaptation plans? Under what institutional conditions do AI-optimised green infrastructure strategies reduce rather than exacerbate spatial inequalities in thermal comfort? Can symbiotic, AI-augmented planning processes demonstrably improve perceptions of procedural justice and trust relative to traditional expert-led or minimally data-driven approaches? In design terms, the book sketches a plausible pathway. In evaluative terms, it invites the planning academy and practice community to generate systematic evidence about when, where, and for whom such approaches work.

Finally, this work raises questions about professional identity and education. If heat is a proving ground for algorithmic urbanism, then planners will need to be as comfortable interrogating model cards and uncertainty bands as they are convening a neighbourhood meeting. They will need to treat every AI output as a hypothesis to be scrutinised: what is the claim? What is the evidence? Who is missing from the data? How are equity and risk quantified? How might this system fail, and what happens then? At the same time, they must deepen their skills in facilitation, conflict management, and ethics: helping communities understand trade-offs, ensuring that technical options remain open to contestation, and keeping human ends at the forefront when the means become increasingly technical.

What *Algorithmic Urbanism for Climate Resilience* offers, in my reading, is not closure on these questions but a structured way of asking them in one critical domain. It shows how AI, green infrastructure, and governance can be assembled to tackle an urgent, unevenly distributed climate risk; it demonstrates the potential and the limits of this integration; and it sketches a direction of travel for adaptation planning under conditions of accelerating urbanisation and warming. The work ahead, for planners, policymakers, researchers, and communities, is to decide what kinds of algorithmic urbanism we are willing to authorise, under what conditions, and with what safeguards.

In that sense, this book is less a verdict than a provocation. It suggests that the tools for more heat-ready, and potentially more just, cities are within reach, but only if we use them under a consciously designed social contract. Whether we do so will depend not only on models and methods but on our collective willingness to govern them.

Professor Zhong-Ren Peng
iAdapt: International Center for Adaptation Planning and Design
School of Landscape Architecture and Planning
University of Florida, Gainesville, Florida, USA

References

Bullard, R. D. (2005). *The quest for environmental justice: Human rights and the politics of pollution*. Sierra Club Books.

Kontokosta, C. E. (2018). Urban informatics in the science and practice of planning. *Journal of Planning Education and Research, 41*(4), 382–395. https://doi.org/10.1177/0739456X18793716

Leszczynski, A. (2019). Glitchy vignettes of platform urbanism. *Environment and Planning D: Society and Space, 38*(2), 189–208. https://journals.sagepub.com/doi/10.1177/0263775819878721

Peng, Z.-R. (2026). *Symbiotic planning for urban futures: Empowering collaborative human-AI planning*. Elsevier.

Peng, Z.-R., Lu, K.-F., Liu, Y., & Zhai, W. (2023). The pathway of urban planning AI: From planning support to plan-making. *Journal of Planning Education and Research, 44*(4), 2263–2279. https://doi.org/10.1177/0739456X231180568

Rittel, H. W. J., & Webber, M. M. (1973). Dilemmas in a general theory of planning. *Policy Sciences, 4*(2), 155–169. https://doi.org/10.1007/BF01405730

Preface

Cities have always been arenas of innovation and adaptation, shaped by the complex interplay of technology, ecological systems, and human decision-making. From the ancient aqueducts of Rome to the high-rise, sensor-enabled cities of today, urban settlements have reflected human ingenuity in managing environmental challenges. Yet, the twenty-first century presents urban areas with challenges of unprecedented scale and intensity. Rising temperatures, intensifying heatwaves, and compounding social and infrastructural vulnerabilities are testing the limits of conventional urban planning and governance. Today, more than half of the world's population resides in cities, and urban growth continues at an accelerating pace. High-density development, impervious surfaces, widespread loss of vegetation, and energy-intensive infrastructures exacerbate urban heat risks while highlighting inequities in who bears the brunt of these impacts.

Urban heat stress is increasingly recognised as a silent but severe threat. Unlike floods, fires, or storms, which often capture immediate attention due to their dramatic impacts, heatwaves unfold slowly, accumulating effects on human health, productivity, and overall urban liveability. Prolonged exposure to high temperatures can exacerbate cardiovascular and respiratory conditions, reduce labour productivity, and disproportionately affect vulnerable populations, including the elderly, children, and low-income communities. Understanding how cities can anticipate, measure, and respond to heat stress is central to the purpose of *Algorithmic Urbanism for Climate Resilience: AI, Green Infrastructure, and Policy Innovation for Heat-Ready Cities*. This book seeks to equip scholars, practitioners, and policymakers with a multidimensional lens to understand, model, and mitigate urban heat in a rapidly urbanising and warming world.

Motivation and Intellectual Journey

The motivation for this book arises from a dual imperative: to address gaps in scholarly understanding and to provide actionable insights for urban adaptation. Research on urban heat, green infrastructure, and artificial intelligence (AI)–assisted planning has grown substantially in recent years. However, these areas often remain siloed. Technological studies frequently emphasise predictive modelling, sensor networks, or machine learning algorithms for detecting and forecasting heat stress but rarely link these tools to policy implementation or urban design strategies. Urban design and landscape architecture research explore ecological interventions, such as tree planting, green roofs, or water-sensitive urban design, yet often lack precise data-driven guidance on optimisation and effectiveness. Policy and governance research, meanwhile, evaluates planning frameworks and regulatory approaches but rarely integrates technological innovation or ecological modelling into decision-making.

Effective urban climate adaptation requires more than isolated solutions. It demands an integrative approach that synthesises technology, ecology, and governance into strategies that are not only scientifically rigorous but also socially equitable and contextually appropriate. This book aims to meet this need by offering a framework that connects computational intelligence, bioclimatic design, and policy innovation, enabling cities to anticipate, mitigate, and adapt to rising heat risks.

My engagement with these challenges began through collaborative research with local governments and urban stakeholders, where heat stress consistently emerged as a critical urban concern. Community workshops, urban field surveys, and climate simulations highlighted the complexity of the problem: heat is not uniform across cities, nor are its impacts evenly distributed. Certain neighbourhoods (often those with fewer trees, higher population density, and limited access to cooling infrastructure) experience disproportionately high temperatures. These experiences raised crucial questions: how can AI support real-time decision-making to improve thermal comfort in public spaces? How can vegetation and green infrastructure be strategically deployed to maximise cooling without compromising other urban priorities? How can governance frameworks facilitate the adoption of technology-driven solutions while avoiding the reinforcement of existing inequities? Over time, these questions shaped my research agenda and inspired the synthesis presented in this book; a combination of theoretical exploration, empirical analysis, and methodological innovation.

At its core, this book introduces the concept of *algorithmic urbanism for climate resilience*. This approach views cities as socio-ecological-technological systems, where urban heat is not only a biophysical phenomenon but also a product of urban form, human behaviour, institutional choices, and technological applications. Algorithmic urbanism emphasises the capacity of AI and computational tools to inform decision-making, optimise ecological interventions, and improve policy implementation. By integrating these three pillars (digital intelligence, green infrastructure, and governance) the book presents a holistic approach to designing heat-ready urban environments that are resilient, efficient, and socially inclusive.

The chapters are organised around three interlinked themes: AI for climate adaptation, green infrastructure and design innovation, and policy and planning perspectives. These dimensions work together to illustrate how cities can leverage technology, ecology, and institutional frameworks to respond effectively to heat stress while enhancing urban liveability.

Positioning the Book within Existing Literature

Although urban heat has been widely studied, most research tends to examine either technological, ecological, or policy responses in isolation. AI and sensor-based studies, for example, often prioritise data acquisition, predictive modelling accuracy, and performance metrics but rarely explore how these insights can be operationalised in real-world urban planning processes. Similarly, green infrastructure research highlights the cooling benefits of urban vegetation, parks, and water-sensitive design, yet frequently lacks evidence-based frameworks for optimising the placement and combination of interventions across complex urban systems. Policy studies, by contrast, focus on regulations, land use planning instruments, and governance models, but seldom engage with emerging technological tools or fine-grained ecological modelling.

This book fills these gaps by explicitly linking AI, green infrastructure, and governance into an integrated, practice-oriented framework. It demonstrates how digital technologies can enhance spatial analysis, predict heat vulnerabilities, and optimise cooling interventions; how green infrastructure can simultaneously provide climate mitigation, biodiversity, and public health benefits; and how governance structures can ensure these solutions are equitably implemented. By combining these domains, the book moves beyond conventional disciplinary boundaries to offer a holistic approach to urban climate resilience.

A further distinctive feature of this book is its global perspective. Cities face heat challenges differently depending on climate, governance structures, socio-economic conditions, and urban morphology. Comparative insights from case studies in Australia, Brazil, and other urban contexts illustrate how context-specific solutions can be designed while identifying common principles for AI-informed, ecologically integrated, and policy-driven adaptation strategies. This approach underscores the need for locally tailored interventions informed by robust analytical methods but guided by broader principles of social equity and ecological sustainability.

Theoretical Foundations and Intellectual Orientation

The theoretical foundation of this book draws on multiple disciplines, reflecting the complexity of urban heat adaptation:

Urban Climate Science informs our understanding of how heat develops and persists in cities. Drivers such as land cover, impervious surfaces, building density, and microclimatic variability shape the spatial distribution of heat. Insights from this field provide a biophysical basis for targeting interventions in areas where heat exposure is highest.

Computational Analytics and Artificial Intelligence provide the tools to monitor, predict, and optimise urban heat responses. Machine learning models, remote sensing, and sensor networks generate fine-grained, real-time insights into urban microclimates, enabling pro-active rather than reactive interventions.

Landscape Ecology and Green Infrastructure Research guide the design and placement of urban vegetation, parks, green roofs, and other cooling interventions. These strategies reduce heat, improve air quality, enhance biodiversity, and foster social well-being. The book emphasises evidence-based optimisation of these interventions using computational tools to maximise benefits relative to urban constraints such as land availability and cost.

Planning Theory and Governance Studies emphasise the institutional and socio-political contexts that shape adaptation outcomes. Policies, regulatory frameworks, and governance arrangements influence whether AI-assisted and green infrastructure interventions are feasible, effective, and socially equitable. Without careful attention to governance, even technically optimal interventions may fail to deliver intended outcomes or may exacerbate existing inequalities.

By integrating these perspectives, the book frames urban heat adaptation as a multidimensional challenge that spans science, technology, design, and governance. It argues that effective solutions require interdisciplinary collaboration, cross-sectoral coordination, and careful attention to social equity.

Organisation of the Book

The book is structured around three thematic parts, guiding readers from foundational knowledge to applied insights.

Part I: Artificial Intelligence for Urban Climate Adaptation

This part explores the role of digital tools in understanding and mitigating urban heat.

- *Chapter 1* reviews the integration of machine learning and remote sensing for urban sustainability, highlighting methods for fine-grained spatial prediction of temperature anomalies.
- *Chapter 2* examines AI-enabled heat mapping and its potential to guide targeted interventions at neighbourhood and street levels.

- *Chapter 3* presents real-time microclimate monitoring platforms that provide actionable information to support vulnerable populations during extreme heat events.
- *Chapter 4* explores the integration of AI-generated insights into planning frameworks to enhance resilience and decision-making at municipal scales.

Part II: Green Infrastructure and Design Innovation

This part focuses on nature-based strategies and their optimisation using computational tools.

- *Chapter 5* investigates vegetation strategies for reducing urban heat while balancing development pressures, including tree species selection, canopy coverage, and urban greening policies.
- *Chapter 6* presents AI-assisted optimisation for tree placement and green infrastructure deployment to improve thermal comfort in microclimates, considering constraints such as land availability, cost, and co-benefits.
- *Chapter 7* synthesises AI-driven green infrastructure optimisation, offering frameworks for efficiently deploying ecological interventions that maximise cooling, enhance biodiversity, and support social well-being.

Part III: Policy, Planning, and Global Perspectives

This part examines governance mechanisms that enable or constrain effective heat adaptation.

- *Chapter 8* investigates climate policy in the Global South, highlighting opportunities for context-sensitive, scalable adaptation strategies in rapidly urbanising regions.
- *Chapter 9* analyses the interaction between transit-oriented development and urban heat, identifying critical trade-offs between mobility, density, and thermal comfort.
- *Chapter 10* examines how policy and technology intersect, proposing integrative approaches to deliver equitable, evidence-based, and sustainable heat mitigation strategies.

Together, these chapters provide a comprehensive, theoretically grounded, and methodologically innovative understanding of algorithmic urbanism for climate resilience.

Rationale and Scope

The scope of this book is deliberately broad, encompassing diverse urban contexts, ranging from high-density metropolises to rapidly growing cities in the Global South. Its focus on urban heat reflects the urgent need to address one of the most pervasive yet underappreciated climate risks. Heatwaves are silent disasters; they cause more fatalities than most other climate hazards and disproportionately affect vulnerable populations. By foregrounding urban heat, this book provides insights into both the biophysical and social dimensions of climate risk, linking ecological design, computational tools, and governance strategies.

Equity is central to the analysis. Vulnerability is unevenly distributed across age, income, ethnicity, and geographic location. Effective adaptation requires interventions that do not exacerbate existing inequalities. The book emphasises methodological rigour, showcasing AI, optimisation algorithms, and spatial modelling to provide robust, replicable, and context-sensitive solutions. It also considers co-benefits, recognising that interventions addressing heat can simultaneously support biodiversity, public health, and social cohesion.

Intended Audience

This book addresses multiple audiences:

- Urban planners and policymakers seeking data-driven, equitable strategies for heat mitigation and climate resilience.
- Researchers in urban studies, environmental science, landscape architecture, and computational analytics interested in interdisciplinary approaches.
- Students and educators exploring the integration of technology, ecology, and governance in urban adaptation.
- Data scientists and geographic information system specialists applying AI and spatial modelling to urban heat challenges.
- Community practitioners advocating for participatory, socially aware climate adaptation initiatives.

Acknowledgements

The ideas and research presented in this book were nurtured through collaboration with colleagues, students, and partners across the world. I am grateful to local governments and research institutions for sharing data, insights, and experiences. I also acknowledge the contributions of early-career researchers and students, particularly at the QUT Urban AI Hub, whose creativity, energy, and curiosity inspired the methodological and conceptual innovations presented in this book. Interdisciplinary discussions with experts in AI, urban planning, landscape ecology, and governance have been invaluable in shaping the book's integrated framework.

Final Reflections

Urban heat is not only a growing environmental challenge but also an opportunity to rethink how cities function, adapt, and thrive. By integrating artificial intelligence, green infrastructure, and governance innovation, this book presents a holistic approach to creating heat-ready cities. It demonstrates that technological intelligence, ecological design, and institutional frameworks are mutually reinforcing tools for climate resilience.

It is my hope that the insights and frameworks offered here will empower researchers, planners, and policymakers to navigate heat risks more effectively, design interventions that are both equitable and efficient, and contribute to urban futures that are sustainable, liveable, and resilient. *Algorithmic Urbanism for Climate Resilience: AI, Green Infrastructure, and Policy Innovation for Heat-Ready Cities* seeks to advance knowledge, inform practice, and inspire interdisciplinary collaboration to address one of the most pressing challenges of our urbanising, warming world.

Part I

Artificial Intelligence for Urban Climate Adaptation

This part of the book explores how artificial intelligence (AI) is reshaping urban climate adaptation strategies, focusing on the integration of machine learning and real-time data for sustainable city planning. It examines AI applications in heat mapping and microclimate monitoring, showing how digital tools provide dynamic insights into urban thermal patterns. The chapters emphasise the potential of AI to support evidence-based decision-making, enabling planners to anticipate, respond to, and mitigate climate risks with greater precision. By linking technological innovation with urban sustainability goals, this part demonstrates how AI can be harnessed to design cities that are both climate-resilient and adaptive to evolving environmental challenges.

DOI: 10.1201/9781003737056-1

1

Integrating Machine Learning for Urban Sustainability

1.1 Introduction and Background

In recent years, rapid urbanisation has resulted in a multitude of urban challenges (Wu et al., 2022; Wurm et al., 2019; Yoo et al., 2018; Zhang et al., 2022a), including disasters, pollution, infrastructure deficits, and unequal living conditions, which pose significant threats to environmental, infrastructure, and societal sustainability. Urban studies are constantly seeking innovative approaches in terms of data and methodologies to address these challenges (Azadbakht et al., 2018; Chi et al., 2022; Vulova et al., 2021; Yigitcanlar et al., 2007, 2008). The advent of machine learning and remote sensing techniques presents a new and promising opportunity for the accurate and efficient monitoring and analysis of urban issues in the pursuit of sustainability (Koumetio Tekouabou et al., 2022; Patino & Duque, 2013; Son et al., 2023; Yigitcanlar, 2009).

Remote sensing images offer an abundance of spatial and spectral information that can provide a macro-level understanding of urban issues. For instance, Demarchi et al. (2014) and Priem et al. (2019) employed hyperspectral remote sensing images to map the spatial distribution of land cover and land use. Furthermore, Chen et al. (2020a) and Geiß et al. (2020) investigated the macro-level spatial attributes of natural and built environmental elements through multispectral satellite imagery. They have become a widely used tool in urban studies due to their readily available data, global coverage, various scales and resolutions, and broad applicability (Patino & Duque, 2013). For example, free global imagery from Landsat, Moderate Resolution Imaging Spectroradiometer (MODIS), and Sentinel satellites/sensors has been provided for decades, with multispectral images and various derivative products that can be applied to urban studies. For instance, the red–green–blue (RGB) and near infrared bands are commonly used in land cover/use classification (Jamali, 2019; Karalas et al., 2016), while thermal bands are effective in heat mapping and soil moisture estimation (Chen et al., 2020b; Das et al., 2021). Additionally, MODIS provides data products that have been widely utilised in various urban studies, such as land surface temperature, vegetation index, and aerosol products (Bogawski et al., 2019; Jato-Espino et al., 2022; Li et al., 2020).

DOI: 10.1201/9781003737056-2

Machine learning has also gained a considerable attention from researchers (Deng et al., 2017; Hay Chung et al., 2021; Li et al., 2020) due to its impressive computational and spatial analysis capabilities. As machine learning is capable of automatically learning from existing data without programming explicitly, it excels in quickly analysing and understanding immense databases to provide valuable insights to users. In the urban studies field, valuable insights refer to meaningful information extracted from complex urban data, aiding decision-making and enhancing urban planning and management.

For example, machine learning can identify patterns in urban growth (Griffiths et al., 2010; Jaad & Abdelghany, 2021), enabling planners to make informed decisions about infrastructure development and resource allocation. There are two main categories of machine learning: supervised and unsupervised. Supervised algorithms involve human supervision in data training, while unsupervised algorithms can identify unknown patterns in data without human supervision (Koumetio Tekouabou et al., 2022). Both types of algorithms can monitor dynamic characteristics, analyse causal relationships, and predict the potential evolution of urban issues.

Numerous urban studies have thus far combined the superior analysis power of machine learning techniques and wide coverage of remote sensing imagery to monitor and mitigate the challenges threatening urban sustainability. For example, urban sustainability can encompass various aspects such as infrastructure allocation optimisation (Dong et al., 2022), disaster risk reduction (Ghaffarian et al., 2021), and ecosystem preservation (Sannigrahi et al., 2019). Machine learning has enabled the analysis and prediction of the impact of urban issues, such as urban heat island (Kafy et al., 2022), land use change (Wang et al., 2016), air pollution (Wei et al., 2021), and demarcation of urban boundary (Mortoja et al., 2020) on a large scale through remote sensing imagery (Martins et al., 2021; Yan et al., 2020). Since these techniques offer superior computational and spatial analysis capacities, vast amounts of remote sensing data can be processed and analysed with minimal human intervention. The application of this integration covers a wide range of research domains, including disaster risk management (Ali et al., 2022; Zhang et al., 2022), land cover and use mapping (Xu et al., 2022; Zhu et al., 2022), air and water pollution monitoring (Guo et al., 2021a; Yang et al., 2022), ecosystem assessment (Liu et al., 2021; Sannigrahi et al., 2019), and infrastructure development (Geiß et al., 2020; Zhang et al., 2018a).

For instance, Kwak et al. (2020) and Mohammad et al. (2022) employed machine learning algorithms to model dynamic characteristics and driving forces of urban heat using Landsat images. Zhang et al. (2018) and Singleton et al. (2022) utilised deep learning algorithms, convolutional neural networks (CNNs), and Sentinel and aerial imagery to classify land covers and uses.

Chen et al. (2021c) and Zhang et al. (2021) utilised support vector machine (SVM) and random forest (RF) algorithms to estimate particulate matter 2.5 ($PM_{2.5}$) concentrations from MODIS aerosol optical depth (AOD) products. Soilán et al. (2016) and Jia and Zhang (2021) employed multiple machine learning algorithms and multi-source remote sensing data to map blue–green and built infrastructures.

Numerous reviews have been published on the applications of machine learning and remote sensing in urban analysis (Alanne & Sierla, 2022; Belgiu & Drăguţ, 2016; Koumetio Tekouabou et al., 2022; Milojevic-Dupont & Creutzig, 2021; Patino & Duque, 2013). However, these studies primarily concentrate on applications of either machine learning or remote sensing in specific domains. For instance, Patino and Duque (2013) solely explore the potential applications of satellite remote sensing in regional urban studies, such as urban growth and population estimation. Furthermore, Milojevic-Dupont and Creutzig (2021), Alanne and Sierla (2022), and Koumetio Tekouabou et al. (2022) exclusively investigate machine learning applications in climate change mitigation, smart buildings, and urban planning decision-making, respectively.

To date, to our knowledge, there has been no comprehensive review examining the integration of machine learning and remote sensing in the studies for urban sustainability. This systematic review aims to address this gap firstly by using a Preferred Reporting Items for Systematic Reviews and Meta-Analyses (PRISMA) methodology. The review provides a comparative analysis of the joint application of remote sensing and machine learning for urban sustainability. Secondly, we propose the machine learning and remote sensing integration framework to explore the research question: "How can urban sustainability studies benefit from advanced spatial analytics integrating both machine learning and remote sensing techniques?" This chapter incorporates the perspectives of general observations of chapter distributions, application areas, remote sensing and machine learning data and techniques, and machine learning and remote sensing integrations.

The remainder of the chapter is organised into four sections: Section 1.2 presents the methodological approach, tools, and criteria employed in this study, thereby enhancing the clarity, replicability, and credibility of this review. Section 1.3 provides a comprehensive overview of the results from the five perspectives mentioned above. We also propose a framework in this section to facilitate understanding of the integration between machine learning and remote sensing, as well as to inform potential directions for their integration in future research. Section 1.4 then analyses and discusses the key findings from each perspective, offering interpretations of the results, identifying limitations, unravelling underlying causality, and examining the broader implications for the field. Finally, Section 1.5 concludes the chapter by summarising the key findings and their implications for theory and practice, as well as avenues for future research to jointly leverage machine learning and remote sensing.

1.2 Research Design

The aim of this study is to explore how advanced spatial analytics, specifically the integration of machine learning and remote sensing techniques, can benefit urban studies in promoting sustainability. A systematic literature review was conducted using the PRISMA approach, which is a recognised methodology for reporting systematic reviews and meta-analyses (Moher et al., 2015). A systematic review allows for a thorough and comprehensive understanding of the existing urban research that integrates machine learning and remote sensing. This helps identify the current state of knowledge, the methodologies used, and the potential gaps in the literature that can be addressed through future urban research. The PRISMA approach ensures that the review process is transparent and reproducible. This helps in establishing the credibility of the study and allows for the findings to be easily verified and updated as new research emerges. The review was conducted in three phases, including planning, review, and reporting, following Li et al.'s (2021) PRISMA review study.

During the planning phase, the research question, objectives, keywords, and inclusion and exclusion criteria were established to ensure that the review would provide valuable insights into the opportunities and challenges of incorporating machine learning and remote sensing techniques in addressing urban sustainability issues. The academic search engine provided by the Queensland University of Technology (QUT) Library, which grants access to 393 different databases, including (a) Directory of Open Access Journals; (b) ScienceDirect; (c) Scopus; (d) TRID; (e) Web of Science, and (f) Wiley Online Library, was utilised to collect related publications. We chose the QUT search engine because it offers access to hundreds of databases and enables comprehensive searches across a wide range of resources, such as books, journal articles, and conference papers. A query was conducted as *((title OR abstract OR keyword) contains ("machine learning" OR "deep learning") AND ("remote sensing" OR "satellite" OR "aerial" OR "spaceborne") AND (urban OR cit*))*. This review left the publication date of the search open, allowing to include as much relevant literature as possible.

Throughout the review phase, a meticulous examination of the selected articles was conducted. Initially, a total of 619 records were obtained through the initial search, as depicted in Figure 1.1. As shown in Table 1.1, the primary inclusion criteria were limited to online, peer-reviewed English journal articles. After a thorough assessment of these records against the established inclusion and exclusion criteria, the number was reduced to 460 articles. Then,

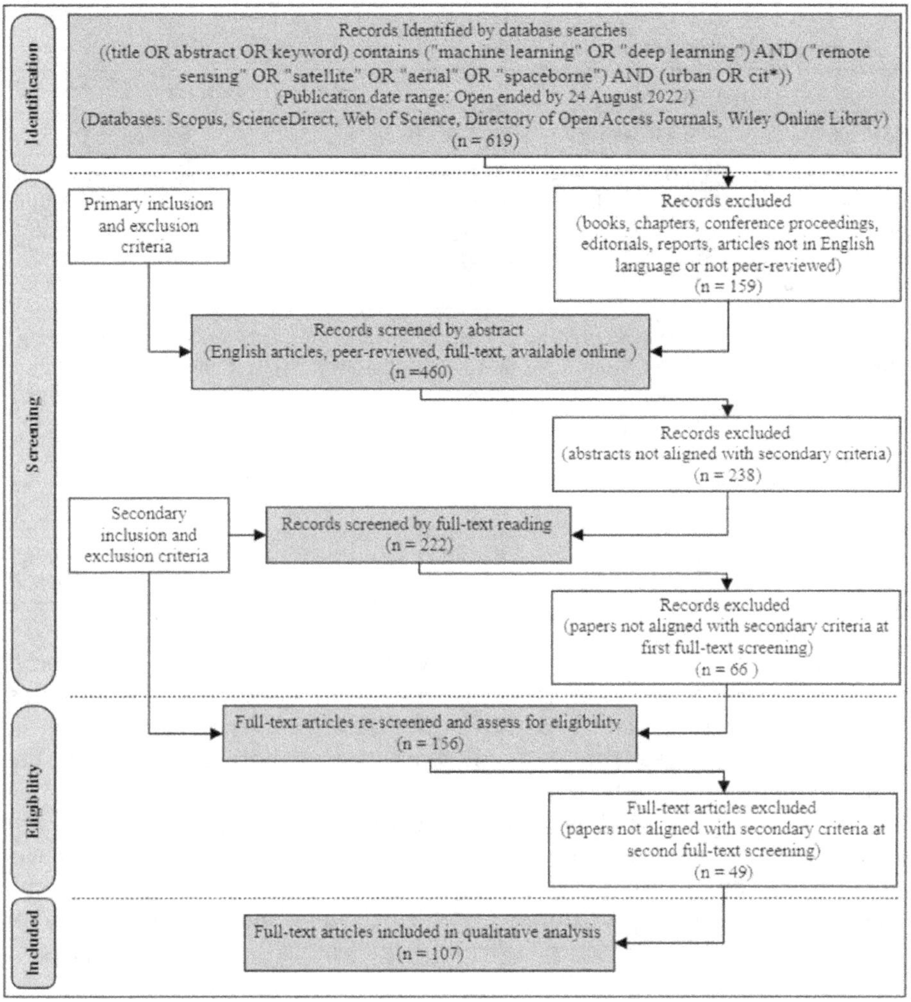

FIGURE 1.1
Literature selection procedure.

these articles underwent a rigorous re-examination process to ensure consistency with the keyword search, and their abstracts were assessed against the secondary criteria. Following this, irrelevant records retrieved from various databases were eliminated, resulting in a final count of 222 articles. After carefully reading the shortlisted articles, we extracted detailed information such as topics, keywords, methodology, research questions, aims, and significance. We then eliminated articles that did not align with the secondary criteria, resulting in a final set of 156 articles. Another round of full-text screening was then conducted, resulting in a further reduction to 107 articles.

TABLE 1.1

Exclusion and Inclusion Criteria

Primary Criteria		Secondary Criteria	
Inclusionary	**Exclusionary**	**Inclusionary**	**Exclusionary**
Academic journal articles Journal articles peer-reviewed Full-text available online Published in English	Duplicate records Journal articles not peer-reviewed Books and chapters Industry reports Government reports	Relevant to (a) machine learning, such as supervised and unsupervised learning, deep learning, neural network, and so on; (b) remote sensing, such as satellites, unmanned aerial vehicles, and LiDAR; (c) studies conducted in urban areas; (d) studies concerning sustainable urban development or urban sustainability	Irrelevant to (a) integration of machine learning and remote sensing integration; (b) studies conducted in urban areas; (c) studies concerning sustainable urban development or urban sustainability
		Relevant to the research question and objectives	Irrelevant to the research question and objective

These articles were analysed and reviewed based on five aspects: (a) general observations, (b) application areas, (c) remote sensing data and techniques, (d) machine learning data and techniques, and (e) integration of machine learning and remote sensing. A detailed description of the categorisation criteria can be found in Table 1.2.

In the final reporting phase, the results and findings from the systematic review of the 107 selected journal articles were presented. The complete list of these papers can be found in Table A1.1. The discussion involves an in-depth examination of the opportunities and limitations of current urban studies that have incorporated machine learning and remote sensing techniques.

TABLE 1.2

Categorisation Criteria

Categorisation Criteria
Select the literature aligning with the research objectives by skimming reading
Identify the literature focusing on machine learning and remote sensing integration in urban studies by full-text reading
Classify similar literature with application areas, data, and technology into broader categories
Refine the broader categories and crosscheck their reliability against other published literature
Final review and corresponding update of category changes
Determine final categories for the classification of literature
Classify the literature selected with the determined categories

1.3 Analysis and Results

1.3.1 General Observations

All 107 eligible articles were categorised by publication year and study area, as shown in Figure 1.2, to analyse and reveal the discrepancy between periods and regions. Figure 1.2 shows related publications fluctuated at a low level from 2010 to 2017. Since 2017, it began to increase and approximately doubled in 2019. Then, 2020 maintains the same level as 2019. In 2021, the publication peaked and nearly tripled. The growth is expected to continue in 2022. Most case studies were in Asia (52%), followed by Europe (20%) and North America (19%). South America, Africa, and Oceania stayed at the same level, below 5%. It is worth noting that China, the USA, and Germany occupied more than half of the case studies, 33%, 16%, and 9%, respectively.

1.3.2 Application Areas

Referring to the categorisation models established by Casali et al. (2022), we focused on application areas that integrate machine learning and remote sensing to address urban issues and categorised the reviewed articles into five categories: land cover/use classification (n = 24, 22%), disaster risk management (n = 22, 21%), pollution monitoring (n = 21, 19%), built infrastructure identification (n = 20, 19%), and ecosystem and environmental assessment (n = 20, 19%). Casali et al.'s (2022) categorisation targeted machine learning applications in urban spatial analysis and established four topical categories: land use and urban form, socio-economic, environmental, and infrastructures.

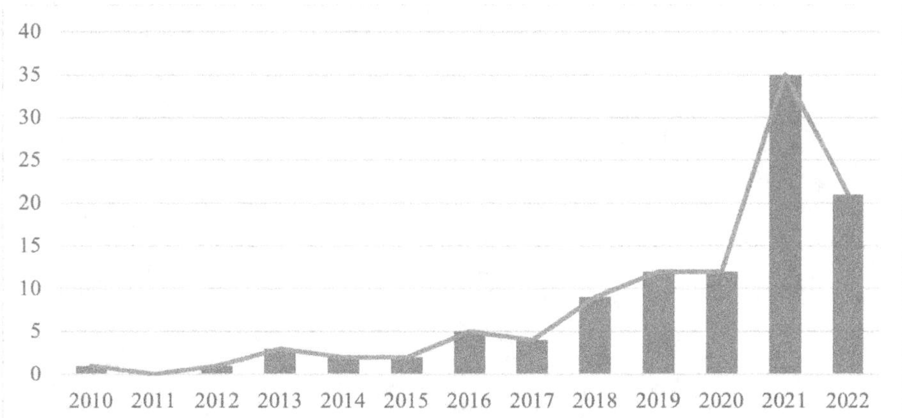

FIGURE 1.2
Publication distribution by year. (Note: the search window is set to end on 24/8/2022.)

Due to the limited scope of urban studies integrating machine learning and remote sensing, these categories were too broad for our review. Therefore, we proposed the categories to focus on application areas that utilise machine learning and remote sensing to address urban issues. We collected research questions, aims, objectives, and the significance of the selected articles through two rounds of full-text reading. Based on the collected information, we investigated their research themes and topics and established the five categories mentioned above. Then, we created subcategories based on the characteristics of each application area. For example, we subcategorised disaster risk management according to the different phases of disaster management and subcategorised pollution monitoring according to pollution types. The following section summarises the research topics and publication distribution of reviewed articles based on these application areas.

1.3.2.1 Land Cover/Use Classification

As illustrated in Figure 1.3, the category of *land cover/use classification* encompassed two topics. *Land cover/use classification* applied various machine learning classification algorithms to process and analyse remote sensing images for producing *single* (n = 11) or *multiple* (n = 13) land cover/use maps. Land cover/use maps serve as critical fundamentals for urban planning policy-making. *Single land cover/use detection* is concentrated on thematic mapping and includes two topics, *urban area extraction* (n = 8) and *neighbourhood environment identification* (n = 3). Urban area extraction is a process that involves isolating and delineating the boundaries of urban regions with geospatial

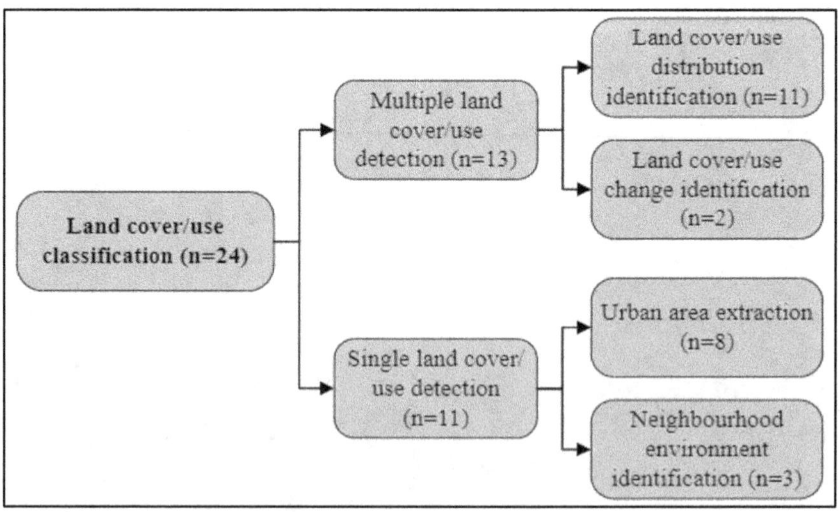

FIGURE 1.3
Research topics and corresponding publication numbers in land cover/use classification.

data or remote sensing imagery (Griffiths et al., 2010; Jaad & Abdelghany, 2021). Urban areas incorporate two similar land cover/use types, built-up areas, and impervious surfaces (Huang et al., 2019a; Khan & Sudheer, 2022), both of which are significant indicators of assessing urban growth during the urbanisation process.

Neighbourhood environment identification addressed the classification and survey of residential settlement types, patterns, and spatial distributions (Wieland & Pittore, 2016; Wurm et al., 2019), which is beneficial to comprehending and improving the dwelling environment. *Land cover/use distribution identification* (n = 11) characterised anthropogenic and natural environmental surfaces spatially (Chen et al., 2021b; Garg et al., 2021). Residential areas, industrial areas, and airports are examples of common human-made land cover/uses, while water, vegetation, and bare soil are among natural land covers. *Land cover/use change identification* (n = 2) employed a long series of remote sensing images and machine learning classification algorithms to identify the historical land cover/use distribution (Schneider, 2012). Furthermore, Wang et al. (2016) investigated the development and evolution of urban areas over time and identified the drivers of land cover and use change using the conditional inference-based regression tree algorithm.

1.3.2.2 Disaster Risk Management

Disaster risk management can be classified into two categories: *disaster risk assessment* (n = 19) and *post-disaster recovery* (n = 3). As depicted in Figure 1.4, *disaster risk assessment* dominated the application domain and aims to monitor and predict potential risks, forcing drivers, and disaster evolutions

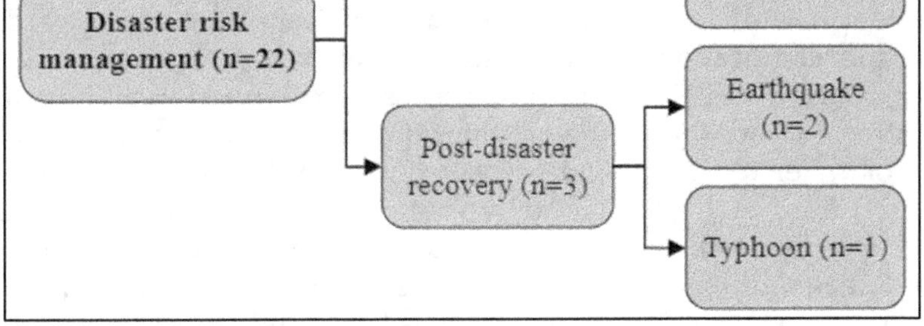

FIGURE 1.4
Research topics and corresponding publication numbers in disaster risk management.

(Guo et al., 2021b; Kafy et al., 2022; Lammers et al., 2021). Most articles focused on the topic of urban heat (n = 15). The most popular solution for assessing and monitoring the urban heat island effect is mapping and predicting the spatiotemporal distribution of land surface temperature (Chen et al., 2022a; Jato-Espino et al., 2022). Additionally, some studies have utilised machine learning regression algorithms to examine the potential impact of land cover surfaces on surface temperatures, such as vegetation covers and built infrastructures (Duncan et al., 2019; Stewart & Kremer, 2022).

Flooding (n = 4) is also a prevalent topic in disaster risk management, with many researchers simulating and predicting flood-prone locations, creating real-time flood maps, assessing comprehensive impacts of floods, and identifying potential driving factors (Lammers et al., 2021; Peng et al., 2019; Santos et al., 2017). *Post-disaster recovery* is a crucial aspect of disaster risk management, including disaster damage assessment and the recovery of normal life for urban residents. Frank et al. (2017) and Wang et al. (2022a) improved the machine learning model to rapidly locate earthquake-damaged buildings and assess post-earthquake damages from high-resolution remote sensing images. Ghaffarian et al. (2021) investigated the impact of the combination of remote sensing data and the eXtreme Gradient Boosting (XGBoost) algorithm in post-typhoon recovery.

1.3.2.3 Pollution Monitoring

Figure 1.5 illustrates the research topics in the field of *pollution monitoring*. A significant area of focus is *air pollution* (n = 15), with most studies dedicated to monitoring its effects and mitigating its impact on public health and human well-being. A key indicator of air pollution is *$PM_{2.5}$ concentration* (n = 12) which has received considerable attention due to its significant impact on respiratory health (Wei et al., 2021; Yan et al., 2021). *Particulate matter 10 (PM_{10}) concentration* (n = 1) is another indicator of air pollution, which measure coarser particulate matter (Tella et al., 2021). Oxynitride, a damaging pollutant, has also been a focus of research, as it not only affects the lives of creatures but also exacerbates global warming. Zhang et al. (2021) and Chi et al. (2022) used public *nitrogen dioxide (NO_2,* n = 2) products and machine learning algorithms to monitor ground-level NO_2 concentrations, an important step towards preventing soil acidification and water eutrophication.

Water pollution (n = 4) is another issue of concern in urban areas. Researchers have extracted and analysed key water quality parameters to monitor the quality of water resources and prevent pollution (Chen et al., 2021a; Zhou et al., 2021). Yang et al. (2021) used hyperspectral data and machine learning to detect black-odorous water, aiming to manage and control water pollution by understanding its spatial distribution and magnitude. *Light pollution* (n = 1), more prevalent in urban areas, has been studied by Xu et al. (2018b) who used multiple machine learning algorithms to detect night ambient lights from high-resolution remote sensing imagery. *Soil heavy metal pollution*

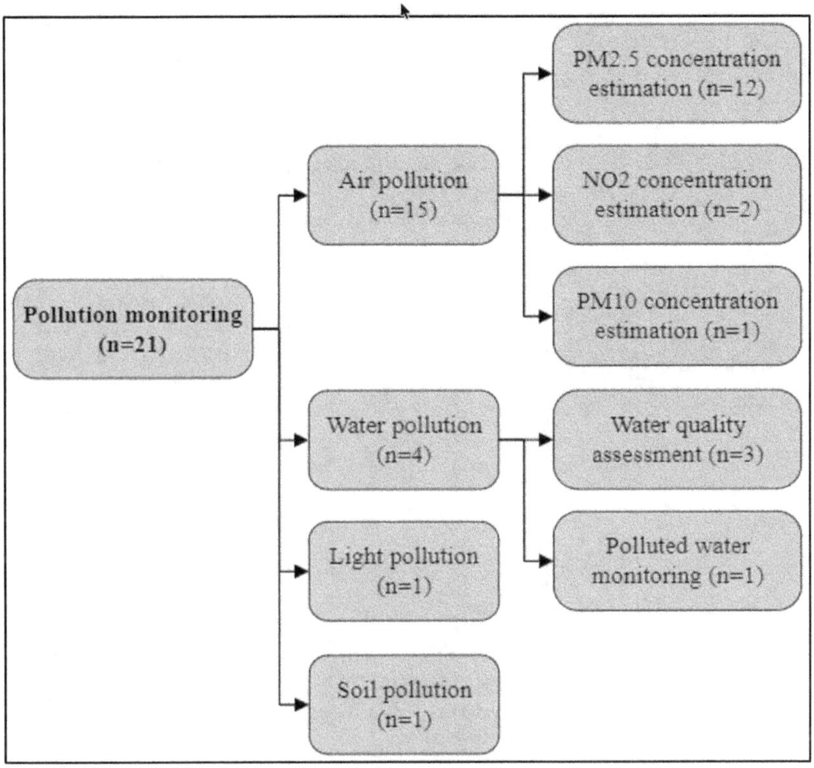

FIGURE 1.5
Research topics and corresponding publication numbers in pollution monitoring.

(n = 1) has also received attention, with Shi et al. (2021) focusing on detecting and analysing the spatial distribution and magnitude of Zn in topsoil.

1.3.2.4 Built Infrastructure Identification

Figure 1.6 illustrates the categorisation of *built infrastructure identification* into two aspects: *building detection* (n = 16) and *traffic object detection* (n = 4). These studies mainly utilised high-resolution aerial images and LiDAR data to depict building structures and a range of urban objects. *Building detection* was further categorised into six subcategories, based on the detected building information. *Building footprints* (n = 6) aimed to provide building outlines with a description of shapes, locations, and dimensions (Janowski et al., 2021; Touzani & Granderson, 2021). Some studies focused on individual building parameters (Frantz et al., 2021; Sanya & Mwebaze, 2020), such as *building height* (n = 3) and *building density* (n = 2), which are important for urban morphology analysis. Other research concentrated on detecting *building structures* (n = 3) (Aissou et al., 2021; Henn et al., 2013; Imam et al., 2021), such as roofs and

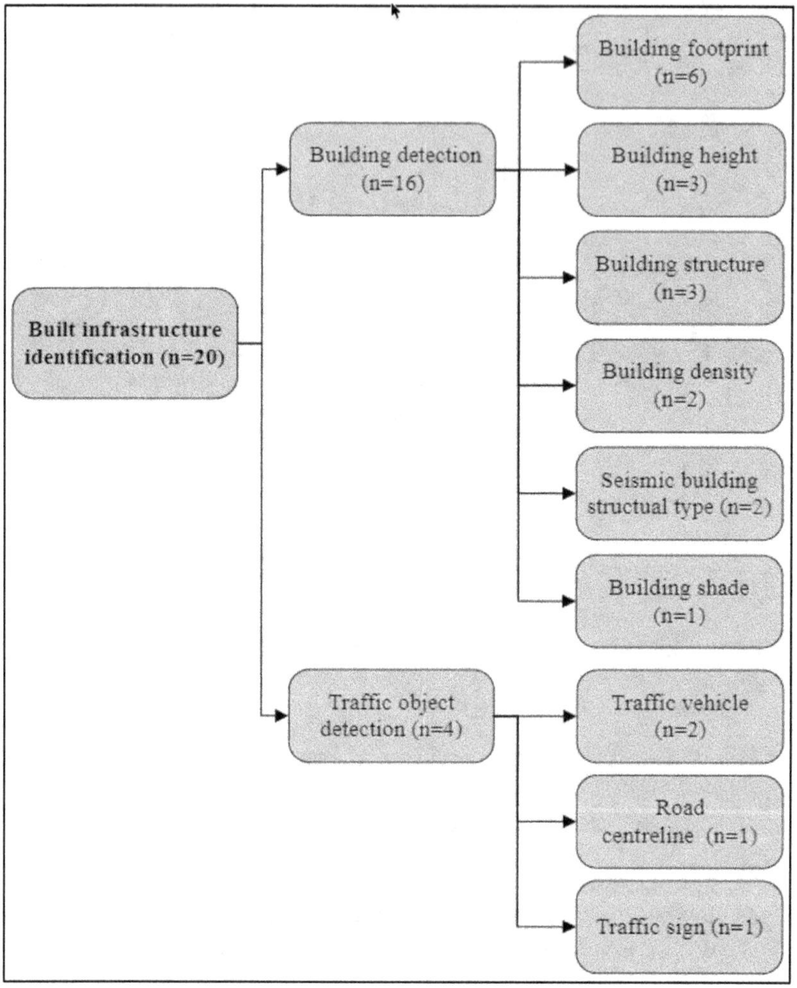

FIGURE 1.6
Research topics and corresponding publication numbers in built infrastructure identification.

fences, as well as *building shades* (n = 1) (Adeline et al., 2013). The estimation of *seismic building structural types* (n = 2) was another crucial aspect of building detection, which assessed the load-bearing capacity of building structures during earthquakes (Geiß et al., 2015; Liuzzi et al., 2019). In addition to *building detection, traffic object detection* is another area of study in urban environments. The most common detection targets in the field include traffic signs (Soilán et al., 2016), vehicles (Han et al., 2020; Schilling et al., 2018), and road centrelines (Zhang et al., 2018). The road centreline is a representation that effectively displays road topology and precisely indicates position. These studies aimed to develop an efficient, precise automated detection and classification methodology for the geometric and semantic properties of traffic objects in urban areas.

1.3.2.5 Ecosystem and Environmental Assessment

Figure 1.7 illustrates the focus of *ecosystem and environmental assessment* in most studies, with a total of 13 studies dedicated to *environment*. The studies were split into three categories: *green infrastructure* (n = 7), *blue infrastructure* (n = 4), and *population distribution* (n = 3). The *green infrastructure* studies analysed the spatial distribution and the economic benefits of green covers, green roofs, and tree species (Uniyal et al., 2022; Wang et al., 2022). The *blue infrastructure* studies extracted and mapped wetlands, canals, and water bodies

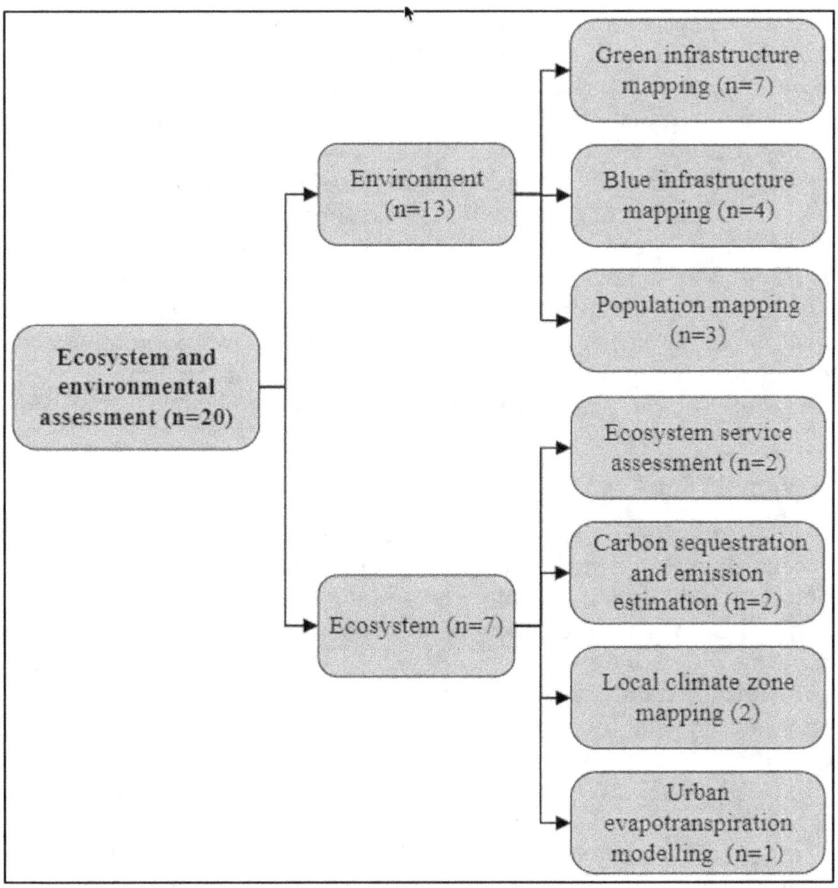

FIGURE 1.7
Research topics and corresponding publication numbers in ecosystem and environmental assessment.

for multiple purposes (Chen et al., 2020; Liao & Wen, 2020; Mahdianpari et al., 2021), such as optimising urban morphology, inspecting water leak, and creating smart city policies. Mahdianpari et al. (2021) indicated that mapping wetlands benefits the creation of smart city policy by providing essential natural infrastructure information for informed decision-making. Furthermore, smart city policies contribute to urban sustainability by promoting resource conservation and biodiversity preservation. Population mapping was included in the environment category because humans are crucial elements of the urban environment. Analysing and comprehending the distribution of urban populations is crucial for improving the living and working environment. In total three *population distribution* studies thus aimed to quantify the magnitude and spatial distribution of the population for more accurate population data (Yang et al., 2019; Ye et al., 2019; Zhao et al., 2021).

Ecosystems are complex anthropogenic ecological communities, featuring distinct characteristics from natural ecosystems. Liu et al. (2021) conducted a systematic analysis of the spatial distribution of *ecosystem services* (n = 2) and the impact of multiple influencing factors, including ecological stock, structure, and circulation of material and energy. Sannigrahi et al. (2019) analysed historical dynamic changes of the ecosystem and quantified the landscape's capacity to provide multiple key ecosystem services. Other studies assessed the function of ecosystem services through specific topics (Milnar & Ramaswami, 2020; Uniyal et al., 2022), such as *carbon sequestration and emission estimation* (n = 2) and evapotranspiration modelling (n = 1). *Local climate zones* (n = 2) classify the entire ecosystem into different types based on similar characteristics, such as surface cover, material, structure, and population activity. Sapena et al. (2021) used local climate zones to examine the bidirectional relationship between the ecosystem and socio-economic society in urban regions.

1.3.3 Remote Sensing Data and Techniques

In 1972, the first Earth observation satellite sensor was launched into space, and since then, hundreds of different types of remote sensing data have become available and accessible worldwide (Osaku et al., 2015). The information retrieved from the reviewed articles are displayed as a heat map in descending order (see Figure 1.8). Based on the method of acquisition, the remote sensing data can be categorised into three subcategories: satellite imaging, aerial imaging, and land imaging (e.g., LiDAR). Figure 1.8 shows that satellite imaging (76%) data was the most dominant, followed by aerial imaging (17%) and land imaging (7%). We found the top five most frequently used satellite data are Landsat (23%), MODIS (15%), Sentinel (11%), nighttime light images (5%), and Satellite Pour l'Observation de la Terre (SPOT, 3%).

Landsat has been in service since 1972 (see Table A1.2), which supports long-term Earth observation. Its moderate spatial resolution and multiple spectral bands, such as multispectral (30 m), panchromatic (15 m), and thermal (60–120 m) bands, cater to a range of applications (Jaad & Abdelghany,

		Land cover/use classification	Disaster risk management	Pollution monitoring	Built infrastructure identification	Ecosystem and environmental assessment
Satellite imaging (76%)	Landsat (23%)	10	10	3	1	7
	MODIS (15%)	2	6	11		1
	Sentinel (11%)	4	2	3	3	3
	Nighttime light (5%)	1	2	1		3
	SPOT (3%)		1	1		2
	Google Earth & Map (2%)	1			1	1
	Pléiades (2%)	2	1			
	WorldView (2%)		1		1	1
	IKONOS (2%)	1			1	
	ASTER (1%)			1		
	Cartosat-1 (1%)					1
	CBERS-2B (1%)	1				
	ERS (1%)	1				
	GeoEye (1%)	1				
	GOES-16 (1%)		1			
	Himawari-8 (1%)			1		
	IKCEST (1%)	1				
	ISS (1%)			1		
	Mapbox (1%)				1	
	NWPU VHR-10 (1%)				1	
	PELICAN (1%)			1		
	Planet Lab (1%)		1			
	QuickBird (1%)		1			
	Rapid-Eye (1%)					1
	xBD (1%)		1			
Aerial imaging (17%)		5	2	3	7	6
Land imaging (Lidar) (7%)		2			6	1

FIGURE 1.8
Heat map of remote sensing images utilised in each application domain. (Note: one paper can be categorised into more than one group. Abbreviations: Advanced Spaceborne Thermal Emission and Reflection Radiometer (ASTER), China-Brazil Earth Resources Satellite 2B (CBERS-2B), European Remote-Sensing Satellite (ERS), Geostationary Operational Environmental Satellites-16 (GOES-16), International Knowledge Centre for Engineering Sciences and Technology (IKCEST), International Space Station (ISS).)

2021; Kafy et al., 2022; Sannigrahi et al., 2019). Although providing free and open-access images, Landsat's shortcomings include a 16-day revisit time and lower spectral resolution compared to advanced hyperspectral sensors. These images have been widely used in urban studies, particularly in land cover/use classification, disaster risk management, and ecosystem and environmental assessment (Kafy et al., 2022; Khan & Sudheer, 2022; Uniyal et al., 2022).

MODIS excels in offering daily observations, diverse spectral bands, and consistent global coverage with free data access (Jato-Espino et al., 2022; Yang et al., 2022). Nevertheless, the low spatial resolution restricts its applicability in finer-scale studies, presenting a trade-off between temporal resolution and spatial granularity. It was frequently used for disaster risk management and pollution monitoring, due to highly available land surface temperature and atmospheric optical depth products. Sentinel satellites provide high spatial resolution (up to 10 m) and diverse sensors, facilitating various monitoring tasks with frequent revisit times (five to ten days) and open data access (Frantz et al., 2021; Hay Chung et al., 2021). However, the relatively short data record (since 2014) and higher data complexity necessitate specialised processing, posing challenges for some users. Sentinel data has been used equally in various application domains, with Sentinel-2 images being the most frequently used for land monitoring, vegetation, soil, and coastal areas, with high-resolution optical imageries (Singleton et al., 2022).

Nighttime light images offer valuable insights into urbanisation, energy consumption, and socio-economic activities (Guo et al., 2021; Wei et al., 2021; Yang et al., 2019), complementing daytime satellite imagery with regular updates. However, limitations include reduced spatial resolution, sensitivity to atmospheric conditions and moonlight, and restriction to nighttime observations. SPOT satellites offer high spatial resolution (up to 1.5 m) and multiple spectral bands for diverse applications (Shi et al., 2021; Yang et al., 2019), backed by a long operational history since 1986. However, they suffer from lower temporal resolution and limited spectral resolution compared to hyperspectral sensors. Data access and cost may also be challenging for some users. Other remote sensing data was used less frequently in urban applications.

Aerial images were the second most popular remote sensing data in urban studies. Unlike satellite images, aerial images were captured by airborne cameras and sensors, and offer high resolutions, but usually cover limited spaces and periods (Dornaika et al., 2016; Xu et al., 2018c). They were often used in land cover/use classification, ecosystem and environmental assessment, and built infrastructure identification, but less frequently in disaster risk management and pollution monitoring. LiDAR data, a kind of point cloud data storing three-dimensional (3D) information, was prominent in the built infrastructure identification due to its ability to provide high-resolution 3D information for depicting urban spatial characteristics (Lu et al., 2014). LiDAR data can be collected by using vehicle-mounted systems for a given geographical area, terrain, building, or space.

All remote sensing images were categorised into three resolution levels: low-, moderate-, and high resolution (see Figure 1.9). The distribution of each application domain was analysed. There are no universally accepted and explicit categorisation criteria for classifying spatial resolutions of remote sensing images (Mushore et al., 2017; Wang et al., 2022). The boundaries

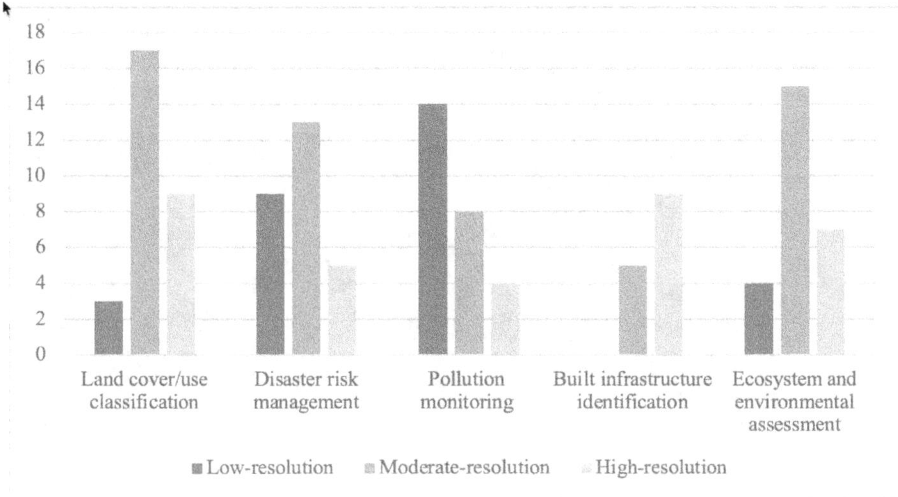

FIGURE 1.9
Remote sensing data with different resolution levels in each application domain.

between categories may vary depending on the specific context, application, or literature source. For example, Remote Sensing Links (2023) categorised remote sensing sensors of coastal and marine environments into low (<10 m), medium (10–100 m), and high (>100 m) levels. However, as technology advances, the categorisation criteria may also evolve to accommodate new sensors with different spatial resolutions. Based on these factors, this chapter categorised spatial resolutions of remote sensing images to cater to the diverse requirements and specific contexts of various urban application domains. In detail, high resolution was defined as below 2 m, moderate resolution as 2–30 m, and low resolution as beyond 30 m. High-resolution images provide detailed spatial information, moderate-resolution images offer a balance between spatial detail and coverage, and low-resolution images cover large areas and are often available at a higher temporal frequency. As shown in Figure 1.9, built infrastructure had a high demand for high spatial resolutions, while public air quality monitoring products were usually limited by low resolutions due to technical constraints. Moderate-resolution data was mainly used in land cover identification, disaster risk management, and ecosystem and environmental assessment.

1.3.4 Machine Learning Data and Techniques

In this section, we compiled and organised the various machine learning algorithms employed in each application domain, accompanied by an illustration and a heat map (see Figure 1.10). Artificial intelligence (AI), the capability of a machine to imitate intelligent human behaviours, is facilitated by

	Land cover/use classification	Disaster risk management	Pollution monitoring	Built infrastructure identification	Ecosystem and environmental assessment
RF (22%)	9	10	8	9	12
SVM (15%)	8	4	5	10	7
CNN (14%)	11	4	2	7	6
ANN (7%)	3	4	4	2	3
Boosting (7%)	3	2	6	1	4
KNN (5%)	3	2	2	2	2
DT (4%)	6	1		2	4
MLP (2%)	4		1		
Cubist (2%)	1	1	2		1
Linear regression (2%)		2	2		
MLC (2%)	2				2
K-means clustering (1%)		1		1	1
Logistic regression (1%)	3				
MARS (1%)	1		2		
NB (1%)	1	1	1		
DBN (1%)	1		1		
FCM clustering (1%)	2				
Lasso regression (1%)			2		
RNN (1%)		1	1		
BR (<1%)	1				
Decision Table (<1%)	1				
DNN (<1%)			1		
GAM (<1%)			1		
GPR (<1%)				1	
ISODATA (<1%)					1
LDA (<1%)					1
LP (<1%)	1				
MDC (<1%)					1
NN ge (<1%)	1				
Optimum-path forest (<1%)	1				
QR (<1%)			1		
Random k-label sets (<1%)	1				
End-to-end GB (<1%)			1		
SGD (<1%)			1		

FIGURE 1.10
Heat map of machine learning algorithms utilised in each application domain. (Note: one paper can be categorised into more than one group. Abbreviations: Binary Relevance (BR), Deep Belief

machine learning, a subfield of AI, which enables a system to automatically learn and enhance its performance through experience (Galaz et al., 2021). The artificial neural network (ANN), invented by McCulloch and Pitts (1943) and developed by Rosenblatt (1958), is a type of supervised machine learning algorithm which simulates human brains using a set of complex algorithms. Deep learning, which encompasses neural networks with three or more hidden node layers, is a particular type of neural network.

As depicted in Figure 1.10, we found the most frequently used machine learning algorithms, in descending order, are RF (22%), SVM (15%), CNN (14%), ANN (7%), and Boosting (7%) algorithms, respectively. These machine learning algorithms are all supervised. The unsupervised machine learning algorithms used in reviewed articles are K-means clustering and Fuzzy C-means clustering. RF, an ensemble learning method of classification and regression, was evenly distributed across all application domains with approximately ten studies each.

SVM, which utilises associated learning algorithms for data analysis in classification and regression analysis, was prominently utilised in land cover/use classification, ecosystem, and environmental assessment, and built infrastructure identification. CNN, a valuable tool for semantic segmentation of visual imagery, was dominant in the land cover/use classification domain and was also commonly used in ecosystem and environmental assessment and built infrastructure identification. Boosting algorithms, such as adaptive boosting, categorical boosting, light gradient boosting machine, and XGBoost, comprise a family of ensemble learning techniques designed to enhance the predictive accuracy of models and create robust classifiers (Shogrkhodaei et al., 2021; Zhao et al., 2021). Although their applicability in the domain of built infrastructure identification is somewhat limited, these algorithms have demonstrated remarkable effectiveness across various application domains, particularly in pollution monitoring, and ecosystem and environmental assessment. ANN algorithms, meanwhile, have been employed with comparable frequency in each of these domains, with roughly three studies conducted in each area.

We classify these machine learning algorithms into four groups: (a) supervised, (b) unsupervised, (c) neural network, and (d) deep learning algorithms (see Figure 1.11). The distribution of the categories across each application domain visually demonstrates that supervised machine learning algorithms dominated in all domains, followed by deep learning algorithms, while

←

Network (DBN), Decision Tree (DT), Deep Neural Network (DNN), Fuzzy C-Means Clustering (FCM Clustering), Generalised Additive Model (GAM), Gaussian Process Regression (GPR), Iterative Self-Organising Data Analysis Technique (ISODATA), K-Nearest Neighbours (KNN), Linear Discriminant Analysis (LDA), Label Powerset (LP), Multivariate Adaptive Regression Splines (MARS), Minimum Description Length Clustering (MDC), Multilayer Perceptron (MLP), Maximum Likelihood Classification (MLC), Naive Bayes (NB), Non-Nested Generalised Exemplars (NN ge), Quantile Regression (QR), End-to-End Gradient Boosting (End-to-end GB), Recurrent Neural Network (RNN), and Stochastic Gradient Descent (SGD).)

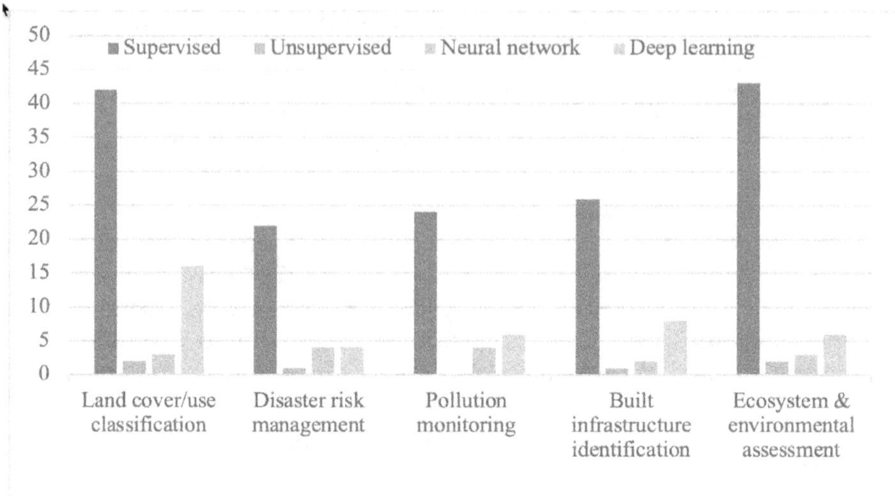

FIGURE 1.11
Machine learning methods in each application domain.

unsupervised machine learning algorithms are the least frequently used with low levels in each domain. Notably, there were no unsupervised algorithms applied in pollution monitoring.

1.3.5 Machine Learning and Remote Sensing Integration

In this section, we propose a framework to better understand the integration of remote sensing and machine learning for urban studies (see Figure 1.12). This framework enables to understand how and where machine learning and remote sensing are integrated to address a myriad of urbanisation issues

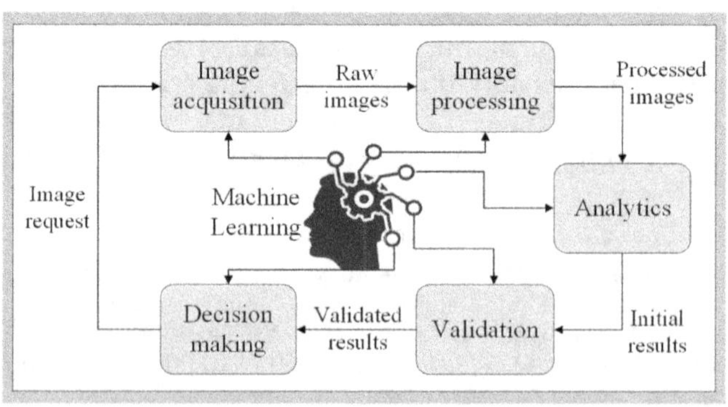

FIGURE 1.12
Machine learning and remote sensing integration framework for urban sustainability.

and sustainability challenges. This framework also allows researchers to compare urban studies and assess the main contributions or novelty of their work/s. The framework has been structured using a recursive loop approach, which effectively demonstrates the iterative nature of the integration process, highlighting the fact that the process of addressing targeted urban issues requires ongoing refinement until a satisfactory solution is achieved.

The proposed framework integrates machine learning and remote sensing through five components: image acquisition, image processing, image analytics, validation, and decision-making. It provides a robust structure for integrating machine learning and remote sensing technologies to address a wide range of urban issues, from land cover/use classification to disaster risk management. By leveraging the strengths of both technologies, the framework can help drive more effective and informed policymaking in various urban applications.

This framework starts with image acquisition. In this stage, machine learning algorithms assist satellites and unmanned aerial vehicles in capturing and downloading as per users' requests. Machine learning has proven pivotal in several aspects of Earth observation data acquisition, such as scheduling satellite tasks, image retrieval, and autonomous navigation of unmanned vehicles (Du et al., 2020; Golbabaei et al., 2020, 2021; Wang et al., 2019).

Image processing focuses on performing operations such as classification, detection, and enhancement on images to extract valuable information or improve their quality. The reviewed studies concentrated on operations like land cover/use classification, building type classification, local climate zone classification, building information detection, and traffic object detection. Some studies focused on enhancing image resolution, for example, Abunnasr and Mhawej (2022) and Chen et al. (2022) deployed multiple machine learning algorithms to downscale land surface temperature images.

Image analytics encompasses the application of descriptive, predictive, and diagnostic analytical techniques to extract insights from processed remote sensing images for understanding and forecasting urban issues. Descriptive analytics analyses real-time and historical data for future insights. Most studies utilised long-term remote sensing images, such as Landsat and nighttime light images, along with machine learning algorithms, to investigate historical evolutions and present status of urban issues, such as extreme heat, urban sprawl, and land cover/use change. Predictive analytics predicts the future evolution of urban issues (Lammers et al., 2021; Mohammad et al., 2022; Yan et al., 2021), and diagnostic analytics determines the potential reasons behind urban problems (Duncan et al., 2019; Stewart & Kremer, 2022; Wang et al., 2016).

Validation is accomplished by utilising credible and dependable datasets as a yardstick to verify research findings. Machine learning algorithms applied to remote sensing images can generate reference data that can corroborate research results. For instance, Priem et al. (2019) classified a 2 m resolution hyperspectral image dataset by employing an SVM, thereby creating ground truth data of high precision. Schilling et al. (2018) compared the outcomes of classification based on RF with those obtained from CNNs to assess the validity and high accuracy of their proposed approach. Many studies in urban areas, such as those by Demarchi et al. (2014), Shafizadeh-Moghadam et al. (2017), and Jamali (2019), have sought to carry out comparative experiments by selecting various machine learning algorithms to evaluate and validate their overall performance.

Finally, decision-making involves utilising the insights and information derived from image analytics to guide informed and effective actions, policy decisions, or automated responses in addressing various urban challenges. The framework involves semi-automatic and automated decision-making. Semi-automatic decision-making is a method for policymakers to make decisions with the support of reliable results achieved by combining machine learning and remote sensing. This kind of decision-making can benefit from all review studies. These studies not only disclose prominent urban issues but also provide validated reliable supporting datasets, such as land cover/use dataset, pollution map, and building information dataset. For instance, Abunnasr and Mhawej (2022) established systems like Ten-Surface Temperature-Google Earth Engine (Ten-ST-GEE) that can automatically retrieve 10-m land surface temperature products using multiple machine learning algorithms and images from Landsat 8 and Sentinel-2. Automated decision-making is a method operated by machine learning to make decisions according to systematically computed and analysed results from remote sensing data. This kind of decision-making is common in some emergency and disaster warning and response systems and unmanned delivery systems. However, there is not one reviewed article involving automated decision-making in its studies. Decision-making in the framework involves identifying locations with prominent issues, which can guide image capturing for long-term observation of urban problems. The reviewed studies revealed high-risk disaster areas, environmentally polluted regions, unreasonable land cover/use situations, and unreasonable land cover/use and distribution of infrastructures (Dong et al., 2022; Hrisko et al., 2021; Jaad & Abdelghany, 2021; Schlosser et al., 2020; Shogrkhodaei et al., 2021), which requires continuous image inputs for improvement. Persistent demand for imagery fosters a recursive integration framework, allowing for continuous improvement in other framework components.

1.4 Findings and Discussion

1.4.1 General Observations

The analysis of urban studies that integrate machine learning and remote sensing techniques reveals that 52% were conducted in Asia, with Europe and North America both having a nearly 20% representation. China, the USA, and Germany had the highest number of reviewed studies. A significant portion of the studies, approximately half, were published post-2020 and utilised ANNs, with a marked emphasis on deep learning algorithms. The superiority of aerospace technology in China, the USA, and Europe, marked by established satellite and sensor resources, provides these regions with a comparative advantage in terms of data accessibility (Avtar et al., 2020), which implies data scarcity might account for the dearth of studies from other regions. However, it is important to note that satellite data from Landsat, MODIS, and SPOT has been publicly accessible globally since the 1980s and 1990s (Patino & Duque, 2013), so there might be other causes.

The advancement in computer science, particularly in the domain of machine learning algorithms, is a crucial driver in the proliferation of machine learning and remote sensing applications in urban studies (Koumetio Tekouabou et al., 2022). Leading nations, such as the USA, Europe, and China, have achieved substantial progress in this field, thereby promoting their widespread use. The results underline the potential to formulate appropriate methodologies for data-scarce settings and to expand successful applications to regions with limited research initiatives (Casali et al., 2022). This can drive the advancement of methodologies and foster the integration of machine learning and remote sensing in urban studies. Regions experiencing rapid urbanisation and facing pressing challenges, such as population growth, environmental degradation, and resource constraints, such as South and Southeast Asia, South America, and Africa, are suggested as ideal candidates for future case studies.

The ethical considerations in remote sensing are just as important as the technical ones and must be addressed before implementing any study. With the advancement of technology, satellite and aerial images capture highly detailed information about our environment, including sensitive data such as personal locations and home addresses. This creates a conflict between technology and privacy, making it difficult to implement studies in some regions. In countries such as Iran, Morocco, and Nicaragua, unmanned aerial vehicles are not allowed due to privacy concerns. It is crucial to develop responsible strategies that avoid privacy invasion and enable the safe and ethical use of remote sensing techniques around the world.

1.4.2 Application Aeras

The literature review has identified five primary urban application domains integrating machine learning and remote sensing, which are land cover/use classification, disaster risk management, pollution monitoring, ecosystem, and environmental assessment, and built infrastructure identification. These domains focus on capturing and analysing physical characteristics and changes in the built and natural environments of cities and do not typically address socio-economic attributes. Although remote sensing data is effective in capturing morphological and spatial features, it is limited in capturing socio-economic information (Sadiq et al., 2022). However, the data achieved from these application domains can serve as a foundation for extended socio-economic application studies. For example, urban heat studies can provide surface temperature maps for heat vulnerability assessments, neighbourhood environment identification studies can provide neighbourhood context and pattern data for urban residential environment assessments, and built infrastructure maps can support equity assessments in social resource allocations (Bogawski et al., 2019; Das et al., 2021; Singleton et al., 2022).

Researchers have proposed the utilisation of multi-source social sensing data, such as statistical data, point of interests (POIs) data, and social media data, to supplement the current lack of socio-economic information. For instance, Chen et al. (2020) integrated POIs and multi-source remote sensing data to estimate anthropogenic heat emissions in energy consumption. These social sensing data contain a wealth of attributes that capture human activities, which are essential to understanding socio-economic characteristics in urban issues. The literature review includes numerous topics in urban sustainability and resilience, including climate change, urban growth, air quality monitoring, green and blue infrastructure, transportation, and resilience from natural disasters (Das et al., 2021; Deng et al., 2017; Ghaffarian et al., 2021; Huang et al., 2021; Jia & Zhang, 2021; Soilán et al., 2016). The resilience to natural disasters refers to the ability of urban areas to withstand, adapt, and recover from such events. This concept encompasses multiple perspectives, including infrastructure resilience, ecosystem resilience, and economic resilience, among others. However, there is a scarcity of interdisciplinary collaborations among these applications. The complexity of urban systems and the numerous issues they face require the integration of machine learning and remote sensing techniques to analyse and manage urban issues in a cross-domain manner (Shen et al., 2018; Xu et al., 2018a; Zhang et al., 2017). For example, Zhang et al. (2023) provided a comprehensive understanding of the impacts of different land use change scenarios on the region's ecosystem service value by integrating a mixed-cell cellular automata model, a system dynamics model, and remote sensing land use images. Additionally, the analysis of urban heat issues encompasses greenhouse gas monitoring, energy consumption, and the built environment, while its management requires collaboration among cooling infrastructure and service allocation, green energy promotion, and resilience enhancement. Resilience enhancement aims to address challenges caused by natural disasters, climate change, socio-economic, and political changes. As integration techniques continue to evolve, cross-field applications are expected to expand in future urban studies.

This chapter adopted Sustainable Development Goals (SDGs) established by the United Nations (2023) to discuss how application domains integrating machine learning and remote sensing contribute to urban sustainability. Land cover/use classification significantly contributes to the pursuit of sustainability, particularly within the context of SDG 11, which aims to foster sustainable cities and communities. By precisely identifying and quantifying diverse urban land attributes, this classification method facilitates sustainable urban planning and management. For instance, it supports the identification of slums, informal settlements, or inadequate housing (Indicator 11.1.1), enabling targeted interventions to enhance living conditions. Additionally, it informs the development of sustainable transportation networks by analysing public transport accessibility (Indicator 11.2.1), thus promoting environmental and social sustainability. Moreover, land cover/use classification assists in monitoring land consumption rates compared to population growth (Indicator 11.3.1), which is crucial for assessing urban development sustainability and optimising land use. Furthermore, it aids in quantifying and analysing public open space distribution within cities (Indicator 11.7.1), ensuring accessibility and inclusivity for all residents.

Disaster risk management is a critical component of urban sustainability, as it enables the development and implementation of effective strategies and interventions to address various disasters, such as urban heat, flooding, earthquakes, and typhoons. Several SDG indicators are closely related to it, which highlight its significance in achieving urban sustainability. For instance, SDG 1 (no poverty) emphasises the need to identify vulnerable populations and improve disaster response efforts in urban areas prone to various disasters. Indicators 1.5.1 and 1.5.2 specifically help monitor the impact of disasters on communities and the economy, highlighting the importance of investing in disaster risk reduction strategies and resilient infrastructure. Moreover, SDG 11 (sustainable cities and communities) underscores the significance of effective disaster risk management in urban areas facing multiple disaster risks. Indicators 11.5.1 and 11.5.2 monitor the impact of disasters on urban populations and infrastructure, encouraging the development of comprehensive disaster risk reduction strategies and resilient infrastructure. Additionally, Indicators 11.b.1 and 11.b.2 promote the adoption and implementation of national and local disaster risk reduction strategies, respectively, specifically tailored to urban areas. SDG 13 (climate action) also highlights the importance of enhancing resilience and adaptive capacity to climate-related hazards, which can exacerbate disaster risks, such as heatwaves, flooding, and extreme weather events.

Pollution monitoring, including air, water, soil, and light pollution, is critical for achieving urban sustainability and improving residents' health by identifying areas that need targeted interventions to improve air and water quality and reduce waste. Effective monitoring of urban pollution is closely related to several SDG indicators. For instance, SDG 3 (good health and well-being) indicates the need for air pollution management. Indicator 3.9.1 Mortality rate attributed to household and ambient air pollution highlights the need for air pollution management in urban areas. Indicator 3.4.2 Mortality rate attributed to unsafe water, unsafe sanitation, and lack of hygiene emphasises the need for clean water and sanitation, and water pollution monitoring can help identify areas with unsafe water sources. Similarly, SDG (clean water and sanitation) emphasises the need for effective water and wastewater management. Indicators 6.3.1 and 6.3.2 highlight the need for effective wastewater and water quality management in urban areas, respectively. The importance of effective waste management in urban areas is highlighted by

Indicator 11.6.1. Additionally, SDG 12 (responsible consumption and production) highlights the need for effective hazardous waste management. Indicators 11.6.2 and 12.4.2 emphasise the importance of monitoring air and hazardous waste pollution, respectively, to reduce related health impacts and protect the environment.

Built infrastructure identification, encompassing building detection and traffic object detection, plays a crucial role in advancing urban sustainability by addressing specific SDG indicators. For example, it is consistent with SDG 9 (industry, innovation, and infrastructure) as it supports the development of resilient, inclusive, and sustainable infrastructure. Road centreline and traffic object detection facilitates urban connectivity and accessibility (Indicator 9.1.1) and fosters sustainable development by providing valuable insights for transportation network planning. Additionally, it promotes sustainable industries (Indicator 9.4.1) by optimising transportation, thus reducing CO_2 emissions. Enhanced public transportation planning (Indicator 11.2.1) ensures accessibility for diverse populations, contributing to social equity. By providing detailed infrastructure data, it encourages civil society participation in urban planning (Indicator 11.3.2) and aids in monitoring public spaces (Indicator 11.7.1), fostering inclusive and sustainable urban environments. Furthermore, accurate infrastructure identification supports resilience to climate-related hazards (Indicator 13.1.1) and informs climate change adaptation and mitigation strategies (Indicator 13.2.1), driving transparent reporting on climate action efforts.

Ecosystem and environmental assessments are essential in promoting urban sustainability. These assessments enable sustainable urban planning by examining ecosystem services and environmental resource distribution, thereby preserving natural ecosystems and efficiently allocating resources (Indicator 11.3.1). Additionally, they facilitate waste management by identifying suitable locations for waste disposal and recycling, consequently minimising environmental impact (Indicator 11.6.1). These assessments also contribute to creating healthier urban environments by mapping green and blue infrastructure, promoting public spaces, and fostering community wellbeing (Indicator 11.7.1). In terms of climate action, local climate zone mapping helps identify vulnerable areas to climate change, improve urban resilience, and reduce disaster risks (Indicator 13.1.1), while carbon sequestration and emission estimation monitor and reduce greenhouse gas emissions (Indicator 13.2.1). Furthermore, green and blue infrastructure mapping supports reforestation and sustainable forest management (Indicators 15.1.1 and 15.2.1, SDG 15: life on land), monitors water-related ecosystems (Indicator 6.6.1), and contributes to improved air quality and public health (Indicator 3.9.1).

1.4.3 Remote Sensing Data and Techniques

The utilisation of satellite imagery has become increasingly prevalent in various urban applications. Among the most frequently employed images are those captured by the Landsat, MODIS, and Sentinel satellites. Aerial images have experienced a significant surge in popularity since 2015 and have been widely employed across various domains within urban environments. The utilisation of LiDAR, a type of land imaging data, has proven to be particularly influential in the identification of built infrastructure due to its ability to capture 3D information. Our analysis of the remote sensing data was focused on evaluating their availability, quality, and costs. All these data types are readily accessible through official channels. Landsat has been providing long-time series satellite images and processed products for free since the 1980s. Similarly, MODIS data products, which encompass fields such as the land, oceans, and atmosphere, have been freely available since 1999, including datasets of land surface temperature, sea ice, and aerosols. The Sentinel satellites, launched by the European Space Agency in 2014, serve as a means of global environment and security monitoring and have been employed in various applications such as land management, monitoring of natural disasters, and assistance in humanitarian operations. These satellite data offer several advantages, including global observations, free accessibility, high spatiotemporal resolution, multispectral imaging, and a wide range of derivative products (Sannigrahi et al., 2019; Wu et al., 2022; Yan et al., 2021). Meanwhile, the advent of new higher-resolution satellites/sensors, particularly because of the proliferation of small satellites such as ICEYE and CubeSat, has opened exciting prospects for more detailed and fine-grained analysis.

The widespread popularity of drones has made the capture of aerial images more easily attainable both in terms of cost and time (Liao & Wen, 2020; Wang et al., 2022). This has enabled individuals and organisations to access aerial images for academic or business purposes with greater ease. Additionally, numerous commercial companies offer aerial data services and historical images to the public. The advancements in airborne camera technology have enabled aerial images to provide multispectral and even hyperspectral information with centimetre-level precision. This improvement in spatial resolution has elevated urban studies to a new level of accuracy and detail. For example, mapping land covers/uses, land surface temperatures, and building footprints at a centimetre level are now possible (Frank et al., 2017; Liuzzi et al., 2019; Zhang et al., 2018). These fundamental maps are critical for modern urban analysis and planning that demands high reliability and accuracy. LiDAR, with its ability to provide reliable detection and ranging over short to long distances, has become a popular method for calculating accurate geospatial measurements. This data can be collected by satellites, aerial vehicles, or even cars in both academic and business settings. LiDAR techniques allow for quick and accurate data collection and the creation of virtual 3D models of the real world, making it a valuable complement to satellite and aerial images (Huang et al., 2019b; Mahdianpari et al., 2021). However, LiDAR is more expensive than aerial images due to the high requirements for equipment and software, which can limit its widespread application. It is important to note that remote sensing data, including satellite, aerial, and LiDAR images, is a type of geospatial data and does not necessarily depict the socio-economic characteristics of urban phenomena. Future urban studies must therefore incorporate social sensing data (Chen et al., 2020; Sadiq et al., 2022), such as social media data, POIs, and demographic data, into research designs to gain a more comprehensive understanding of urban environments.

1.4.4 Machine Learning Data and Techniques

The dominance of certain machine learning algorithms in specific urban application domains can be attributed to various factors. One of the most significant factors is the outstanding capabilities of machine learning algorithms. For instance, CNNs are especially effective in processing visual data, such as satellite imagery, and are therefore extensively used in domains like land cover/use classification (Singleton et al., 2022; Wurm et al., 2019), where visual features play a crucial role in classification. Additionally, the popularity of certain algorithms may be due to their demonstrated performance and accuracy in previous applications. For example, RF and SVM have been extensively used in various machine learning applications and have proven to be effective in many scenarios (Aissou et al., 2021; Deng et al., 2017; Milnar & Ramaswami, 2020; Shi et al., 2021). Therefore, researchers might opt for these popular algorithms as a default option for their own research. Lastly, the availability of tools and libraries that implement these algorithms may also influence their usage (Abunnasr & Mhawej, 2022). In sum, several factors like algorithm capabilities, previous performance, and tool availability play a critical role in determining the dominance of specific machine learning algorithms in urban application domains.

Over 90% of the reviewed articles employed supervised machine learning techniques to address various challenges. The RF and SVM algorithms were the most widely used in traditional supervised learning (Belgiu & Drăguţ, 2016; Griffiths et al., 2010), while deep learning, specifically CNNs, was utilised to analyse urban issues and provide efficient solutions. The prevalence of supervised algorithms in reviewed urban studies is due to the fundamental distinction from unsupervised algorithms, as supervised algorithms can leverage labelled datasets, whereas unsupervised algorithms do not (Demarchi et al., 2014; Jamali, 2019). Labelled datasets which contain pre-collected pairs of inputs and corresponding outputs are used to explicitly train supervised learning models to learn the underlying mapping function from the inputs to the expected outputs. When a large number of labelled datasets exist, supervised learning models are able to achieve high performance and good generalisation to new unseen data, which makes them popular in urban studies (Jamali, 2019). Supervised learning algorithms also provide accurate and more interpretable results (Casali et al., 2022), which are essential in urban studies to support decision-making processes. With the growing availability of open-source labelled datasets and pre-trained models, more researchers tend to apply supervised machine learning in their studies. Supervised algorithms employ labelled data in two domains: classification and regression. Classification algorithms are trained with labelled data to classify inputs into categories, while regression algorithms gradually fit models to labelled data to estimate continuous values of outputs (Hartling et al., 2021; Jato-Espino et al., 2022). These two capabilities align with the requirements of reviewed urban studies utilising remote sensing data, thus contributing to their popularity.

Supervised machine learning has gained popularity in urban studies due to its capacity to handle abundant, diverse data sources and model complex interactions (Casali et al., 2022; Jamali, 2019). Its suitability for predictive modelling, continuous improvement, and scalability facilitates analysing data across various urban contexts (Casali et al., 2022). Furthermore, it supports decision-making by providing actionable insights for stakeholders and boasts interdisciplinary applicability (Koumetio Tekouabou et al., 2022). The utilisation of supervised machine learning in urban studies has enabled researchers to better understand and address challenges related to urban development, infrastructure, and policy interventions. However, it requires large-labelled datasets that can be time-consuming, intensive, and expensive to label (Alanne & Sierla, 2022; Jamali, 2019). This is where unsupervised learning comes in. Unsupervised learning is a machine learning approach that doesn't require labels for data (Demarchi et al., 2014). It discovers patterns and structures in data without any external guidance. Unsupervised learning has the potential to help address the need for large-labelled datasets, especially for rich datasets that don't have labels. By utilising unsupervised learning, we can extract valuable insights from data without the need for labelling, which can save time, money, and resources. Although unsupervised learning is not yet as widely used as supervised learning, it has been gaining momentum in recent years. With the increasing availability of data and advancements in unsupervised learning algorithms, we can expect to see more work in urban field. For instance, capitalising on its capacity to explore and identify undiscovered patterns or connections within urban data (Koumetio Tekouabou et al., 2022), unsupervised machine learning could be employed for the examination of urban morphology, the identification of gentrification, and the uncovering of urban mobility trends. Unlike supervised machine learning, which depends on predefined categories, unsupervised machine learning operates without such constraints, potentially yielding a more comprehensive understanding of the urban environment.

Nevertheless, it is notable that most reviewed articles lack clear explanations regarding the selection of machine learning algorithms, and there are limited systematic comparisons between different algorithms. Further efforts are necessary to enhance the interpretability of algorithms prior to case studies, which would benefit both the introduction of machine learning algorithms and the extension of their application in the urban domain (Casali et al., 2022), such as the hot topic called eXplainable AI (XAI). Land cover/use classification is the most common application domain of machine learning algorithms, while the utilisation of algorithms in built infrastructure is limited. Efforts are required to introduce successful algorithms from one domain to another within the urban field.

1.4.5 Machine Learning and Remote Sensing Integration

In recent years, numerous studies have been conducted to explore the application of remote sensing images and machine learning methods to optimise traditional models in urban studies. Traditional models have been useful for urban planning and management; however, they often rely on static, outdated, or incomplete data, limiting their accuracy and adaptability (Touzani & Granderson, 2021; Zhang et al., 2018). By integrating remote sensing images with machine learning techniques, researchers have been able to create dynamic, up-to-date, and comprehensive models that can better grasp changes and developments in urban areas (Kwak et al., 2020; Wu et al., 2022).

For instance, machine learning algorithms such as SVM, RF, and CNNs have been employed in conjunction with remote sensing images to classify land use and land cover, as well as detect changes in urban areas (Jamali, 2019; Shafizadeh-Moghadam et al., 2017; Wieland & Pittore, 2016). These approaches have demonstrated improved accuracy and efficiency compared to traditional methods, enabling more effective monitoring of urban growth, residential environments, and other essential urban elements. Furthermore, in the context of urban climate and environmental studies, the integration of remote sensing and machine learning has facilitated a better understanding of urban heat islands, air quality, and other environmental factors that directly affect urban residents' well-being (Jato-Espino et al., 2022; Jia & Zhang, 2021; Zhang et al., 2022). By using these advanced models, urban planners and policymakers can make more informed decisions regarding sustainable urban development and environmental protection.

Despite the progress made thus far, there is still ample room for further research and development in the integration of remote sensing and machine learning to optimise traditional models in urban studies. As technology and algorithms continue to evolve, and as more comprehensive and high-resolution remote sensing datasets become available, the potential for significant advancements in urban planning, management, and sustainability will only continue to grow. In this context, the proposed machine learning and remote sensing integration framework can provide a guideline for future directions with a comprehensive analysis of current integration.

The integration of machine learning and remote sensing in the field of urban studies has brought high automation and thorough understanding and analysis in the image processing and analytics processes. This is due to the combination of high-quality remote sensing images and machine learning algorithms with exceptional computational capacities. The advancement in remote sensing techniques has resulted in the availability of high-resolution remote sensing images that capture a wealth of spatial characteristics

(Patino & Duque, 2013). Image processing is focused on automatically classifying and detecting information that is useful for urban analysis. As the computational capacity of algorithms continues to improve, machine learning is increasingly being applied in urban studies to automate the analysis of causes, evolutions, and strategies of urban issues (Koumetio Tekouabou et al., 2022). For example, the analytics process leverages extracted information to predict changes in urban features, identify potential drivers of urban issues, and evaluate potential solutions.

The process of image acquisition and decision-making was traditionally performed manually, with support from machine learning and remote sensing techniques. In the reviewed literature, the emphasis was not on automatic image acquisition, as manual operations were sufficient for case studies that involved a limited amount of data. Researchers could manage the selection of data sources and screening of data (Chen et al., 2020c; Correa Martins et al., 2021). However, when the results and methods are extended to a larger geographical area, there may be a need to implement automatic remote sensing image acquisition using machine learning (Abunnasr & Mhawej, 2022). The limited adoption of machine learning for decision-making in the reviewed articles could be attributed to the inherent risks associated with such systems, including algorithmic bias, unequal distribution of benefits and access, cascading failures, external disruptions, and trade-offs between efficiency and resilience (Galaz et al., 2021).

Urban studies are complex compositions that encompass various components, including social, economic, climate, environmental, and cultural aspects. It is challenging to quantify indices from these aspects, which are critical for informed decision-making (Koumetio Tekouabou et al., 2022). The absence of adequate quantified parameters hinders, to some extent, the application of machine learning in automatic decision-making. Therefore, more research is needed to enhance the automation level of the image acquisition and decision-making processes. Swift and targeted image acquisition and decision support can promote the efficiency of addressing urban challenges. Maintaining a high level of automation in each process of the integration framework can lead to the development of a time- and resource-efficient advanced spatial analytics system for urban sustainability. In the image acquisition process, reinforcement learning algorithms can be utilised for task planning and scheduling of satellites, autonomous navigation of unmanned aerial vehicles, and automatic retrieval and downloading of remote sensing datasets (Du et al., 2020; Wang et al., 2019). For the decision-making process, remote sensing and machine learning can be leveraged to simulate the effectiveness of potential solutions in addressing urban issues (Koumetio Tekouabou et al., 2022).

We have discussed how application domains integrating machine learning and remote sensing contribute to urban sustainability in line with SDGs. Furthermore, the integration of machine learning and remote sensing technologies also contributes significantly to urban sustainability compared to traditional models in three key aspects: enhanced precision and accuracy (Lammers et al., 2021; Xu et al., 2018c; Yan et al., 2021), real-time monitoring capabilities (Peng et al., 2019; Yan et al., 2020; Zhao et al., 2021), and improved adaptability to dynamic urban environments (Janowski et al., 2021; Kwak et al., 2020; Liu et al., 2021). Firstly, machine learning algorithms, when combined with high-resolution remote sensing images, provide superior precision and accuracy in identifying and quantifying urban land attributes, environmental factors, and infrastructure (Chen et al., 2021b; Han et al., 2020; Xu et al., 2018c). This enables informed decision-making and effective interventions in areas such as land use planning, pollution management, and resource allocation, ultimately promoting sustainable urban development and reducing the environmental impact of urbanisation.

Secondly, the real-time monitoring capabilities of machine learning and remote sensing technologies provide a substantial advantage over traditional models, which often rely on static, outdated, or incomplete data. The ability to continuously monitor urban environments enables urban planners and policymakers to detect and respond to changes more effectively, such as population growth (Zhao et al., 2021), air conditions (Yan et al., 2020), and disaster impact (Peng et al., 2019), and consequently develop relevant policy actions (Arbolino et al., 2018). This real-time information facilitates proactive and timely interventions, ensuring urban growth remains sustainable and adaptable to emerging challenges.

Lastly, the combination of machine learning and remote sensing technologies offers improved adaptability to the ever-changing nature of urban environments (Kwak et al., 2020). Traditional models may struggle to accommodate rapid urban growth, technological advancements, or evolving environmental conditions. However, the integration of remote sensing and machine learning allows for dynamic, up-to-date, and comprehensive models that can adapt to these changes and provide accurate insights for urban planning and management. This adaptability ensures that urban development remains sustainable and resilient, even in the face of complex and evolving challenges (Sannigrahi et al., 2019).

1.5 Conclusion

This chapter reviewed the integration of machine learning and remote sensing in urban studies for urban sustainability. It employed the PRISMA approach to analyse the joint utilisation of machine learning and remote sensing in urban studies. The chapter proposed a framework to better understand the benefits of advanced spatial analytics through the integration of machine learning and remote sensing. The framework is analysed from five perspectives: application areas, remote sensing and machine learning data and techniques, and machine learning and remote sensing integration. The major findings and their implications are summarised below.

Firstly, most of the reviewed urban studies (over 90%) were conducted in Asia, Europe, and North America, with China, the USA, and Europe having most of the articles. Technical and ethical factors contribute to this concentration. Machine learning and remote sensing provide technical advantages in these regions, and ethical restrictions limit data acquisition in data-scarce regions. This highlights the need for responsible strategies to develop suitable methodologies for data-sparse environments and extend successful applications while avoiding privacy invasion.

Secondly, the reviewed urban studies focused on physical characteristics, ignoring society and economy. This is due to remote sensing data being better at capturing physical features, not humanistic ones. The chapter suggests supplementing with multi-source social sensing data to address this gap. The reviewed articles have covered many aspects of urban sustainability and resilience, but future cross-domain research is needed for a comprehensive analysis and targeted policymaking.

Thirdly, the most used satellite images are Landsat, MODIS, and Sentinel. Aerial images have become widespread since 2015 in urban applications. LiDAR point cloud data dominates infrastructure identification due to its excellent capacity for capturing 3D information. The availability, high quality, and low cost make remote sensing data highly popular. Remote sensing provides not only preprocessed images but also various derivative products for the public. It holds promise for incorporating social sensing data, which reflects socio-economic characteristics, to analyse and investigate urban issues.

Fourthly, over 90% of reviewed articles employed supervised machine learning algorithms, which excel at classification and regression but rely on labelled datasets for algorithm training. Unsupervised machine learning has the potential to help address the need for a large number of labelled datasets and is expected to be extensively applied for exploring and identifying undiscovered patterns or connections in urban areas. Moreover, a paradigm for selecting approaches is urgently needed, as many of the reviewed studies lacked clear explanations for algorithm selection and extension.

Finally, so far, the integration in urban studies remains highly automated in image processing and analytics procedures, which provides a better understanding and analysis of urban issues. It benefits from the improvements in both remote sensing image quality and machine learning computational capacity. To meet potential application demands, there is a need to improve automation in the image acquisition process. We suggest the extensive application of reinforcement learning in task scheduling, autonomous navigation, and data retrieval and downloading. Despite the difficulties in quantifying parameters and algorithm bias, we advocate for further attempts to analyse and investigate urban policy feasibility and reliability through the joint use of machine learning and remote sensing techniques.

A systematic review of literature provides valuable insights into the integration of machine learning and remote sensing in urban studies for urban sustainability. However, the focus on current urban research may limit the coverage of all possible machine learning and remote sensing data and techniques, and urban application domains. As a result, the findings from the review may not be comprehensive enough to inform all possible applications of machine learning and remote sensing integration in urban studies. This limitation could impact the selection of appropriate data and techniques for addressing urban issues effectively.

To address this limitation, it is essential to explore machine learning and remote sensing integration in other research fields, such as environmental monitoring, public health, and agriculture, to gain a more comprehensive understanding of the benefits and challenges associated with the integration of machine learning and remote sensing in various domains. Ultimately, this knowledge can inform urban studies and promote sustainable urban development. The insights gained from exploring other research fields can help researchers and practitioners select appropriate data and techniques for urban studies, leading to more informed decisions for sustainable urban development.

In addition to exploring more possibilities in other research fields, this study intends to further refine and test the framework for understanding the benefits of advanced spatial analytics through the integration of machine learning and remote sensing in empirical research. Additionally, the framework could be applied to different application areas, such as transportation planning, urban ecology, and disaster management, to examine the potential benefits of machine learning and remote sensing integration in these domains. Lastly, future research could focus on developing more advanced machine learning and remote sensing techniques to enhance the accuracy and efficiency of data analysis in urban studies.

This study provides useful guidelines and a framework for researchers and policymakers in using remote sensing data, machine learning approaches, and their integration in future urban studies. The key contribution of this study is to show how to integrate machine learning and remote sensing techniques to promote urban sustainability in an explicable and efficient manner.

Acknowledgements

This chapter, with permission from the copyright holder, is a reproduced version of the following journal article: Li, F., Yigitcanlar, T., Nepal, M., Nguyen, K., & Dur, F. (2023). Machine learning and remote sensing integration for leveraging urban sustainability: A review and framework. *Sustainable Cities and Society, 96*(1), 104653.

Appendix

TABLE A1.1

Analysis Highlights of the Reviewed Literature

Author	Journal	Title	Aim	Region	Domain	Sub-Domain	Machine Learning Technique	Remote Sensing Data
Khan and Sudheer (2022)	*The Egyptian Journal of Remote Sensing and Space Sciences*	"Machine learning-based monitoring and modelling for spatio-temporal urban growth of Islamabad."	To predict Islamabad's urban growth and inform policymakers of the urbanisation trend.	Islamabad, Pakistanw	Land cover/use classification	Urban area extraction	Artificial neural network (ANN)–based cellular automata (CA)	Landsat 5, 8
Singleton et al. (2022)	*Computers, Environment and Urban Systems*	"Estimating generalized measures of local neighbourhood context from multispectral satellite images using a convolutional neural network."	To extract urban context from multispectral satellite imagery with a convolutional autoencoder.	Great Britain, UK	Land cover/use classification	Neighbourhood environment identification	Convolutional neural network (CNN)	Sentinel-2
Xu et al. (2022)	*IEEE Transactions on Intelligent Transportation Systems*	"MM-UrbanFAC: Urban Functional Area Classification Model Based on Multimodal Machine Learning."	To provide a multimodal machine learning–based classification model for urban functional areas.	Not mentioned	Land cover/use classification	Land cover/use distribution identification	Dual-branch neural network combining SE-ResNeXt and Dual Path Network (DPN), Gradient Boosting Decision Tree (GBDT), and light gradient boosting machine (LightGBM)	IKCEST

(Continued)

TABLE A1.1 *(Continued)*

Analysis Highlights of the Reviewed Literature

Author	Journal	Title	Aim	Region	Domain	Sub-Domain	Machine Learning Technique	Remote Sensing Data
Zhu et al. (2022)	*IEEE Transactions on Geoscience and Remote Sensing*	"A Weakly Pseudo-Supervised Decorrelated Subdomain Adaptation Framework for Cross-Domain Land-Use Classification."	To develop a comprehensive approach to scene-level land use classification in remote sensing images with high spatial resolutions.	Wuhan and Shanghai, China	Land cover/use classification	Land cover/use distribution identification	CNN	Aerial images and remote sensing scene images
Chen et al. (2021b)	*ISPRS Journal of Photogrammetry and Remote Sensing*	"Mapping essential urban land use categories with open big data: Results for five metropolitan areas in the United States of America."	To investigate urban land use classification in five US metropolitan areas and propose a cost-effective framework for accurate results.	San Francisco Bay Area, Denver, New Orleans, Chicago, and New York City, USA	Land cover/use classification	Land cover/use distribution identification	Random forest (RF), categorical boosting (CatBoost), LightGBM, and ANN	Sentinel-1, 2 and NPP/VIIRS
Garg et al. (2021)	*Scientific Reports*	"Semantic segmentation of PolSAR image data using advanced deep learning model."	To develop a deep learning solution to reduce misclassification of vegetated urban targets and release a new LULC dataset.	Houston, USA	Land cover/use classification	Land cover/use distribution identification	RF, K-Nearest Neighbour (KNN), support vector machine (SVM), and DeepLabV3+	Aerial images
Jaad and Abdelghany (2021)	*Cities*	"The story of five MENA cities: Urban growth prediction modelling using remote sensing and video analytics."	To predict urban growth with limited data by using machine learning-based framework integrating remote sensing and computer vision.	Dubai, UAE; Cairo, Egypt; Doha, Qatar; Casablanca, Morocco; and Riyadh, KSA	Land cover/use classification	Urban area extraction	CNN	Landsat 5, 7, 8

Huang et al. (2019a)	*Journal of Visual Communication and Image Representation*	"Automatic extraction of impervious surfaces from high resolution remote sensing images based on deep learning."	To develop a deep learning model to extract impervious surfaces from high-resolution remote sensing images and reduce spectral interference and shadows.	Beijing and Ningxia, China	Land cover/use classification	Urban area extraction	CNN, Fuzzy C-Means (FCM) clustering	WorldView-2 and Pléiades-1A
Huang et al. (2019b)	*Journal of Visual Communication and Image Representation*	"Automatic extraction of urban impervious surfaces based on deep learning and multi-source remote sensing data."	To develop an automated impervious surface extraction method using deep learning and multi-source remote sensing data.	Ningxia, China	Land cover/use classification	Urban area extraction	FCM, CNN, Deep Belief Network (DBN), and extreme learning machine (ELM)	Pléiades-1A, Landsat 8, and LiDAR
Jamali (2019)	*SN Applied Sciences*	"Evaluation and comparison of eight machine learning models in land use/land cover mapping using Landsat 8 OLI: A case study of the northern region of Iran."	To assess the effectiveness of eight machine learning algorithms for image classification using WEKA and R programming languages.	Sari, Iran	Land cover/use classification	Land cover/use distribution identification	RF, decision table, DTNB, J48, Lazy IBK, multilayer perceptron (MLP), non-nested generalised exemplars (NN ge), and simple logistic	Landsat 8
Priem et al. (2019)	*International Journal of Applied Earth Observation and Geoinformation*	"Comparing map-based and library-based training approaches for urban land-cover fraction mapping from Sentinel-2 imagery."	To assess the effectiveness of a model building scheme for VIS fraction mapping on Sentinel-2 using a hyperspectral library.	Brussels, Belgium	Land cover/use classification	Land cover/use distribution identification	SVM	Sentinel-2
Wurm et al. (2019)	*ISPRS Journal of Photogrammetry and Remote Sensing*	"Semantic segmentation of slums in satellite images using transfer learning on fully convolutional neural networks."	To investigate the transfer learning potential of FCNs for slum mapping across different satellite images.	Mumbai, India	Land cover/use classification	Neighbourhood environment identification	Fully convolutional network (FCN)	QuickBird and Sentinel-2

(*Continued*)

TABLE A1.1 *(Continued)*

Analysis Highlights of the Reviewed Literature

Author	Journal	Title	Aim	Region	Domain	Sub-Domain	Machine Learning Technique	Remote Sensing Data
Azadbakht et al. (2018)	*International Journal of Applied Earth Observation and Geoinformation*	"Synergy of sampling techniques and ensemble classifiers for classification of urban environments using full-waveform LiDAR data."	To suggest a sampling technique synergy with ensemble classifiers to tackle data imbalance issues in training datasets for data mining.	Woodridge and Stretton, Australia	Land cover/use classification	Land cover/use distribution identification	Ensemble classifiers with bagging, RF and boosting	LiDAR
Zhang et al. (2018a)	*ISPRS Journal of Photogrammetry and Remote Sensing*	"A hybrid MLP-CNN classifier for very fine resolution remotely sensed image classification."	To combine CNN and MLP with a rule-based decision fusion approach for classifying very high-resolution remotely sensed imagery.	Southampton, UK	Land cover/use classification	Land cover/use distribution identification	MLP–CNN	Aerial images
Deng et al. (2017)	*ISPRS Journal of Photogrammetry and Remote Sensing*	"Subpixel urban impervious surface mapping: The impact of input Landsat images."	To explore the impact of atmospheric correction and seasonality on urban impervious surface estimation using Landsat imagery.	Binghamton and Los Angeles, USA; San Juan, Puerto Rico	Land cover/use classification	Urban area extraction	RF	Landsat 7, 8; aerial images, and Google Earth images
Shafizadeh-Moghadam et al. (2017)	*Computers, Environment and Urban Systems*	"Coupling machine learning, tree-based and statistical models with cellular automata to simulate urban growth."	To model urban growth in the Tehran Metropolitan Area using six distinct models.	Tehran, Iran	Land cover/use classification	Urban area extraction	ANN, SVM, RF, classification and regression trees (CART), logistic regression (LR), and multivariate adaptive regression splines (MARS)	Landsat 5, 7, 8

Karalas et al. (2016)	*IEEE Transactions on Geoscience and Remote Sensing*	"Land Classification Using Remotely Sensed Data: Going Multilabel."	To obtain an up-to-date high-resolution description of land cover by applying the multilabel classification.	New York, USA	Land cover/use classification	Land cover/use distribution identification	Binary relevance (BR), multilabel KNN (ML-KNN), ensemble of classifier chains (ECC), label powerset (LP), instance-based logistic regression (IBLR), random K-labelsets (RAKEL)	MODIS
Wang et al. (2016)	*Agriculture, Ecosystems & Environment*	"Identification and apportionment of the drivers of land use change on a regional scale: Unbiased recursive partitioning-based stochastic model application."	To propose a new framework using remote sensing, GIS, and machine learning identifies factors responsible for regional land use change.	Jiangxi, China	Land cover/use classification	Land cover/use change identification	Conditional inference tree (CIT) and RF	Not mentioned
Wieland and Pittore (2016)	*ISPRS Journal of Photogrammetry and Remote Sensing*	"Large-area settlement pattern recognition from Landsat-8 data."	To generate a multilayered settlement product from Landsat 8 data for a vast region in Central Asia.	Southern Kazakhstan, Kyrgyzstan, Tajikistan and Eastern Uzbekistan, Asia	Land cover/use classification	Neighbourhood environment identification	SVM	Landsat 8
Osaku et al. (2015)	*Information Sciences*	"Improving land cover classification through contextual-based optimum-path forest."	To improve land cover image classification using a contextual approach called OPF–MRF.	Itatinga, Duque de Caxias	Land cover/use classification	Land cover/use distribution identification	Optimum-path forest, SVM	CBERS-2B, Landsat 5, IKONOS-2 MS, and GeoEye
Demarchi et al. (2014)	*ISPRS Journal of Photogrammetry and Remote Sensing*	"Assessing the performance of two unsupervised dimensionality reduction techniques on hyperspectral APEX data for high resolution urban land-cover mapping."	To evaluate the efficacy of AANN and BandClust in large-scale hyperspectral urban land-cover mapping.	Brussels, Belgium	Land cover/use classification	Land cover/use distribution identification	RF, AdaBoost (ADB), MLP, and SVM.	Aerial images

(Continued)

TABLE A1.1 *(Continued)*

Analysis Highlights of the Reviewed Literature

Author	Journal	Title	Aim	Region	Domain	Sub-Domain	Machine Learning Technique	Remote Sensing Data
Deng and Wu (2013)	*ISPRS Journal of Photogrammetry and Remote Sensing*	"The use of single-date MODIS imagery for estimating large-scale urban impervious surface fraction with spectral mixture analysis and machine learning techniques."	To obtain SMA end member spectra via a least-squares solution and compare SMA to machine learning techniques.	Ohio and Virginia, USA	Land cover/use classification	Urban area extraction	Cubist regression tree and RF	MODIS
Schneider (2012)	*Remote Sensing of Environment*	"Monitoring land cover change in urban and peri-urban areas using dense time stacks of Landsat satellite data and a data mining approach."	To develop multi-date change detection technique for complex urban environments.	Chengdu, Xi'an, and Kunming, China	Land cover/use classification	Land cover/use change identification	Maximum likelihood classification (MLC), boosted decision trees, and SVM	Landsat 5, 7
Griffiths et al. (2010)	*Remote Sensing of Environment*	"Mapping megacity growth with multi-sensor data."	To enhance the mapping of landscape-level urban growth and land use conversion in extended megacity regions.	Dhaka, Bangladesh	Land cover/use classification	Urban area extraction	SVM	Landsat 5, 7, and European Remote Sensing Satellite (ERS) 1 and 2
Abunnasr and Mhawej (2022)	*Environmental Modelling & Software*	"Towards a combined Landsat-8 and Sentinel-2 for 10-m land surface temperature products: The Google Earth Engine monthly Ten-ST-GEE system."	To introduce Ten-ST-GEE, a 10-m LST system based on Landsat 8 and Sentinel-2 bands and ML, using Google Earth Engine.	Two agricultural lands and two urban agglo-merations in both the USA and in Lebanon	Disaster risk management	Urban heat	Ordinary least square (OLS), robust least squares (RLS), DisTrad, RF, and SVM	Landsat 8 and Sentinel-2

Ali et al. (2022)	*Remote Sensing*	"Assessment Analysis of Flood Susceptibility in Tropical Desert Area: A Case Study of Yemen."	To identify flood-prone areas and investigate the factors responsible for flooding in Tarim city.	Tarim, Yemen	Disaster risk management	Flooding	KNN, Naïve Bayes (NB), RF, and eXtreme Gradient Boosting (XGBoost)	Sentinel-1
Chen et al. (2022b)	*Atmospheric Measurement Techniques*	"A high-resolution monitoring approach of canopy urban heat island using a random forest model and multi-platform observations."	To analyse the pattern and causes of CUHIs with RF model and local environment and background weather data.	Nanjing, China	Disaster risk management	Urban heat	RF	Landsat 8
Chen et al. (2022a)	*International Journal of Applied Earth Observation and Geoinformation*	"DeepUrbanDownscale: A physics informed deep learningframeworkforhigh-resolution urban surface temperature estimation via 3D point clouds."	To estimate high-resolution urban surface temperature by using DUD.	Zhangzhou Harbor, China	Disaster risk management	Urban heat	Physics-informed neural network (PINN)	Landsat 8
Jato-Espino et al. (2022)	*Urban Climate*	"ArcUHI: A GIS add-in for automated modelling of the Urban Heat Island effect through machine learning."	To develop a highly applicable and interpretable ArcGIS Pro add-in (ArcUHI) for automated modelling of the UHI effect.	Madrid, Spain	Disaster risk management	Urban heat	Multiple linear regression (MLR), SVM and RF	MODIS
Kafy et al. (2022)	*Building and Environment*	"Predicting the impacts of land use/land cover changes on seasonal urban thermal characteristics using machine learning algorithms."	To evaluate thermal characteristics of Bangladesh by assessing and predicting seasonal UTFVI patterns.	Sylhet City, Bangladesh	Disaster risk management	Urban heat	ANN and CA	Landsat 4, 5, 8

(Continued)

TABLE A1.1 *(Continued)*

Analysis Highlights of the Reviewed Literature

Author	Journal	Title	Aim	Region	Domain	Sub-Domain	Machine Learning Technique	Remote Sensing Data
Mohammad et al. (2022)	*Urban Climate*	"Machine learning algorithm based prediction of land use land cover and land surface temperature changes to characterize the surface urban heat island phenomena over Ahmedabad city, India."	To investigate and predict the spatial and temporal distribution of change in LULC, seasonal LST and UTFVI.	Ahmedabad, India	Disaster risk management	Urban heat	ANN–CA	Landsat 5, 8
Stewart and Kremer (2022)	*Environment and Planning. B, Urban Analytics and City Science*	"Temporal change in relationships between urban structure and surface temperature."	To identify how green space and the built environment influence surface temperature.	New York, USA	Disaster risk management	Urban heat	RF	Landsat 7
Wang et al. (2022c)	*Remote Sensing*	"A Two-Stage Seismic Damage Assessment Method for Small, Dense, and Imbalanced Buildings in Remote Sensing Images."	To perform post-earthquake building location and damage assessment with the improved YOLOv4 and SVM classification.	Mexico City, Mexico	Disaster risk management	Earthquake	CNN, SVM	xBD database
Zhang et al. (2022b)	*Atmospheric Environment*	"Construction of cloud-free MODIS-like land surface temperatures coupled with a regional weather research and forecasting (WRF) model."	To demonstratethe feasibility of using RS-WRF coupled model and ML to generate cloud-free MODIS-like LST images.	Beijing, China	Disaster risk management	Urban heat	Linearregression model, ANN and RF	MODIS

Das et al. (2021)	*GeoJournal*	"Spatio-temporal pattern of land use and land cover and its effects on land surface temperature using remote sensing and GIS techniques: A case study of Bhubaneswar city, Eastern India (1991-2021)."	To investigate the impact of LULCC on LST and predict future trends using a novel approach.	Bhubaneswar, India	Disaster risk management	Urban heat	SVM	Landsat 5, 7, 8
Ghaffarian et al. (2021)	*International Journal of Disaster Risk Reduction*	"Agent-based modelling of post-disaster recovery with remote sensing data."	To explore the potential of combining remote sensing data and agent-based model to simulate recovery processes and test real-world scenarios.	Tacloban, Philippines	Disaster risk management	Typhoon	XGBoost	WorldView-2, Pleiades
Guo et al. (2021b)	*Frontiers in Environmental Science*	"Dual Effects of Synoptic Weather Patterns and Urbanization on Summer Diurnal Temperature Range in an Urban Agglomeration of East China."	To analyse the different characteristics and influencing factors of diurnal temperature range (DTR) and DTR (urban–rural) in YRDUA under different SWPs.	Yangtze River Delta region, China	Disaster risk management	Urban heat	RF	MODIS, DMSP/OLS
Hrisko et al. (2021)	*Remote Sensing of Environment*	"Estimating heat storage in urban areas using multispectral satellite data and machine learning."	To propose a multispectral hysteresis model to predict heat storage flux in urban areas.	New York, USA	Disaster risk management	Urban heat	Gradient-boosted regression tree (GBRT)	GOES-16
Lammers et al. (2021)	*Journal of Hydrology*	"Prediction models for urban flood evolution for satellite remote sensing."	To propose alternative flood models to reduce computational expenses and use WRF-Hydro to validate the prediction accuracy.	Atlanta, USA	Disaster risk management	Flooding	Long short-term memory (LSTM) models	Not mentioned

(Continued)

TABLE A1.1 *(Continued)*

Analysis Highlights of the Reviewed Literature

Author	Journal	Title	Aim	Region	Domain	Sub-Domain	Machine Learning Technique	Remote Sensing Data
Chen et al. (2020b)	*Environmental Pollution*	"Estimation of anthropogenic heat emissions in China using Cubist with points-of-interest and multisource remote sensing data."	To provide an updated view of anthropogenic heat emission (AHE) across China to meet the demand for a national database of AHE profiles.	Mainland of China	Disaster risk management	Urban heat	Cubist model	MODIS: MOD11A2, SPOT 5, and DMSP/OLS
Kwak et al. (2020)	*Sustainable Cities and Society*	"Discerning the success of sustainable planning: A comparative analysis of urban heat island dynamics in Korean new towns."	To evaluate sustainability planning by comparing UHIs' characteristics with machine learning and statistical methods.	Bundang, Ilsan, Gwanggyo, and Pangyo, Korea	Disaster risk management	Urban heat	K-means clustering	Landsat 5, 7, 8
Duncan et al. (2019)	*The Science of the Total Environment*	"Turning down the heat: An enhanced understanding of the relationship between urban vegetation and surface temperature at the city scale."	To assess how urban vegetation type influenced urban LST.	Perth, Australia	Disaster risk management	Urban heat	RF	MODIS and Landsat 5, 7, 8
Peng et al. (2019)	*Remote Sensing*	"Patch similarity convolutional neural network for urban flood extent mapping using bi-temporal satellite multispectral imagery."	To map floods in near real time over urban areas by using patch similarity estimation.	West Houston and Lumberton, USA	Disaster risk management	Flooding	CNN	Multispectral images from Planet Lab

Yoo et al. (2018)	*ISPRS Journal of Photogrammetry and Remote Sensing*	"Estimation of daily maximum and minimum air temperatures in urban landscapes using MODIS time series satellite data."	To understand the relationship between LST and T_{max}/T_{min} in urban regions.	Los Angeles, USA, and Seoul, Korea	Disaster risk management	Urban heat	RF	MODIS and Landsat 5, 7
Frank et al. (2017)	*Remote Sensing*	"Effect of label noise on the machine-learned classification of earthquake damage."	To investigate the impact of label noise on rubble classification in high-resolution remotely sensed images.	Urban area in downtown Christchurch, New Zealand	Disaster risk management	Earthquake	RF	Aerial images
Santos et al. (2017)	*IEEE Geoscience and Remote Sensing Letters*	"An RS-GIS-Based Comprehensive Impact Assessment of Floods-A Case Study in Madeira River, Western Brazilian Amazon."	To develop an RS–GIS–based comprehensive impact assessment of floods.	Porto Velho, Brazil	Disaster risk management	Flooding	KNN	Aerial images
Chi et al. (2022)	*The Science of the Total Environment*	"Machine learning-based estimation of ground-level NO_2 concentrations over China."	To estimate ground-level NO_2 concentrations using the XGBoost model.	Mainland China	Pollution monitoring	NO_2 concentration estimation	XGBoost	TROPOMI NO_2 products
Yang et al. (2022)	*Remote Sensing of Environment*	"Geographical and temporal encoding for improving the estimation of $PM_{2.5}$ concentrations in China using end-to-end gradient boosting."	To estimate the concentration of $PM_{2.5}$ over a large-scale area using remote sensing data.	China	Pollution monitoring	$PM_{2.5}$ concentration estimation	End-to-end gradient boosting	MODIS
Zhang et al. (2022a)	*Environmental Pollution*	"Spatiotemporal neural network for estimating surface NO_2 concentrations over north China and their human health impact."	To estimate surface NO_2 from multiple sources of information by using STNN.	Beijing, Tianjin, Shijiazhuang, and Jinan, China	Pollution monitoring	NO_2 concentration estimation	Spatiotemporal Neural Network (STNN)	TROPOMI NO_2 products

(Continued)

TABLE A1.1 *(Continued)*

Analysis Highlights of the Reviewed Literature

Author	Journal	Title	Aim	Region	Domain	Sub-Domain	Machine Learning Technique	Remote Sensing Data
Chen et al. (2021c)	*Environmental Pollution*	"Estimating monthly $PM_{2.5}$ concentrations from satellite remote sensing data, meteorological variables, and land use data using ensemble statistical modelling and a random forest approach."	To estimate monthly $PM_{2.5}$ concentrations by using a two-stage ensemble GAM and RF modelling method with multi-source data.	Taiwan, China	Pollution monitoring	$PM_{2.5}$ concentration estimation	RF	MODIS
Chen et al. (2021a)	*Ecological Indicators*	"Machine learning-based inversion of water quality parameters in typical reach of the urban river by UAV multispectral data."	To develop a highly reliable urban river water quality monitoring system by combining UAV multispectral images and machine learning.	Hefei, China	Pollution monitoring	Water quality assessment	Genetic algorithm (GA)–XGBoost	Aerial images
Guo et al. (2021a)	*International Journal of Remote Sensing*	"A machine learning-based strategy for estimating non-optically active water quality parameters using Sentinel-2 imagery."	To extract non-optically active parameters from Sentinel-2 data and apply to small water bodies for analysis.	Tianjin, China	Pollution monitoring	Water quality assessment	RF, SVM, and ANN	Sentinel-2
Huang et al. (2021)	*Atmospheric Research*	"$PM_{2.5}$ concentration estimation with 1-km resolution at high coverage over urban agglomerations in China using the BPNN-KED approach and potential application."	To investigate the potential applications of high-resolution $PM_{2.5}$ concentration products with broad spatial coverage.	Beijing-Tianjin-Hebei, Central China, Yangtze River Delta, and Pearl River Delta, China	Pollution monitoring	$PM_{2.5}$ concentration estimation	Back-propagation neural network coupled kriging with external drifting approach (BPNN-KED)	MODIS

Yang et al. (2021)	*Environmental Pollution*	"Monitoring urban black-odorous water by using hyperspectral data and machine learning."	To identify primary indicators of urban black-odorous water through their spectral characteristics and differentiation.	Guangzhou, China	Pollution monitoring	Polluted water monitoring	Lasso regression	Hyperspectral images
Shi et al. (2021)	*The Science of the Total Environment*	"Digital mapping of zinc in urban topsoil using multisource geospatial data and random forest."	To produce a high-resolution map of Zn concentration in urban topsoil using machine learning algorithms and multi-source geospatial data.	Shenzhen, China	Pollution monitoring	Soil heavy metal measurement	RF	ASTER, Landsat 8, and SPOT 5
Shogrkhodaei et al. (2021)	*Environmental Pollution*	"Spatio-temporal modelling of $PM_{2.5}$ risk mapping using three machine learning algorithms."	To prepare a $PM_{2.5}$ risk map using meteorological, environmental, and remote sensing variables through spatiotemporal modelling with three machine learning algorithms.	Tehran, Iran	Pollution monitoring	$PM_{2.5}$ concentration estimation	RF, AdaBoost, and stochastic gradient descent (SGD)	Landsat 7
Tella et al. (2021)	*Atmospheric Pollution Research*	"Spatial assessment of PM_{10} hotspots using Random Forest, K-Nearest Neighbour and Naïve Bayes."	To predict spatial PM_{10} hazard hotspot, determine the variable importance of the predictors, and compare the models in the prediction of PM_{10} hotspot.	Selangor, Malaysia	Pollution monitoring	PM_{10} concentration estimation	RF, KNN, NB	Landsat 8

(Continued)

TABLE A1.1 *(Continued)*

Analysis Highlights of the Reviewed Literature

Author	Journal	Title	Aim	Region	Domain	Sub-Domain	Machine Learning Technique	Remote Sensing Data
Wei et al. (2021)	*Environmental Impact Assessment Review*	"Delineating the spatial-temporal variation of air pollution with urbanization in the Belt and Road Initiative area."	To identify the factors responsible for $PM_{2.5}$ accumulation or dissipation.	73 countries in 7 regions of East Asia, Southeast Asia, Central Asia, West Asia, South Asia, Europe, and North Africa	Pollution monitoring	$PM_{2.5}$ concentration estimation	Bi-LSTM	DMSP/OLS, NPP/VIIRS
Yan et al. (2021)	*Environmental Pollution*	"A Spatial-Temporal Interpretable Deep Learning Model for improving interpretability and predictive accuracy of satellite-based $PM_{2.5}$."	To improve the interpretability and predictive accuracy of satellite-based $PM_{2.5}$ measurements.	Beijing, China	Pollution monitoring	$PM_{2.5}$ concentration estimation	Spatial-temporal interpretable deep learning model (SIDLM)	MODIS
Zhang et al. (2021)	*Ecotoxicology and Environmental Safety*	"Estimating $PM_{2.5}$ concentration using the machine learning GA-SVM method to improve the land use regression model in Shaanxi, China."	To estimate $PM_{2.5}$ concentration with an improved LUR model by using the GA-SVM method and multi-source data.	Shanxi, China	Pollution monitoring	$PM_{2.5}$ concentration estimation	GA-SVM	MODIS
Zhou et al. (2021)	*Remote Sensing*	"Spectral and spatial feature integrated ensemble learning method for grading urban river network water quality."	To develop a universal inversion model for key water quality parameters using feature extraction for accurate water quality monitoring.	Shanghai, China	Pollution monitoring	Water quality assessment	SVM, MLP, and XGBoost	Aerial images

Chen et al. (2020c)	*The Science of the Total Environment*	"Estimating $PM_{2.5}$ with high-resolution 1-km AOD data and an improved machine learning model over Shenzhen, China."	To validate the performance of the IRF model in estimating $PM_{2.5}$ concentrations over Shenzhen, China.	Shenzhen, China	Pollution monitoring	$PM_{2.5}$ concentration estimation	RF	MODIS
Li et al. (2020)	*Environment International*	"Ensemble-based deep learning for estimating $PM_{2.5}$ over California with multisource big data including wildfire smoke."	To develop an ensemble deep learning method for high spatiotemporal resolution estimation and prediction of $PM_{2.5}$ concentrations and uncertainties.	California, USA	Pollution monitoring	$PM_{2.5}$ concentration estimation	Bagging of the base residual deep network	MODIS
Yan et al. (2020)	*Environment International*	"New interpretable deep learning model to monitor real-time $PM_{2.5}$ concentrations from satellite data."	To obtain a real-time knowledge of $PM_{2.5}$ with a new deep learning model—EntityDenseNet.	China	Pollution monitoring	$PM_{2.5}$ concentration estimation	EntityDenseNet	Himawari-8
Shen et al. (2018)	*Journal of Geophysical Research. Atmospheres*	"Estimating Regional Ground-Level $PM_{2.5}$ Directly From Satellite Top-Of-Atmosphere Reflectance Using Deep Belief Networks."	To develop a novel data–driven approach to estimate ground-level $PM_{2.5}$ concentration using satellite TOA reflectance.	Wuhan, China	Pollution monitoring	$PM_{2.5}$ concentration estimation	DBN	MODIS
Xu et al. (2018a)	*Environmental Pollution*	"Evaluation of machine learning techniques with multiple remote sensing datasets in estimating monthly concentrations of ground-level $PM_{2.5}$."	To propose a model utilising remote sensing, re-analysis, and air quality data for $PM_{2.5}$ concentration prediction and mapping.	British Columbia, Canada	Pollution monitoring	$PM_{2.5}$ concentration estimation	MLR, Bayesian Regularised Neural Networks (BRNN), SVM with radial basis function kernel, least absolute shrinkage and Selection Operator (LASSO), MARS, RF, XGBoost, and Cubist	MODIS

(Continued)

TABLE A1.1 *(Continued)*

Analysis Highlights of the Reviewed Literature

Author	Journal	Title	Aim	Region	Domain	Sub-Domain	Machine Learning Technique	Remote Sensing Data
Xu et al. (2018b)	*Building and Environment*	"Mapping ambient light at night using field observations and high-resolution remote sensing imagery for studies of urban environments."	To map ambient light at night using field observations and high-resolution remote sensing imagery.	NCR, Canada	Pollution monitoring	Light pollution	MLR, quantile regression (QR), BRNN, KNN, generalised additive model (GAM), SVM, MARS, RF, XGBoost, Cubist	ISS and aerial images
Wu et al. (2022)	*Sustainable Cities and Society*	"Developing a data-fusing method for mapping fine-scale urban three-dimensional building structure."	To estimate building height information with a data-fusing methodology utilising multi-source spatial variables.	Shenzhen, China	Built infrastructure identification	Building height	XGBoost, RF, and ANN	Sentinel-1, 2
Aissou et al. (2021)	*IEEE Sensors Journal*	"Building Roof Superstructures Classification From Imbalanced and Low Density Airborne LiDAR Point Cloud."	To classify roof superstructures using LiDAR data with low density and imbalanced classes for an urban area.	Vaihingen, Germany	Built infrastructure identification	Building structure	SVM	LiDAR
Frantz et al. (2021)	*Remote Sensing of Environment*	"National-scale mapping of building height using Sentinel-1 and Sentinel-2 time series."	To develop a machine learning regression–based methodology for predicting building height using free and open image archives.	Germany	Built infrastructure identification	Building height	SVM, RF	Sentinel-1, 2

Imam et al. (2021)	*International Journal of Low Carbon Technologies*	"An automatic approach for detecting building fences from high-resolution images: The case study of Makkah, Saudi Arabia."	To extract building edges and identify illegal building fences with remote sensing, GIS, and machine learning techniques.	Al-Hudaybiyah region, Saudi Arabia	Built infrastructure identification	Building structure	CNN	WorldView-3
Janowski et al. (2021)	*Land Use Policy*	"Remote measurement of building usable floor area – Algorithms fusion."	To elaborate precise, complete, and detailed property information with the use of remote sensing observations in a suitable numerical algorithm.	Olsztyn, Poland	Built infrastructure identification	Building footprint	YOLO	LiDAR
Touzani and Granderson (2021)	*Remote Sensing*	"Open data and deep semantic segmentation for automated extraction of building footprints."	To automatically generate a larger training dataset using openly available data for improving deep learning models' accuracy.	15 cities and 5 counties in the USA	Built infrastructure identification	Building footprint	DeepLabV3+	Mapbox
Geiß et al. (2020)	*ISPRS Journal of Photogrammetry and Remote Sensing*	"Multistrategy ensemble regression for mapping of built-up density and height with Sentinel-2 data."	To extract built-up density and height information from multispectral Sentinel-2 data through machine learning algorithms.	Berlin, Hamburg, Munich, and Cologne, German	Built infrastructure identification	Building density and height	RF, SVM, Gaussian process regression (GPR), and ANN	Sentinel-2
Han et al. (2020)	*Engineering Applications of Artificial Intelligence*	"Precise object detection using adversarially augmented local/global feature fusion."	To develop a novel deep learning framework for accurately detecting objects of interest in high spatial resolution remote sensing imagery.	Vaihingen, Germany	Built infrastructure identification	Traffic vehicle	CNN	NWPU VHR-10 dataset

(*Continued*)

TABLE A1.1 *(Continued)*

Analysis Highlights of the Reviewed Literature

Author	Journal	Title	Aim	Region	Domain	Sub-Domain	Machine Learning Technique	Remote Sensing Data
Sanya and Mwebaze (2020)	*Heliyon*	"Identifying patterns in urban housing density in developing countries using convolutional networks and satellite imagery."	To explore the use of CNNs in predicting housing density in developing countries through high-level semantic scene classification.	Uganda	Built infrastructure identification	Building density	CNN	Google map images
Schlosser et al. (2020)	*Remote Sensing*	"Building extraction using orthophotos and dense point cloud derived from visual band aerial imagery based on machine learning and segmentation."	To investigate building identification using only aerial survey–based visual bands orthoimages and photogrammetric point clouds.	Debrecen, Hungary	Built infrastructure identification	Building footprint	RF, SVM, and KNN	Aerial images
Liuzzi et al. (2019)	*Bulletin of Earthquake Engineering*	"A transferable remote sensing approach to classify building structural types for seismic risk analyses: The case of Val d'Agri area (Italy)."	To evaluate seismic vulnerability of buildings via efficient classification of structural types using multispectral remote sensing.	Val d'Agri area, Italy	Built infrastructure identification	Seismic building structural type	RF, KNN	Aerial images
Schilling et al. (2018)	*ISPRS Journal of Photogrammetry and Remote Sensing*	"Object-based detection of vehicles using combined optical and elevation data."	To develop a rapid system for detecting stationary vehicles using optical and elevation data in combination.	Vaihingen, Germany	Built infrastructure identification	Traffic vehicle	RF and CNN	Aerial images

Xu et al. (2018c)	*Remote Sensing*	"Building extraction in very high resolution remote sensing imagery using deep learning and guided filters."	To create a network that enhances building extraction accuracy and utilises a guided filter for post-processing improvement.	Vaihingen and Potsdam, Germany	Built infrastructure identification	Building footprint	ResNet	Aerial images
Zhang et al. (2018b)	*Remote Sensing*	"Road centerline extraction from very-high-resolution aerial image and LiDAR data based on road connectivity."	To enhance the extraction of road centrelines by accounting for road connectivity, utilising both very-high-resolution aerial images and LiDAR data.	New York, USA; Vaihingen, Germany; Guangzhou, China	Built infrastructure identification	Road centreline	RF	Aerial images and LiDAR
Dornaika et al. (2016)	*Expert Systems with Applications*	"Building detection from orthophotos using a machine learning approach: An empirical study on image segmentation and descriptors."	To develop a precise and automated framework for detecting buildings in aerial orthophotos using a generic approach.	Belfort, France	Built infrastructure identification	Building footprint	SVM	Aerial images
Soilán et al. (2016)	*ISPRS Journal of Photogrammetry and Remote Sensing*	"Traffic sign detection in MLS acquired point clouds for geometric and image-based semantic inventory."	To identify geometric and semantic characteristics of traffic signs in urban and highway areas using advanced techniques.	Galicia, Spain	Built infrastructure identification	Traffic sign	SVM	LiDAR
Geiß et al. (2015)	*ISPRS Journal of Photogrammetry and Remote Sensing*	"Estimation of seismic building structural types using multi-sensor remote sensing and machine learning techniques."	To evaluate the suitability of a remote sensing–based approach in estimating seismic building structural types.	Padang, Indonesia	Built infrastructure identification	Seismic building structural type	SVM, RF	IKONOS, Landsat 5

(Continued)

TABLE A1.1 *(Continued)*

Analysis Highlights of the Reviewed Literature

Author	Journal	Title	Aim	Region	Domain	Sub-Domain	Machine Learning Technique	Remote Sensing Data
Lu et al. (2014)	*Landscape and Urban Planning*	"Building type classification using spatial and landscape attributes derived from LiDAR remote sensing data."	To obtain building information, generate environmental landscape attributes, and classify building types by using LiDAR data and machine learning methods.	Denver, USA	Built infrastructure identification	Building footprint	DT, aggregated decision tree (ADT), RF, and SVM	LiDAR
Adeline et al. (2013)	*ISPRS Journal of Photogrammetry and Remote Sensing*	"Shadow detection in very high spatial resolution aerial images: A comparative study."	To better understand the radiative impact of shadows.	Toulouse and Paris, France	Built infrastructure identification	Building shade	K-means, SVM	PELICAN, CASI, and DMC
Henn et al. (2013)	*ISPRS Journal of Photogrammetry and Remote Sensing*	"Model driven reconstruction of roofs from sparse LiDAR point clouds."	To develop a fully automatic approach for 3D building reconstruction using LiDAR data and building footprints.	Cologne, Oberhausen, Germany	Built infrastructure identification	Building structure	SVM	LiDAR
Dong et al. (2022)	*The Science of the Total Environment*	"Optimization of green infrastructure networks based on potential green roof integration in a high-density urban area—A case study of Beijing, China."	To apply green roofs to GIN optimisation.	Beijing, China	Ecosystem and environmental assessment	Green infrastructure mapping	CNN	Google map images and LUC products

Uniyal et al. (2022)	*Urban Forestry & Urban Greening*	"Quantification of carbon sequestration by urban forest using Landsat 8 OLI and machine learning algorithms in Jodhpur, India."	To estimate the aboveground biomass of urban forests and assess their role in carbon sequestration by using ML-based regression algorithms.	Jodhpur, India	Ecosystem and environmental assessment	Carbon sequestration estimation	SVM, RF, KNN, and XGBoost	Landsat 8
Wang et al. (2022b)	*Computers and Electronics in Agriculture*	"UAV remote sensing based estimation of green cover during turfgrass establishment."	To compare methods for estimating green cover during turfgrass establishment using UAV data and assess its effectiveness against ground-level data.	Dallas, USA	Ecosystem and environmental assessment	Green infrastructure mapping	SVM, MLC, minimum distance classification (MDC), K-means, and ISODATA	Aerial images
Correa Martins et al. (2021)	*Ecological Informatics*	"Machine learning and SLIC for Tree Canopies segmentation in urban areas."	To detect trees in a typical urban environment with high-resolution images by combining SLIC and CNN.	The municipality of Campo Grande, Brazil	Ecosystem and environmental assessment	Green infrastructure mapping	CNN	Aerial images
Hartling et al. (2021)	*GIScience and Remote Sensing*	"Urban tree species classification using UAV-based multi-sensor data fusion and machine learning."	To improve urban tree species classification with limited training samples, a multi-sensor data fusion technique was developed.	St. Louis, USA	Ecosystem and environmental assessment	Green infrastructure mapping	RF and SVM	Aerial images
Hay Chung et al. (2021)	*Building and Environment*	"Improved machine-learning mapping of local climate zones in metropolitan areas using composite Earth observation data in Google Earth Engine."	To improve LCZ classification with a workflow using the GEE platform and ML algorithms by identifying optimal input sources, investigating variable contributions, and comparing results with "elite variables."	Guangdong, Macao, and Hong Kong, China	Ecosystem and environmental assessment	Local climate zone mapping	RF, SVM, and CART	Landsat 8 and Sentinel-2

(Continued)

TABLE A1.1 *(Continued)*

Analysis Highlights of the Reviewed Literature

Author	Journal	Title	Aim	Region	Domain	Sub-Domain	Machine Learning Technique	Remote Sensing Data
Jia and Zhang (2021)	*Journal of Cleaner Production*	"A human-scale investigation into economic benefits of urban green and blue infrastructure based on big data and machine learning: A case study of Wuhan."	To investigate the economic benefits of green and blue infrastructure by combining big data and machine learning.	Wuhan, China	Ecosystem and environmental assessment	Green and blue infrastructure mapping	FCN	Landsat 8
Liu et al. (2021)	*Journal of Cleaner Production*	"Factors of ecosystem service values in a fast-developing region in China: Insights from the joint impacts of human activities and natural conditions."	To quantify the spatiotemporal variation of ecosystem services in the GBA from 2000 to 2015 and analyse the combined effects of multiple factors.	Guangdong–Hong Kong–Macao Greater Bay Area (GBA), China	Ecosystem and environmental assessment	Ecosystem service assessment	DT	Not mentioned
Mahdianpari et al. (2021)	*Journal of Environmental Management*	"Smart solutions for smart cities: Urban wetland mapping using very-high resolution satellite imagery and airborne LiDAR data in the City of St. John's, NL, Canada."	To produce a highly accurate wetland map using VHR and LiDAR elevation data for the City of St. John's.	St. John's, Canada	Ecosystem and environmental assessment	Blue infrastructure mapping	RF	WorldView-4, GeoEye-1, and LiDAR
Martins et al. (2021)	*Urban Forestry & Urban Greening*	"Deep learning-based tree species mapping in a highly diverse tropical urban setting."	To extract tree types, use a new multi-task CNN on VHR aerial photos for species composition.	Rio de Janeiro, Brazil	Ecosystem and environmental assessment	Green infrastructure mapping	CNN	Aerial images

Sapena et al. (2021)	*Computers, Environment and Urban Systems*	"Estimating quality of life dimensions from urban spatial pattern metrics."	To assess the interplay between urban form and socio-economic status, quantify their spatial relationship in cities.	North Rhine-Westphalia, Germany	Ecosystem and environmental assessment	Local climate zone mapping	RF	Rapid-Eye and Cartosat-1
Vulova et al. (2021)	*The Science of the Total Environment*	"Modelling urban evapotranspiration using remote sensing, flux footprints, and artificial intelligence."	To model urban evapotranspiration and identify drivers of evapotranspiration with remote sensing and GIS data and ML and DL algorithms.	Berlin, Germany	Ecosystem and environmental assessment	Evapotranspiration modelling	CNN and RF	Landsat 8, Sentinel-2
Zhao et al. (2021)	*IEEE Journal of Selected Topics in Applied Earth Observations and Remote Sensing*	"Mapping Population Distribution Based on XGBoost Using Multisource Data."	To explore the combination of multi-source datasets and multidimensional features for population estimation and compare different models.	Shenzhen, China	Ecosystem and environmental assessment	Population mapping	RF, GBDT, XGBoost, LightGBM, and CatBoost	Landsat 8, Luojia 1-01
Chen et al. (2020a)	*Advanced Engineering Informatics*	"Augmenting a deep-learning algorithm with canal inspection knowledge for reliable water leak detection from multispectral satellite images."	To enhance visibility, reliability, and extensibility by using deep learning augmented by domain knowledge about model physical parameter patterns.	Arizona Canal, South Canal, and Western Canal, USA	Ecosystem and environmental assessment	Blue infrastructure mapping	Physics-guided neural network (PGNN) and CNN	Landsat 8
Liao and Wen (2020)	*International Journal of Applied Earth Observation and Geoinformation*	"Extracting urban water bodies from high-resolution radar images: Measuring the urban surface morphology to control for radar's double-bounce effect."	To develop a dependable urban water extraction method for SAR imagery with incorporating surface morphology to regulate radar multi-bouncing.	Taipei, China	Ecosystem and environmental assessment	Blue infrastructure mapping	RF	Aerial images and Sentinel-1

(Continued)

TABLE A1.1 *(Continued)*

Analysis Highlights of the Reviewed Literature

Author	Journal	Title	Aim	Region	Domain	Sub-Domain	Machine Learning Technique	Remote Sensing Data
Milnar and Ramaswami (2020)	*Environmental Science & Technology*	"Impact of Urban Expansion and In Situ Greenery on Community-Wide Carbon Emissions: Method Development and Insights from 11 US Cities."	To evaluate the effect of biogenic CO_2 discharges municipally with a method that employs fine-grained remote sensing data.	11 cities in the USA	Ecosystem and environmental assessment	Carbon emission estimation	SVM	Aerial images
Bogawski et al. (2019)	*The Science of the Total Environment*	"Predicting the onset of Betula pendula flowering in Poznań (Poland) using remote sensing thermal data."	To model the spatial pattern of flowering onset in silver birch in Poznań using land surface temperature data and in situ phenological observations.	Poznań, Poland	Ecosystem and environmental assessment	Green infrastructure mapping	SVM, RF	MODIS
Sannigrahi et al. (2019)	*Ocean & Coastal Management*	"Evaluating landscape capacity to provide spatially explicit valued ecosystem services for sustainable coastal resource management."	To assess the capability of landscapes to offer various ecosystem services and major eco-regions' spatiotemporal trends from 1973 to 2018.	SBR, India	Ecosystem and environmental assessment	Ecosystem service assessment	ANN, DT, NB, GBDT, linear discriminant analysis (LDA), KNN, MLC, SVM, radial basis function (RBF), and RF	Landsat 4, 5, 7, 8

Yang et al. (2019)	*International Journal of Environmental Research and Public Health*	"Improved estimates of population exposure in low-elevation coastal zones of China."	To quantify the magnitude and spatial distribution of population in China's low-elevation coastal zones.	Low-elevation coastal zones of China	Ecosystem and environmental assessment	Population mapping	Cubist and RF	DMSP/OLS, SPOT 5
Ye et al. (2019)	*The Science of the Total Environment*	"Improved population mapping for China using remotely sensed and points-of-interest data within a random forests model."	To improve China's population map beyond the WorldPop dataset by integrating multi-source remote sensing data and POIs.	China	Ecosystem and environmental assessment	Population mapping	RF	DMSP/OLS, SPOT 5

TABLE A1.2

Specifications of Satellites and Sensors Utilised in the Selected Articles

Remote Sensing Data		Spatial Resolution	Temporal Resolution	Data Available Since	Swath Width
Landsat (31)	Landsat 4 TM (1)	30 m/120 m (Thermal band)	16 days	1982–1993	185 km
	Landsat 4 MSS (1)	60 m	16 days	1982–1992	185 km
	Landsat 5 TM (11)	30 m/120 m (Thermal band)	16 days	1984–2012	185 km
	Landsat 5 MSS (11)	60 m	16 days	1984–2013	185 km
	Landsat 7 ETM+ (12)	30 m/60 m (Thermal band)	16 days	1999	185 km
	Landsat 8 OLI (24)	15 m (Pan)/30 m	16 days	2013	185 km
	Landsat 8 TIRS (24)	100 m	16 days	2013	185 km
MOIDS (17)		250 m (bands 1–2)/ 500 m (bands 3–7)/ 1 km (bands 8–36)	1–2 days	2000	2330 km
Sentinel (15)	Sentinel-1 (5)	10 m (VNIR)/20 m (RE and SWIR) and 60 m (AC).	12 days	2014	250 km
	Sentinel-2 (11)	10 m (VNIR)/20 m (RE and SWIR) and 60 m (AC).	5–10 days	2015	290 km
	Sentinel-5 (2)	7 km	16 days	2017	2600 km
Nighttime light (6)	DMSP/OLS (5)	1 km	24 hours	1992–2013	3000 km
	NPP/VIIRS (2)	500 m	12 hours	2012	3000 km
	Luojia 1-01 (1)	130 m	15 days	2018	250 km
WorldView (4)	WorldView-2 (2)	0.46 m (Pan)/1.8 m	1.1 days	2009	16.4 km
	WorldView-3 (1)	0.31 m (Pan)/1.24 m (MS)/3.7 m (SWIR)/ 30 m (CAVIS)	1–4.5 days	2014	13.1 km
	WorldView-4 (1)	0.31 m (Pan)/1.24 m (MS)	1–4.5 days	2016	13.1 km
SPOT-5 (4)		2.5–5 m (Pan)/10 m (MS)/20 m (SWIR)	26 days	2002	60 km
Pléiades-1A (3)		0.5 m (Pan)/2 m (MS)	26 days	2011	20 km
IKONOS (2)		1 m (PAN)/4 m (MS)	14 days	1999–2015	13.8 km
ASTER (1)		15 m (VNIR)/30 m (SWIR)/90 m (TIR)	16 days	1999	60 km
Cartosat-1 (1)		2.5 m	5 days	2005–2019	30 km
CBERS-2B (1)		2.5 m (Pan)/20 m (MS)/260 m (NIR)	26 days	2007–2010	27 km (Pan)/ 113 km (MS)/ 890 km (NIR)
DMC-1 (1)		32 m	1 day	2002	600 km

(Continued)

TABLE A1.2 *(Continued)*

Specifications of Satellites and Sensors Utilised in the Selected Articles

Remote Sensing Data		Spatial Resolution	Temporal Resolution	Data Available Since	Swath Width
ERS (1)	ERS-1 (1)	10–30 m	35 days	1991–2000	100 km
	ERS-2 (1)	30 m	35 days	1995–2011	100 km
GeoEye (1)		0.41 m (Pan)/ 1.64 m (MS)	1.7–4.6 days	2008	15.2 km
GOES-16 (1)		0.5–2 km	10 minutes	2016	\
Himawari-8 (1)		0.5 km	10 minutes	2014	500 km
ISS (1)		4 km	4 days	2017	550 km
QuickBird (1)		0.61 m (Pan)/ 2.44 m (MS)	1.5–2.8 days	2001–2015	16.5 km
Rapid-Eye (1)		5 m	1–5.5 days	2008–2020	77 km

Note: One paper can be categorised into more than one group.

References

Abunnasr, Y., & Mhawej, M. (2022). Towards a combined Landsat-8 and Sentinel-2 for 10-m land surface temperature products: The Google Earth Engine monthly Ten-ST-GEE system. *Environmental Modelling & Software, 155*. https://doi.org/10.1016/j.envsoft.2022.105456

Adeline, K. R. M., Chen, M., Briottet, X., Pang, S. K., & Paparoditis, N. (2013). Shadow detection in very high spatial resolution aerial images: A comparative study. *ISPRS Journal of Photogrammetry and Remote Sensing, 80*, 21–38. https://doi.org/10.1016/j.isprsjprs.2013.02.003

Aissou, B. E., Aissa, A. B., Dairi, A., Harrou, F., Wichmann, A., & Kada, M. (2021). Building roof superstructures classification from imbalanced and low density airborne LiDAR point cloud. *IEEE Sensors Journal, 21*(13), 14960–14976. https://doi.org/10.1109/JSEN.2021.3073535

Alanne, K., & Sierla, S. (2022). An overview of machine learning applications for smart buildings. *Sustainable Cities and Society, 76*. https://doi.org/10.1016/j.scs.2021.103445

Ali, R. A.-A., Yousef, A. A.-M., Ali, A., Jiquan, Z., Kashif, U., Abu, R. M., Towfiqul, I., Tayyiba, H., Dawuda Usman, K., Jean Claude, N., Bazel, A.-S., Yasser, M. K., Wafaa, M. M. A. L. H., & Xingpeng, L. (2022). Assessment analysis of flood susceptibility in tropical desert area: A case study of Yemen. *Remote Sensing (Basel, Switzerland), 14*(4050), 4050. https://doi.org/10.3390/rs14164050

Arbolino, R., Carlucci, F., De Simone, L., Ioppolo, G., & Yigitcanlar, T. (2018). The policy diffusion of environmental performance in the European countries. *Ecological Indicators, 89*, 130–138. https://doi.org/10.1016/j.ecolind.2018.01.062

Avtar, R., Komolafe, A. A., Kouser, A., Singh, D., Yunus, A. P., Dou, J., Kumar, P., Gupta, R. D., Johnson, B. A., Thu Minh, H. V., Aggarwal, A. K., & Kurniawan, T. A. (2020). Assessing sustainable development prospects through remote sensing: A review. *Remote Sensing Applications, 20*, 100402. https://doi.org/10.1016/j.rsase.2020.100402

Azadbakht, M., Fraser, C. S., & Khoshelham, K. (2018). Synergy of sampling techniques and ensemble classifiers for classification of urban environments using full-waveform LiDAR data. *International Journal of Applied Earth Observation and Geoinformation*, *73*, 277–291. https://doi.org/10.1016/j.jag.2018.06.009

Belgiu, M., & Drăguţ, L. (2016). Random forest in remote sensing: A review of applications and future directions. *ISPRS Journal of Photogrammetry and Remote Sensing*, *114*, 24–31. https://doi.org/10.1016/j.isprsjprs.2016.01.011

Bogawski, P., Grewling, Ł., & Jackowiak, B. (2019). Predicting the onset of *Betula pendula* flowering in Poznań (Poland) using remote sensing thermal data. *The Science of the Total Environment*, *658*, 1485–1499. https://doi.org/10.1016/j.scitotenv.2018.12.295

Casali, Y., Aydin, N. Y., & Comes, T. (2022). Machine learning for spatial analyses in urban areas: A scoping review. *Sustainable Cities and Society*, *85*. https://doi.org/10.1016/j.scs.2022.104050

Chen, B., Mu, X., Chen, P., Wang, B., Choi, J., Park, H., Xu, S., Wu, Y., & Yang, H. (2021a). Machine learning-based inversion of water quality parameters in typical reach of the urban river by UAV multispectral data. *Ecological Indicators*, *133*, 108434–108442. https://doi.org/10.1016/j.ecolind.2021.108434

Chen, B., Tu, Y., Song, Y., Theobald, D. M., Zhang, T., Ren, Z., Li, X., Yang, J., Wang, J., Wang, X., Gong, P., Bai, Y., & Xu, B. (2021b). Mapping essential urban land use categories with open big data: Results for five metropolitan areas in the United States of America. *ISPRS Journal of Photogrammetry and Remote Sensing*, *178*, 203–218. https://doi.org/10.1016/j.isprsjprs.2021.06.010

Chen, C.-C., Wang, Y.-R., Yeh, H.-Y., Lin, T.-H., Huang, C.-S., & Wu, C.-F. (2021c). Estimating monthly $PM_{2.5}$ concentrations from satellite remote sensing data, meteorological variables, and land use data using ensemble statistical modeling and a random forest approach. *Environmental Pollution*, *291*. https://doi.org/10.1016/j.envpol.2021.118159

Chen, J., Tang, P., Rakstad, T., Patrick, M., & Zhou, X. (2020a). Augmenting a deep-learning algorithm with canal inspection knowledge for reliable water leak detection from multispectral satellite images. *Advanced Engineering Informatics*, *46*, 101161. https://doi.org/10.1016/j.aei.2020.101161

Chen, L., Fang, B., Zhao, L., Zang, Y., Liu, W., Chen, Y., Wang, C., & Li, J. (2022a). DeepUrbanDownscale: A physics informed deep learning framework for high-resolution urban surface temperature estimation via 3D point clouds. *International Journal of Applied Earth Observation and Geoinformation*, *106*, 102650. https://doi.org/10.1016/j.jag.2021.102650

Chen, Q., Yang, X., Ouyang, Z., Zhao, N., Jiang, Q., Ye, T., Qi, J., & Yue, W. (2020b). Estimation of anthropogenic heat emissions in China using Cubist with points-of-interest and multisource remote sensing data. *Environmental Pollution*, *266*, 115183. https://doi.org/10.1016/j.envpol.2020.115183

Chen, S., Yang, Y., Deng, F., Zhang, Y., Liu, D., Liu, C., & Gao, Z. (2022b). A high-resolution monitoring approach of canopy urban heat island using a random forest model and multi-platform observations. *Atmospheric Measurement Techniques*, *15*(3), 735–756. https://doi.org/10.5194/amt-15-735-2022

Chen, W., Ran, H., Cao, X., Wang, J., Teng, D., Chen, J., & Zheng, X. (2020c). Estimating $PM_{2.5}$ with high-resolution 1-km AOD data and an improved machine learning model over Shenzhen, China. *The Science of the Total Environment*, *746*, 141093. https://www.sciencedirect.com/science/article/pii/S0048969720346222?via%3Dihub

Chi, Y., Fan, M., Zhao, C., Yang, Y., Fan, H., Yang, X., Yang, J., & Tao, J. (2022). Machine learning-based estimation of ground-level NO_2 concentrations over China. *The Science of the Total Environment, 807*(Pt 1), 150721. https://www.sciencedirect.com/science/article/pii/S0048969721057995?via%3Dihub

Correa Martins, J. A., Menezes, G., Gonçalves, W., Sant'Ana, D. A., Osco, L. P., Liesenberg, V., Li, J., Ma, L., Oliveira, P. T., Astolfi, G., Pistori, H., & Junior, J. M. (2021). Machine learning and SLIC for Tree Canopies segmentation in urban areas. *Ecological Informatics, 66,* 101465. https://doi.org/10.1016/j.ecoinf.2021.101465

Das, T., Jana, A., Mandal, B., & Sutradhar, A. (2021). Spatio-temporal pattern of land use and land cover and its effects on land surface temperature using remote sensing and GIS techniques: A case study of Bhubaneswar city, Eastern India (1991-2021). *GeoJournal.* https://doi.org/10.1007/s10708-021-10541-z

Demarchi, L., Canters, F., Cariou, C., Licciardi, G., & Chan, J. C.-W. (2014). Assessing the performance of two unsupervised dimensionality reduction techniques on hyperspectral APEX data for high resolution urban land-cover mapping. *ISPRS Journal of Photogrammetry and Remote Sensing, 87,* 166–179. https://doi.org/10.1016/j.isprsjprs.2013.10.012

Deng, C., Li, C., Zhu, Z., Lin, W., & Xi, L. (2017). Subpixel urban impervious surface mapping: The impact of input Landsat images. *ISPRS Journal of Photogrammetry and Remote Sensing, 133,* 89–103. https://doi.org/10.1016/j.isprsjprs.2017.09.015

Deng, C., & Wu, C. (2013). A spatially adaptive spectral mixture analysis for mapping subpixel urban impervious surface distribution. *Remote Sensing of Environment, 133,* 62–70.

Dong, J., Guo, F., Lin, M., Zhang, H., & Zhu, P. (2022). Optimization of green infrastructure networks based on potential green roof integration in a high-density urban area—A case study of Beijing, China. *The Science of the Total Environment, 834,* 155307–155307. https://doi.org/10.1016/j.scitotenv.2022.155307

Dornaika, F., Moujahid, A., El Merabet, Y., & Ruichek, Y. (2016). Building detection from orthophotos using a machine learning approach: An empirical study on image segmentation and descriptors. *Expert Systems With Applications, 58,* 130–142. https://doi.org/10.1016/j.eswa.2016.03.024

Du, Y., Wang, T., Xin, B., Wang, L., Chen, Y., & Xing, L. (2020). A data-driven parallel scheduling approach for multiple agile earth observation satellites. *IEEE Transactions on Evolutionary Computation, 24*(4), 679–693. https://doi.org/10.1109/tevc.2019.2934148

Duncan, J. M. A., Boruff, B., Saunders, A., Sun, Q., Hurley, J., & Amati, M. (2019). Turning down the heat: An enhanced understanding of the relationship between urban vegetation and surface temperature at the city scale. *The Science of the Total Environment, 656,* 118–128. https://doi.org/10.1016/j.scitotenv.2018.11.223

Frank, J., Rebbapragada, U., Bialas, J., Oommen, T., & Havens, T. C. (2017). Effect of label noise on the machine-learned classification of earthquake damage. *Remote Sensing (Basel, Switzerland), 9*(8), 803. https://doi.org/10.3390/rs9080803

Frantz, D., Schug, F., Okujeni, A., Navacchi, C., Wagner, W., van der Linden, S., & Hostert, P. (2021). National-scale mapping of building height using Sentinel-1 and Sentinel-2 time series. *Remote Sensing of Environment, 252,* 112128–112128. https://doi.org/10.1016/j.rse.2020.112128

Galaz, V., Centeno, M. A., Callahan, P. W., Causevic, A., Patterson, T., Brass, I., Baum, S., Farber, D., Fischer, J., Garcia, D., McPhearson, T., Jimenez, D., King, B., Larcey, P., & Levy, K. (2021). Artificial intelligence, systemic risks, and sustainability. *Technology in Society, 67.* https://doi.org/10.1016/j.techsoc.2021.101741

Garg, R., Kumar, A., Bansal, N., Prateek, M., & Kumar, S. (2021). Semantic segmentation of PolSAR image data using advanced deep learning model. *Scientific Reports, 11*(1), 15365–15365. https://doi.org/10.1038/s41598-021-94422-y

Geiß, C., Aravena Pelizari, P., Marconcini, M., Sengara, W., Edwards, M., Lakes, T., & Taubenböck, H. (2015). Estimation of seismic building structural types using multi-sensor remote sensing and machine learning techniques. *ISPRS Journal of Photogrammetry and Remote Sensing, 104,* 175–188. https://doi.org/10.1016/j.isprsjprs.2014.07.016

Geiß, C., Schrade, H., Aravena Pelizari, P., & Taubenböck, H. (2020). Multistrategy ensemble regression for mapping of built-up density and height with Sentinel-2 data. *ISPRS Journal of Photogrammetry and Remote Sensing, 170,* 57–71. https://doi.org/10.1016/j.isprsjprs.2020.10.004

Ghaffarian, S., Roy, D., Filatova, T., & Kerle, N. (2021). Agent-based modelling of post-disaster recovery with remote sensing data. *International Journal of Disaster Risk Reduction, 60,* 102285.

Golbabaei, F., Yigitcanlar, T., & Bunker, J. (2021). The role of shared autonomous vehicle systems in delivering smart urban mobility: A systematic review of the literature. *International Journal of Sustainable Transportation, 15*(10), 731–748. https://doi.org/10.1080/15568318.2020.1798571

Golbabaei, F., Yigitcanlar, T., Paz, A., & Bunker, J. (2020). Individual predictors of autonomous vehicle public acceptance and intention to use: A systematic review of the literature. *Journal of Open Innovation: Technology, Market, and Complexity, 6*(4), 106. https://doi.org/10.3390/joitmc6040106

Griffiths, P., Hostert, P., Gruebner, O., & van der Linden, S. (2010). Mapping megacity growth with multi-sensor data. *Remote Sensing of Environment, 114*(2), 426–439. https://doi.org/10.1016/j.rse.2009.09.012

Guo, H., Huang, J. J., Chen, B., Guo, X., & Singh, V. P. (2021a). A machine learning-based strategy for estimating non-optically active water quality parameters using Sentinel-2 imagery. *International Journal of Remote Sensing, 42*(5), 1841–1866. https://doi.org/10.1080/01431161.2020.1846222

Guo, M., Zhang, M., Wang, H., Wang, L., Liu, S., Zong, L., Zhang, Y., & Li, Y. (2021b). Dual effects of synoptic weather patterns and urbanization on summer diurnal temperature range in an urban agglomeration of East China. *Frontiers in Environmental Science, 9.* https://doi.org/10.3389/fenvs.2021.672295

Han, X., He, T., Ong, Y.-S., & Zhong, Y. (2020). Precise object detection using adversarially augmented local/global feature fusion. *Engineering Applications of Artificial Intelligence, 94,* 103710. https://doi.org/10.1016/j.engappai.2020.103710

Hartling, S., Sagan, V., & Maimaitijiang, M. (2021). Urban tree species classification using UAV-based multi-sensor data fusion and machine learning. *GIScience and Remote Sensing, 58*(8), 1250–1275. https://doi.org/10.1080/15481603.2021.1974275

Hay Chung, L. C., Xie, J., & Ren, C. (2021). Improved machine-learning mapping of local climate zones in metropolitan areas using composite Earth observation data in Google Earth Engine. *Building and Environment, 199,* 107879. https://doi.org/10.1016/j.buildenv.2021.107879

Henn, A., Gröger, G., Stroh, V., & Plümer, L. (2013). Model driven reconstruction of roofs from sparse LIDAR point clouds. *ISPRS Journal of Photogrammetry and Remote Sensing, 76*, 17–29. https://doi.org/10.1016/j.isprsjprs.2012.11.004

Hrisko, J., Ramamurthy, P., & Gonzalez, J. E. (2021). Estimating heat storage in urban areas using multispectral satellite data and machine learning. *Remote Sensing of Environment, 252*, 112125. https://doi.org/10.1016/j.rse.2020.112125

Huang, F., Yu, Y., & Feng, T. (2019a). Automatic extraction of impervious surfaces from high resolution remote sensing images based on deep learning. *Journal of Visual Communication and Image Representation, 58*, 453–461. https://doi.org/10.1016/j.jvcir.2018.11.041

Huang, F., Yu, Y., & Feng, T. (2019b). Automatic extraction of urban impervious surfaces based on deep learning and multi-source remote sensing data. *Journal of Visual Communication and Image Representation, 60*, 16–27. https://doi.org/10.1016/j.jvcir.2018.12.051

Huang, Y., Zhang, T., Zhu, Z., Gong, W., & Xia, X. (2021). $PM_{2.5}$ concentration estimation with 1-km resolution at high coverage over urban agglomerations in China using the BPNN-KED approach and potential application. *Atmospheric Research, 258*, 105628. https://doi.org/10.1016/j.atmosres.2021.105628

Imam, A., Faisal, K., Majrashi, A., & Hegazy, I. (2021). An automatic approach for detecting building fences from high-resolution images: The case study of Makkah, Saudi Arabia. *International Journal of Low Carbon Technologies, 16*(3), 1087–1097. https://doi.org/10.1093/ijlct/ctab039

Jaad, A., & Abdelghany, K. (2021). The story of five MENA cities: Urban growth prediction modeling using remote sensing and video analytics. *Cities, 118*, 103393. https://doi.org/10.1016/j.cities.2021.103393

Jamali, A. (2019). Evaluation and comparison of eight machine learning models in land use/land cover mapping using Landsat 8 OLI: A case study of the northern region of Iran. *SN Applied Sciences, 1*(11). https://doi.org/10.1007/s42452-019-1527-8

Janowski, A., Renigier-Biłozor, M., Walacik, M., & Chmielewska, A. (2021). Remote measurement of building usable floor area – Algorithms fusion. *Land Use Policy, 100*, 104938. https://doi.org/10.1016/j.landusepol.2020.104938

Jato-Espino, D., Manchado, C., Roldán-Valcarce, A., & Moscardó, V. (2022). ArcUHI: A GIS add-in for automated modelling of the Urban Heat Island effect through machine learning. *Urban Climate, 44*, 101203. https://doi.org/10.1016/j.uclim.2022.101203

Jia, J., & Zhang, X. (2021). A human-scale investigation into economic benefits of urban green and blue infrastructure based on big data and machine learning: A case study of Wuhan. *Journal of Cleaner Production, 316*, 128321. https://doi.org/10.1016/j.jclepro.2021.128321

Kafy, A. A., Saha, M., Faisal, A.-A., Rahaman, Z. A., Rahman, M. T., Liu, D., Fattah, M. A., Al Rakib, A., AlDousari, A. E., Rahaman, S. N., Hasan, M. Z., & Ahasan, M. A. K. (2022). Predicting the impacts of land use/land cover changes on seasonal urban thermal characteristics using machine learning algorithms. *Building and Environment, 217*, 109066. https://doi.org/10.1016/j.buildenv.2022.109066

Karalas, K., Tsagkatakis, G., Zervakis, M., & Tsakalides, P. (2016). Land classification using remotely sensed data: Going multilabel. *IEEE Transactions on Geoscience and Remote Sensing, 54*(6), 3548–3563. https://doi.org/10.1109/TGRS.2016.2520203

Khan, A., & Sudheer, M. (2022). Machine learning-based monitoring and modeling for spatio-temporal urban growth of Islamabad. *The Egyptian Journal of Remote Sensing and Space Sciences, 25*(2), 541–550. https://doi.org/10.1016/j.ejrs.2022.03.012

Koumetio Tekouabou, S. C., Diop, E. B., Azmi, R., Jaligot, R., & Chenal, J. (2022). Reviewing the application of machine learning methods to model urban form indicators in planning decision support systems: Potential, issues and challenges. *Journal of King Saud University – Computer and Information Sciences, 34*(8), 5943–5967. https://doi.org/10.1016/j.jksuci.2021.08.007

Kwak, Y., Park, C., & Deal, B. (2020). Discerning the success of sustainable planning: A comparative analysis of urban heat island dynamics in Korean new towns. *Sustainable Cities and Society, 61*, 102341. https://doi.org/10.1016/j.scs.2020.102341

Lammers, R., Li, A., Nag, S., & Ravindra, V. (2021). Prediction models for urban flood evolution for satellite remote sensing. *Journal of Hydrology, 603*, 127175. https://doi.org/10.1016/j.jhydrol.2021.127175

Li, L., Girguis, M., Lurmann, F., Pavlovic, N., McClure, C., Franklin, M., Wu, J., Oman, L. D., Breton, C., Gilliland, F., & Habre, R. (2020). Ensemble-based deep learning for estimating $PM_{2.5}$ over California with multisource big data including wildfire smoke. *Environment International, 145*. https://doi.org/10.1016/j.envint.2020.106143

Li, W., Yigitcanlar, T., Erol, I., & Liu, A. (2021). Motivations, barriers and risks of smart home adoption: From systematic literature review to conceptual framework. *Energy Research & Social Science, 80*. https://doi.org/10.1016/j.erss.2021.102211

Liao, H.-Y., & Wen, T.-H. (2020). Extracting urban water bodies from high-resolution radar images: Measuring the urban surface morphology to control for radar's double-bounce effect. *International Journal of Applied Earth Observation and Geoinformation, 85*, 102003. https://doi.org/10.1016/j.jag.2019.102003

Liu, Z., Wu, R., Chen, Y., Fang, C., & Wang, S. (2021). Factors of ecosystem service values in a fast-developing region in China: Insights from the joint impacts of human activities and natural conditions. *Journal of Cleaner Production, 297*, 126588. https://doi.org/10.1016/j.jclepro.2021.126588

Liuzzi, M., Aravena Pelizari, P., Geiß, C., Masi, A., Tramutoli, V., & Taubenböck, H. (2019). A transferable remote sensing approach to classify building structural types for seismic risk analyses: The case of Val d'Agri area (Italy). *Bulletin of Earthquake Engineering, 17*(9), 4825–4853. https://doi.org/10.1007/s10518-019-00648-7

Lu, Z., Im, J., Rhee, J., & Hodgson, M. (2014). Building type classification using spatial and landscape attributes derived from LiDAR remote sensing data. *Landscape and Urban Planning, 130*(1), 134–148. https://doi.org/10.1016/j.landurbplan.2014.07.005

Mahdianpari, M., Granger, J. E., Mohammadimanesh, F., Warren, S., Puestow, T., Salehi, B., & Brisco, B. (2021). Smart solutions for smart cities: Urban wetland mapping using very-high resolution satellite imagery and airborne LiDAR data in the City of St. John's, NL, Canada. *Journal of Environmental Management, 280*, 111676–111676. https://doi.org/10.1016/j.jenvman.2020.111676

Martins, G. B., La Rosa, L. E. C., Happ, P. N., Filho, L. C. T. C., Santos, C. J. F., Feitosa, R. Q., & Ferreira, M. P. (2021). Deep learning-based tree species mapping in a highly diverse tropical urban setting. *Urban Forestry & Urban Greening, 64*, 127241. https://doi.org/10.1016/j.ufug.2021.127241

McCulloch, W. S., & Pitts, W. (1943). A logical calculus of the ideas immanent in nervous activity. *Bulletin of Mathematical Biophysics, 5,* 115–133. https://doi.org/10.1007/BF02478259

Milnar, M., & Ramaswami, A. (2020). Impact of urban expansion and in situ greenery on community-wide carbon emissions: Method development and insights from 11 US cities. *Environmental Science & Technology, 54*(24), 16086–16096. https://doi.org/10.1021/acs.est.0c02723

Milojevic-Dupont, N., & Creutzig, F. (2021). Machine learning for geographically differentiated climate change mitigation in urban areas. *Sustainable Cities and Society, 64,* 102526. https://doi.org/10.1016/j.scs.2020.102526

Mohammad, P., Goswami, A., Chauhan, S., & Nayak, S. (2022). Machine learning algorithm based prediction of land use land cover and land surface temperature changes to characterize the surface urban heat island phenomena over Ahmedabad city, India. *Urban Climate, 42.* https://doi.org/10.1016/j.uclim.2022.101116

Moher, D., Shamseer, L., Clarke, M., Ghersi, D., Liberati, A., Petticrew, M., Shekelle, P., Stewart, L. A., & Group, P.-P. (2015). Preferred reporting items for systematic review and meta-analysis protocols (PRISMA-P) 2015 statement. *Systematic Reviews, 4*(1), 1. https://doi.org/10.1186/2046-4053-4-1

Mortoja, M., Yigitcanlar, T., & Mayere, S. (2020). What is the most suitable methodological approach to demarcate peri-urban areas? A systematic review of the literature. *Land Use Policy, 95,* 104601. https://doi.org/10.1016/j.landusepol.2020.104601

Mushore, T. D., Odindi, J., Dube, T., Matongera, T. N., & Mutanga, O. (2017). Remote sensing applications in monitoring urban growth impacts on in-and-out door thermal conditions: A review. *Remote Sensing Applications: Society and Environment, 8,* 83–93. http://dx.doi.org/10.1016/j.rsase.2017.08.001

Osaku, D., Nakamura, R. Y. M., Pereira, L. A. M., Pisani, R. J., Levada, A. L. M., Cappabianco, F. A. M., Falcão, A. X., & Papa, J. P. (2015). Improving land cover classification through contextual-based optimum-path forest. *Information Sciences, 324,* 60–87. https://doi.org/10.1016/j.ins.2015.06.020

Patino, J. E., & Duque, J. C. (2013). A review of regional science applications of satellite remote sensing in urban settings. *Computers, Environment and Urban Systems, 37,* 1–17. https://doi.org/10.1016/j.compenvurbsys.2012.06.003

Peng, B., Meng, Z., Huang, Q., & Wang, C. (2019). Patch similarity convolutional neural network for urban flood extent mapping using bi-temporal satellite multispectral imagery. *Remote Sensing (Basel, Switzerland), 11*(21), 2492. https://doi.org/10.3390/rs11212492

Priem, F., Okujeni, A., van der Linden, S., & Canters, F. (2019). Comparing map-based and library-based training approaches for urban land-cover fraction mapping from Sentinel-2 imagery. *International Journal of Applied Earth Observation and Geoinformation, 78,* 295–305. https://doi.org/10.1016/j.jag.2019.02.003

Remote Sensing Links. (2023). Remote sensing internet resources directory. Available from http://www.ncl.ac.uk/tcmweb/msc_tcm/rs.htm.

Rosenblatt, F. (1958). The perceptron: A probabilistic model for information storage and organization in the brain. *Psychological Review, 65*(6), 386. https://doi.org/10.1037/h0042519

Sadiq, R., Akhtar, Z., Imran, M., & Ofli, F. (2022). Integrating remote sensing and social sensing for flood mapping. *Remote Sensing Applications: Society and Environment, 25.* https://doi.org/10.1016/j.rsase.2022.100697

Sannigrahi, S., Joshi, P. K., Keesstra, S., Paul, S. K., Sen, S., Roy, P. S., Chakraborti, S., & Bhatt, S. (2019). Evaluating landscape capacity to provide spatially explicit valued ecosystem services for sustainable coastal resource management. *Ocean & Coastal Management, 182,* 104918. https://doi.org/10.1016/j.ocecoaman.2019.104918

Santos, L., Carvalho, B. L., Anderson, T., Rudorff, L. O., Marchezini, C. M., Londe, V., & Saito, L. R. (2017). An RS-GIS-based comprehensive impact assessment of floods – A case study in Madeira River, Western Brazilian Amazon. *IEEE Geoscience and Remote Sensing Letters, 14*(9), 1614–1617. https://doi.org/10.1109/LGRS.2017.2726524

Sanya, R., & Mwebaze, E. (2020). Identifying patterns in urban housing density in developing countries using convolutional networks and satellite imagery. *Heliyon, 6*(12), e05617–e05617. https://doi.org/10.1016/j.heliyon.2020.e05617

Sapena, M., Wurm, M., Taubenböck, H., Tuia, D., & Ruiz, L. A. (2021). Estimating quality of life dimensions from urban spatial pattern metrics. *Computers, Environment and Urban Systems, 85,* 101549. https://doi.org/10.1016/j.compenvurbsys.2020.101549

Schilling, H., Bulatov, D., & Middelmann, W. (2018). Object-based detection of vehicles using combined optical and elevation data. *ISPRS Journal of Photogrammetry and Remote Sensing, 136,* 85–105. https://doi.org/10.1016/j.isprsjprs.2017.11.023

Schlosser, A. D., Szabó, G., Bertalan, L., Varga, Z., Enyedi, P., & Szabó, S. (2020). Building extraction using orthophotos and dense point cloud derived from visual band aerial imagery based on machine learning and segmentation. *Remote Sensing (Basel, Switzerland), 12*(15), 2397. https://doi.org/10.3390/RS12152397

Schneider, A. (2012). Monitoring land cover change in urban and peri-urban areas using dense time stacks of Landsat satellite data and a data mining approach. *Remote Sensing of Environment, 124,* 689–704. https://doi.org/10.1016/j.rse.2012.06.006

Shafizadeh-Moghadam, H., Asghari, A., Tayyebi, A., & Taleai, M. (2017). Coupling machine learning, tree-based and statistical models with cellular automata to simulate urban growth. *Computers, Environment and Urban Systems, 64,* 297–308. https://doi.org/10.1016/j.compenvurbsys.2017.04.002

Shen, H., Li, T., Yuan, Q., & Zhang, L. (2018). Estimating regional ground-level $PM_{2.5}$ directly from satellite top-of-atmosphere reflectance using deep belief networks. *Journal of Geophysical Research. Atmospheres, 123*(24), 13,875–13,886. https://doi.org/10.1029/2018JD028759

Shi, T., Hu, X., Guo, L., Su, F., Tu, W., Hu, Z., Liu, H., Yang, C., Wang, J., Zhang, J., & Wu, G. (2021). Digital mapping of zinc in urban topsoil using multisource geospatial data and random forest. *The Science of the Total Environment, 792,* 148455–148455. https://doi.org/10.1016/j.scitotenv.2021.148455

Shogrkhodaei, S. Z., Razavi-Termeh, S. V., & Fathnia, A. (2021). Spatio-temporal modeling of $PM_{2.5}$ risk mapping using three machine learning algorithms. *Environmental Pollution, 289,* 117859–117859. https://doi.org/10.1016/j.envpol.2021.117859

Singleton, A., Arribas-Bel, D., Murray, J., & Fleischmann, M. (2022). Estimating generalized measures of local neighbourhood context from multispectral satellite images using a convolutional neural network. *Computers, Environment and Urban Systems, 95,* 101802. https://doi.org/10.1016/j.compenvurbsys.2022.101802

Soilán, M., Riveiro, B., Martínez-Sánchez, J., & Arias, P. (2016). Traffic sign detection in MLS acquired point clouds for geometric and image-based semantic inventory. *ISPRS Journal of Photogrammetry and Remote Sensing, 114,* 92–101. https://doi.org/10.1016/j.isprsjprs.2016.01.019

Son, T., Weedon, Z., Yigitcanlar, T., Sanchez, T., Corchado, J., & Mehmood, R. (2023). Algorithmic urban planning for smart and sustainable development: Systematic review of the literature. *Sustainable Cities and Society, 94,* 104562. https://doi.org/10.1016/j.scs.2023.104562

Stewart, J. D., & Kremer, P. (2022). Temporal change in relationships between urban structure and surface temperature. *Environment and Planning. B, Urban Analytics and City Science.* https://doi.org/10.1177/23998083221083677

Tella, A., Balogun, A.-L., Adebisi, N., & Abdullah, S. (2021). Spatial assessment of PM_{10} hotspots using Random Forest, K-Nearest Neighbour and Naïve Bayes. *Atmospheric Pollution Research, 12*(10), 101202. https://doi.org/10.1016/j.apr.2021.101202

Touzani, S., & Granderson, J. (2021). Open data and deep semantic segmentation for automated extraction of building footprints. *Remote Sensing (Basel, Switzerland), 13*(13), 2578. https://doi.org/10.3390/rs13132578

United Nations. (2023). *Global indicator framework for the Sustainable Development Goals and targets of the 2030 Agenda for Sustainable Development.* https://unstats.un.org/sdgs/indicators/Global%20Indicator%20Framework%20after%202023%20refinement_Eng.pdf

Uniyal, S., Purohit, S., Chaurasia, K., Rao, S. S., & Amminedu, E. (2022). Quantification of carbon sequestration by urban forest using Landsat 8 OLI and machine learning algorithms in Jodhpur, India. *Urban Forestry & Urban Greening, 67,* 127445. https://doi.org/10.1016/j.ufug.2021.127445

Vulova, S., Meier, F., Rocha, A. D., Quanz, J., Nouri, H., & Kleinschmit, B. (2021). Modeling urban evapotranspiration using remote sensing, flux footprints, and artificial intelligence. *The Science of the Total Environment, 786,* 147293–147293. https://doi.org/10.1016/j.scitotenv.2021.147293

Wang, C., Wang, J., Shen, Y., & Zhang, X. (2019). Autonomous navigation of UAVs in large-scale complex environments: A deep reinforcement learning approach. *IEEE Transactions on Vehicular Technology, 68*(3), 2124–2136. https://doi.org/10.1109/tvt.2018.2890773

Wang, P., Bayram, B., & Sertel, E. (2022a). A comprehensive review on deep learning based remote sensing image super-resolution methods. *Earth-Science Reviews,* 104110. https://doi.org/10.1016/j.earscirev.2022.104110

Wang, Q., Ren, Q., & Liu, J. (2016). Identification and apportionment of the drivers of land use change on a regional scale: Unbiased recursive partitioning-based stochastic model application. *Agriculture, Ecosystems & Environment, 217,* 99–110. https://doi.org/10.1016/j.agee.2015.11.004

Wang, T., Chandra, A., Jung, J., & Chang, A. (2022b). UAV remote sensing based estimation of green cover during turfgrass establishment. *Computers and Electronics in Agriculture, 194,* 106721. https://doi.org/10.1016/j.compag.2022.106721

Wang, Y., Cui, L., Zhang, C., Chen, W., Xu, Y., & Zhang, Q. (2022c). A two-stage seismic damage assessment method for small, dense, and imbalanced buildings in remote sensing images. *Remote Sensing (Basel, Switzerland), 14*(4), 1012. https://doi.org/10.3390/rs14041012

Wei, G., Zhang, Z., Ouyang, X., Shen, Y., Jiang, S., Liu, B., & He, B.-J. (2021). Delineating the spatial-temporal variation of air pollution with urbanization in the Belt and Road Initiative area. *Environmental Impact Assessment Review, 91*, 106646. https://doi.org/10.1016/j.eiar.2021.106646

Wieland, M., & Pittore, M. (2016). Large-area settlement pattern recognition from Landsat-8 data. *ISPRS Journal of Photogrammetry and Remote Sensing, 119*, 294–308. https://doi.org/10.1016/j.isprsjprs.2016.06.010

Wu, X., Ou, J., Wen, Y., Liu, X., He, J., & Zhang, J. (2022). Developing a data-fusing method for mapping fine-scale urban three-dimensional building structure. *Sustainable Cities and Society, 80*, 103716. https://doi.org/10.1016/j.scs.2022.103716

Wurm, M., Stark, T., Zhu, X. X., Weigand, M., & Taubenböck, H. (2019). Semantic segmentation of slums in satellite images using transfer learning on fully convolutional neural networks. *ISPRS Journal of Photogrammetry and Remote Sensing, 150*, 59–69. https://doi.org/10.1016/j.isprsjprs.2019.02.006

Xu, X., Bai, Y., Liu, Y., Zhao, X., & Sun, Y. (2022). MM-UrbanFAC: Urban functional area classification model based on multimodal machine learning. *IEEE Transactions on Intelligent Transportation Systems, 23*(7), 8488–8497. https://doi.org/10.1109/TITS.2021.3083486

Xu, Y., Ho, H. C., Wong, M. S., Deng, C., Shi, Y., Chan, T. C., & Knudby, A. (2018a). Evaluation of machine learning techniques with multiple remote sensing datasets in estimating monthly concentrations of ground-level $PM_{2.5}$. *Environmental Pollution, 242*, 1417–1426. https://doi.org/10.1016/j.envpol.2018.08.029

Xu, Y., Knudby, A., & Côté-Lussier, C. (2018b). Mapping ambient light at night using field observations and high-resolution remote sensing imagery for studies of urban environments. *Building and Environment, 145*, 104–114. https://doi.org/10.1016/j.buildenv.2018.09.002

Xu, Y., Wu, L., Xie, Z., & Chen, Z. (2018c). Building extraction in very high resolution remote sensing imagery using deep learning and guided filters. *Remote Sensing (Basel, Switzerland), 10*(1), 144. https://doi.org/10.3390/rs10010144

Yan, X., Zang, Z., Jiang, Y., Shi, W., Guo, Y., Li, D., Zhao, C., & Husi, L. (2021). A spatial-temporal interpretable deep learning model for improving interpretability and predictive accuracy of satellite-based $PM_{2.5}$. *Environmental Pollution, 273*. https://doi.org/10.1016/j.envpol.2021.116459

Yan, X., Zang, Z., Luo, N., Jiang, Y., & Li, Z. (2020). New interpretable deep learning model to monitor real-time $PM_{2.5}$ concentrations from satellite data. *Environment International, 144*. https://doi.org/10.1016/j.envint.2020.106060

Yang, J., Zhou, A., Han, L., Li, Y., & Xie, Y. (2021). Monitoring urban black-odorous water by using hyperspectral data and machine learning. *Environmental Pollution, 269*, 116166. https://doi.org/10.1016/j.envpol.2020.116166

Yang, N., Shi, H., Tang, H., & Yang, X. (2022). Geographical and temporal encoding for improving the estimation of $PM_{2.5}$ concentrations in China using end-to-end gradient boosting. *Remote Sensing of Environment, 269*. https://doi.org/10.1016/j.rse.2021.112828

Yang, X., Yao, C., Chen, Q., Ye, T., & Jin, C. (2019). Improved estimates of population exposure in low-elevation coastal zones of China. *International Journal of Environmental Research and Public Health, 16*(20), 4012. https://doi.org/10.3390/ijerph16204012

Ye, T., Zhao, N., Yang, X., Ouyang, Z., Liu, X., Chen, Q., Hu, K., Yue, W., Qi, J., Li, Z., & Jia, P. (2019). Improved population mapping for China using remotely sensed and points-of-interest data within a random forests model. *The Science of the Total Environment, 658*, 936–946. https://doi.org/10.1016/j.scitotenv.2018.12.276

Yigitcanlar, T. (2009). Planning for smart urban ecosystems: Information technology applications for capacity building in environmental decision making. *Theoretical and Empirical Researches in Urban Management*, 4(3 (12)), 5–21.

Yigitcanlar, T., Dodson, J., Gleeson, B., & Sipe, N. (2007). Travel self-containment in master planned estates: Analysis of recent Australian trends. *Urban Policy and Research, 25*(1), 129–149. https://doi.org/10.1080/08111140701255823

Yigitcanlar, T., Fabian, L., & Coiacetto, E. (2008). Challenges to urban transport sustainability and smart transport in a tourist city: The Gold Coast, Australia. *Open Transportation Journal, 2*, 29–46. https://doi.org/10.2174/1874447800802010029

Yoo, C., Im, J., Park, S., & Quackenbush, L. J. (2018). Estimation of daily maximum and minimum air temperatures in urban landscapes using MODIS time series satellite data. *ISPRS Journal of Photogrammetry and Remote Sensing, 137*, 149–162. https://doi.org/10.1016/j.isprsjprs.2018.01.018

Zhang, C., Liu, C., Li, B., Zhao, F., & Zhao, C. (2022a). Spatiotemporal neural network for estimating surface NO_2 concentrations over north China and their human health impact. *Environmental Pollution, 307*. https://doi.org/10.1016/j.envpol.2022.119510

Zhang, C., Pan, X., Li, H., Gardiner, A., Sargent, I., Hare, J., & Atkinson, P. M. (2018a). A hybrid MLP-CNN classifier for very fine resolution remotely sensed image classification. *ISPRS Journal of Photogrammetry and Remote Sensing, 140*, 133–144. https://doi.org/10.1016/j.isprsjprs.2017.07.014

Zhang, C., Pan, X., Zhang, S. Q., Li, H. P., & Atkinson, P. M. (2017). A rough set decision tree based MLP-CNN for very high resolution remotely sensed image classification. *International Archives of the Photogrammetry, Remote Sensing and Spatial Information Sciences – ISPRS Archives, 42*(2), 1451–1454. https://doi.org/10.5194/isprs-archives-XLII-2-W7-1451-2017

Zhang, P., Liu, L., Yang, L., Zhao, J., Li, Y., Qi, Y., Ma, X., & Cao, L. (2023). Exploring the response of ecosystem service value to land use changes under multiple scenarios coupling a mixed-cell cellular automata model and system dynamics model in Xi'an, China. *Ecological Indicators, 147*, 110009. https://doi.org/10.1016/j.ecolind.2023.110009

Zhang, P., Ma, W., Wen, F., Liu, L., Yang, L., Song, J., Wang, N., & Liu, Q. (2021). Estimating $PM_{2.5}$ concentration using the machine learning GA-SVM method to improve the land use regression model in Shaanxi, China. *Ecotoxicology and Environmental Safety, 225*, 112772–112772. https://doi.org/10.1016/j.ecoenv.2021.112772

Zhang, X., Chen, W., Chen, Z., Yang, F., Meng, C., Gou, P., Zhang, F., Feng, J., Li, G., & Wang, Z. (2022b). Construction of cloud-free MODIS-like land surface temperatures coupled with a regional weather research and forecasting (WRF) model. *Atmospheric Environment, 283*, 119190. https://doi.org/10.1016/j.atmosenv.2022.119190

Zhang, Z., Zhang, X., Sun, Y., & Zhang, P. (2018b). Road centerline extraction from very-high-resolution aerial image and LiDAR data based on road connectivity. *Remote Sensing (Basel, Switzerland), 10*(8), 1284. https://doi.org/10.3390/rs10081284

Zhao, X., Xia, N., Xu, Y., Huang, X., & Li, M. (2021). Mapping population distribution based on XGBoost using multisource data. *IEEE Journal of Selected Topics in Applied Earth Observations and Remote Sensing, 14*, 11567–11580. https://doi.org/10.1109/JSTARS.2021.3125197

Zhou, X., Liu, C., Akbar, A., Xue, Y., & Zhou, Y. (2021). Spectral and spatial feature integrated ensemble learning method for grading urban river network water quality. *Remote Sensing (Basel, Switzerland)*, 13(22), 4591. https://doi.org/10.3390/rs13224591

Zhu, Q., Sun, Y., Guan, Q., Wang, L., & Lin, W. (2022). A weakly pseudo-supervised decorrelated subdomain adaptation framework for cross-domain land-use classification. *IEEE Transactions on Geoscience and Remote Sensing, 60*, 1–13. https://doi.org/10.1109/TGRS.2022.3170335

2

Urban AI Applications in Heat Mapping

2.1 Introduction

The rapid pace of urbanisation has intensified urban heat challenges, posing serious risks to public health and well-being (Shivanna, 2022) and threatening the achievement of Sustainable Development Goals (Kamruzzaman et al., 2018). These risks are further exacerbated by climate change, necessitating a reassessment of urban infrastructure to mitigate adverse effects (Yin et al., 2023). Microclimates, where design and material choices create significant localised climatic variations, often result in notably higher temperatures and humidity compared to rural areas (Degirmenci et al., 2021; Vinayak et al., 2022). Managing such conditions demands a nuanced understanding and practical strategies (Cheval et al., 2024). Urban heat mapping is crucial in this context, providing reliable, high-resolution data to support diverse stakeholders in addressing these challenges (Dimitrov et al., 2024).

Despite extensive research, accurately predicting urban thermal environments remains complex due to the dynamic nature of variables and the technological limitations of traditional methods (Wang et al., 2022). Generating precise, georeferenced, and high-resolution data is challenging, as traditional techniques, grounded in empirical data and environmental simulation tools, often fail to process large-scale data efficiently or adapt to rapidly changing environmental conditions (Li et al., 2022; Shi et al., 2021). These shortcomings underscore the limitations of conventional heat mapping approaches and highlight the urgent need for advanced computational tools that can manage spatial complexity, improve accuracy, and support real-time decision-making in urban settings.

Urban artificial intelligence (Urban AI) offers a transformative approach to addressing the limitations of traditional geospatial analysis by leveraging advanced machine learning (ML) and deep learning (DL) techniques (Yigitcanlar et al., 2023b, 2024c, 2024d). By integrating large-scale urban datasets, including satellite imagery, sensor networks, and real-time environmental data, Urban AI enhances spatial accuracy, computational efficiency, and adaptability in complex urban environments (Yigitcanlar, 2024a, 2024b). This study introduces a novel U-Net convolutional neural network (CNN) model as an Urban AI agent to enhance urban heat mapping and microclimate analysis. Using high-resolution satellite imagery, the model effectively captures fine-scale spatial temperature variations, offering a robust tool for identifying urban heat island (UHI) effects. The U-Net architecture significantly reduces computational time and complexity, processing heat maps rapidly while maintaining high spatial accuracy. This enables the development of targeted mitigation strategies, such as optimising green infrastructure and managing anthropogenic heat sources (Irfeey et al., 2023; Mathew et al., 2019).

This study introduces a novel U-Net CNN model as an Urban AI agent for high-resolution urban heat mapping. Trained on RGB satellite imagery, the model captures fine-scale spatial temperature variations, enabling accurate detection of UHI effects. The encoder–decoder architecture of U-Net enhances spatial accuracy while maintaining computational efficiency, supporting timely and localised heat mitigation strategies such as optimising green infrastructure or reducing anthropogenic heat sources.

Local governments have long aimed to integrate urban technologies into planning and decision-making (Yigitcanlar, 2006; Yigitcanlar et al., 2023a, 2023b). To support these efforts, the proposed model of near-real-time heat maps can guide policy and emergency response. Its high-resolution outputs are valuable for identifying heat-prone zones at street and neighbourhood levels, informing targeted interventions like tree placement and surface treatments. The model's efficiency and low hardware requirements also make it suitable for widespread use, including in resource-limited contexts.

Overall, the study demonstrates how Urban AI can bridge the gap between geospatial modelling and actionable planning. By offering a scalable and accessible tool for heat monitoring, the model contributes to more adaptive, resilient, and liveable urban environments in the face of intensifying climate risks.

2.2 Literature Background

The rapid growth of urban populations is driving the spatial expansion of cities, both horizontally and vertically (Wang et al., 2023). This urban sprawl is widely recognised as a contributing factor to numerous environmental and social challenges, including public health concerns (Mohan & Kandya, 2015), elevated pollutant concentrations (Ge et al., 2024; Sarrat et al., 2006), and increased energy consumption (Akbari et al., 2016). A significant consequence of this growth is the replacement of natural landscapes with impervious surfaces, leading to altered urban climates (McCarthy et al., 2010; Zhang et al., 2013) and notably higher temperatures, especially in the late afternoon and evening. This UHI effect, where urban areas are warmer than their rural counterparts (Voogt & Oke, 2003), is further intensified by global temperature increases. Classified as an atmospheric hazard impacting a large proportion of the population (Wang et al., 2023), addressing the UHI effect requires coordinated local and national government strategies.

Urban heat mapping is an essential tool for informing these strategies and promoting sustainable urban growth. Traditional approaches to heat mapping include ground-based measurements from weather stations (Fabrizi et al., 2010), remote sensing to derive land surface temperatures (LSTs) (Zhou et al., 2018), and empirical models based on land use or topographic variables. However, these methods often face limitations, including low spatial resolution, limited coverage, time constraints, and inefficient data processing (Elmes et al., 2017). Additionally, their reliance on simplistic algorithms fails to capture the complexity of urban thermal climates (Shi et al., 2021; Sidiqui et al., 2022). The advent of high-resolution Earth observation satellites has improved this process, enabling more granular analyses of air temperature and microclimate variations (Coutts et al., 2016). Despite these advancements, traditional techniques still lack the adaptability and precision required for dynamic urban environments (Hung et al., 2006; Nichol et al., 2009).

In parallel, AI has increasingly been employed to address diverse urban challenges, from transportation to climate risk management (Faisal et al., 2021; Ye et al., 2025). Urban AI offers a transformative approach to urban heat mapping by integrating ML with geospatial data, enabling more accurate and efficient analysis. Recent studies have used algorithms such as artificial neural networks (ANN), random forests (RF), and eXtreme Gradient Boosting (XGB) for microclimate prediction (Guo et al., 2024; Li et al., 2023). However, these models often struggle to capture spatial hierarchies and contextual dependencies in complex urban environments (Bhakare et al., 2024; Li et al., 2024). In contrast, convolutional neural networks (CNNs) offer pixel-level precision and spatial awareness, making them more suitable for geospatial applications (Nguyen et al., 2022; Ronneberger et al., 2015).

Building on this foundation, the U-Net model has gained prominence for its encoder–decoder structure and skip connections that retain both high- and low-level features. Originally introduced for biomedical segmentation (Ronneberger et al., 2015), U-Net is now widely adopted in urban applications for its ability to preserve spatial detail and delineate boundaries effectively (Ramani et al., 2024). These features are particularly important for urban heat mapping, where small-scale thermal gradients must be detected across varied urban morphologies. Comparative studies show U-Net's consistent superiority over FCN, SegNet, and even advanced variants like Attention U-Net and Residual U-Net (Supritha & Murthy, 2023; Wu et al., 2022). Rampal et al. (2022) further confirm that simpler encoder–decoder architectures like U-Net offer more interpretable, flexible, and stable learning, especially in geospatial contexts (Afshari et al., 2023).

Recent studies validated the feasibility of using RGB satellite imagery for estimating LST via DL models. Although RGB data lacks thermal bands, it provides rich spatial and morphological features correlated with surface heat distribution. Zhang et al. (2013) demonstrated that CNNs can infer LST from RGB images by learning urban structural patterns. Similarly, Zhao et al. (2023) achieved high spatial accuracy in predicting LST using only visible bands.

More recently, Abunnasr and Mhawej (2023) introduced a high-resolution downscaling model (HSR-LST) that predicted LST using RGB data with a mean absolute error of 0.88°C when validated against airborne thermal imagery. These studies support the use of RGB as a practical and scalable alternative, especially in data-constrained urban environments.

This study leverages the U-Net architecture within an Urban AI framework to generate high-resolution heat maps from RGB imagery, eliminating the need for direct thermal inputs. Traditional urban heat mapping depends heavily on thermal remote sensing, which often suffers from low spatial resolution, high acquisition costs, and limited frequency (Elmes et al., 2017). In contrast, RGB data is widely available and, when combined with U-Net's capabilities, offers a scalable, cost-effective, and practical alternative for real-time urban heat analysis.

The model effectively captures fine-scale variations across urban typologies, including dense city centres, suburban areas, and vegetated zones, yielding consistent and interpretable results (Morgan et al., 2024). In addition to its accuracy, U-Net is computationally efficient and accessible for implementation in cities with limited technical resources. It supports urban planning by informing interventions such as green infrastructure, reflective surfaces, and cooling corridors.

Furthermore, its ability to generate heat maps in near real-time makes it valuable for emergency response during heatwaves. By integrating Urban AI with planning systems, this study demonstrates a scalable and adaptive approach to urban climate resilience. As cities face escalating thermal risks, AI-driven heat mapping represents a critical step towards more sustainable, liveable, and climate-responsive urban development.

2.3 Research Design

2.3.1 Study Area and Data Collection

This study focused on urban heat mapping within the metropolitan area of Adelaide, South Australia (Figure 2A.1). Adelaide's urban landscape, comprising densely populated city centres, suburban neighbourhoods, and surrounding green spaces, offers a dynamic and varied environment for analysing urban heat distribution. This diversity provides an ideal case study for assessing how different urban forms and infrastructure impact surface temperatures, making Adelaide a valuable model for understanding UHI effects in cities with similarly mixed landscapes. The selection of Adelaide allows for a detailed analysis of how densely built-up areas, green spaces, and suburban layouts each contribute uniquely to localised temperature patterns.

2.3.2 Data Sources and Preparation

High-resolution thermal and spatial data were obtained from Data SA, the South Australian Government Data Directory (SA Dataset, 2024). This dataset provides advanced satellite imagery capable of capturing detailed surface temperature variations across both urban and non-urban areas. The thermal data were analysed at a spatial resolution of 2 m, where each pixel value corresponds to LST. This fine spatial resolution enables the precise identification of localised thermal variations, a key requirement for understanding UHI patterns and their relationship with urban morphology and vegetation. High-resolution imagery was specifically selected to capture fine-scale variations in surface heat intensity, thereby enabling a precise examination of temperature distribution. This level of detail is essential for identifying UHI patterns, understanding how temperature gradients shift across different urban layouts, and supporting strategies for urban heat mitigation.

The dataset used in this study contained spatial and thermal information at a resolution of 1024 × 1024 pixels per image, offering a balanced combination of spatial granularity and computational feasibility, which is particularly suitable for DL applications. The dataset captures a single temporal snapshot during summer, with recorded temperatures ranging from a maximum of 33°C to a minimum of 16°C. While these conditions provide critical insights into peak urban heat scenarios, the static nature of the dataset limits its ability to account for seasonal variability and dynamic changes in urban heat patterns.

The dataset includes critical spatial and thermal details that allow the model to learn the impacts of urban morphology and vegetation on UHI patterns. High-density urban areas with impervious surfaces, such as roads and rooftops, retain more heat, resulting in localised hotspots. Conversely, green spaces, including parks and vegetation, mitigate urban heat through natural cooling processes such as evapotranspiration. The fine spatial resolution of 2 m enables the model to detect subtle thermal gradients and distinguish between densely built and vegetated areas. By leveraging these capabilities, the U-Net model captures complex interactions between urban structure, vegetation, and thermal dynamics, delivering actionable insights for targeted UHI mitigation strategies.

This study focused on high-resolution thermal mapping as the primary approach for urban heat assessment. Although variables such as humidity and wind influence urban heat dynamics, they were not included in this model due to constraints in data availability, computational efficiency, and the study's focus on LST mapping.

2.3.3 Data Normalisation

As a preparatory step, the values of each pixel in the satellite imagery were normalised to a continuous range [0, 1]. Normalising pixel values is particularly beneficial in DL, as it helps prevent issues such as vanishing or exploding gradients, thereby promoting stable training and faster convergence. This step is essential for achieving consistent results and improving the overall effectiveness of the model training process.

2.3.4 Ground Truth Heat Maps

Urban heat maps representing ground truth heat intensity were converted into greyscale for single-channel processing, where pixel intensity represents relative heat levels across the urban landscape. This simplified format focuses the model's learning on heat intensity distribution without colour channel complexities. Following prediction, the greyscale output was colour-mapped through a predefined lookup table (LUT) for enhanced interpretability. This colourisation aligns heat intensities with visually distinct colour bands, which is crucial for visual assessments.

2.3.5 Training–Validation Split

The dataset was split into training (80%) and validation (20%) subsets to evaluate the model's generalisation capability. This ratio was selected to maximise the training set size, ensuring the model could learn robust spatial and thermal patterns while retaining sufficient validation data to assess performance on unseen samples.

To ensure the model generalised effectively, validation error was used as the primary indicator. The objective was to minimise training errors without compromising validation performance. If validation error began increasing while training error continued decreasing, it indicated overfitting, where the model memorised training data rather than learning meaningful patterns. To mitigate this, validation loss was continuously monitored, and early stopping was applied, halting training when validation performance began to deteriorate.

2.3.6 U-Net Model Architecture

U-Net CNN architecture was employed owing to its strengths in pixel-wise image-segmentation tasks. Originally developed for biomedical image segmentation, U-Net's encoder–decoder structure has proven effective in diverse applications where both fine and broad spatial details are essential to evaluate the output (Ronneberger et al., 2015). Its architecture allows the model to retain the spatial details of the input image, making it highly suitable for mapping tasks (Figure 2A.2). The U-Net model facilitates the detection of urban heat patterns by learning both local and global features, effectively balancing spatial accuracy with computational efficiency.

2.3.6.1 Encoder Path

The encoder consists of four convolutional blocks, each performing two convolutional operations with ReLU activations and batch normalisation to standardise outputs. Each block includes a max-pooling layer to downsample the feature maps, successively capturing broader and more abstract spatial features as the resolution decreases. This hierarchical feature extraction process enables the model to capture complex urban thermal landscape patterns, including isolated hotspots and expansive heat-affected areas.

2.3.6.2 Bottleneck Layer

The bottleneck layer, located at the centre of the model, contains 1024 filters that capture high-level feature representations of urban heat intensity. This dense Feature Layer compresses and abstracts the learned patterns from the encoder, enabling the model to recognise global spatial dependencies. This plays a crucial role in bridging the encoded feature representations with the decoder, where these abstracted features are used to generate the final output, such as identifying heat-prone zones.

2.3.6.3 Decoder Path with Skip Connections

The decoder path restores the spatial resolution of the encoded features, gradually transforming them back to the original image dimensions. Upsampling is performed through transposed convolutions, which learn to map lower-resolution feature maps to higher resolutions. Skip connections play a crucial role by linking each encoder layer with its corresponding decoder layer, allowing the decoder to access high-resolution features from the encoder. These connections reintegrate fine-grained spatial details, which enhances the ability of the model to accurately localise high-heat areas within urban grids. Skip connections help prevent information loss during downsampling, ensuring that critical details such as edges and small features vital for UHI detection are preserved across the network layers.

2.3.6.4 Output Layer

The output layer consists of a single convolution with a sigmoid activation function, producing a greyscale heat map where each pixel value indicates relative heat intensity. Post-processing applies a colour LUT to transform greyscale predictions into a coloured heat map, making the model output visually accessible for planners and decision-makers. This intuitive colour coding facilitates quick identification of areas most affected by UHI.

2.3.7 Model Training and Optimisation

The U-Net model was trained using a GPU-based environment, which is necessary for handling large-scale data and complex architecture efficiently. Model training was guided by a range of hyperparameters optimised to ensure effective convergence and robust performance on unseen data.

2.3.7.1 Loss Function

The mean squared error (MSE) loss function was selected, aligning with the continuous nature of the heat intensity target variable. Unlike binary segmentation tasks, in which the objective is to classify discrete classes, UHI mapping requires precise intensity prediction. MSE measures the squared difference between predicted and actual pixel values, penalising larger deviations and ensuring a focus on pixel-wise accuracy across the entire heat map.

2.3.7.2 Optimisation Algorithm and Learning Rate

The Adam optimiser was applied for its adaptability in handling large-scale, sparse gradients, which is an advantage in DL with high-dimensional data. An initial learning rate of 1e-4 was selected to balance convergence speed with stability. An adaptive learning rate scheduler further adjusted this rate, reducing it by a factor of 0.1 if validation loss did not improve after eight consecutive epochs, effectively fine-tuning the learning process, allowing the model to refine its solution while avoiding overshooting.

2.3.7.3 Training Duration and Batch Size

To capture detailed patterns within urban heat maps, the model was trained over 400 epochs with a batch size of 2. This batch size was selected to balance memory constraints imposed by high-resolution imagery and help avoid local minima. Prolonged training allows the model to learn intricate spatial and thermal patterns, with careful monitoring to prevent overfitting. Regularisation techniques and validation performance checks help ensure that the learned patterns generalise well to diverse urban areas, thereby improving the ability of the model to generate accurate predictions on unseen data.

2.3.7.4 Training–Validation Split

A training–validation split of 80/20 was adopted in this study based on a sensitivity analysis conducted using different ratios, including 85/15 and 70/30. The 80/20 configuration provided the best balance between training data sufficiency and model generalisation performance. This setup aligns with typical practices in DL and ensures the robustness of the model's predictive capabilities on unseen data.

2.3.7.5 Evaluation Metrics

The efficacy of the model was assessed using a combination of quantitative and qualitative metrics, specifically designed to align with the requirements of urban heat mapping.

Quantitative Metrics—MSE, Peak Signal-to-Noise Ratio (PSNR), and Structural Similarity Index (SSIM): MSE served as the primary quantitative metric, reflecting the pixel-wise accuracy of the heat intensity predictions. PSNR was used to evaluate the visual fidelity of the predicted heat maps in comparison to the ground truth, while SSIM assessed the model's ability to preserve spatial patterns and structural consistency. Together, these metrics provide a comprehensive evaluation of the model's effectiveness in replicating real urban heat distributions, where lower MSE and higher PSNR and SSIM values signify more substantial alignment with the reference data.

Qualitative Analysis: In addition to quantitative metrics, qualitative visual inspections were performed to evaluate the model's predictions. The predicted heat maps were visually compared against actual heat distributions, providing a practical assessment of spatial alignment and the model's accuracy in localising high-heat areas. This step is essential in applications such as UHI mapping, where real-world relevance and spatial coherence are critical for decision-making.

2.4 Analysis and Results

The following section presents a comprehensive evaluation of the U-Net model's performance in predicting urban heat distribution, detailing quantitative accuracy metrics, visual consistency with ground truth data, robustness across diverse urban environments, and error sources.

2.4.1 Model Performance and Training Metrics

The U-Net model was trained on high-resolution satellite imagery of urban areas, utilising an 85/15 split between training and validation data. The model's training process captured both fine-scale and broad thermal patterns within urban heat distribution. Over 400 epochs, the model's training performance was monitored using MSE as the primary loss function (Figure 2A.3). The training loss consistently decreased from an initial MSE of approximately 0.0013 to a final value of 0.0007, indicating effective convergence. The validation loss followed a similar decreasing trend, confirming the ability of the model to generalise to unseen data.

The model was optimised using the Adam algorithm with an initial learning rate of 1e−4. A learning rate scheduler dynamically adjusted the learning rate by a factor of 0.1 when the validation loss plateaued for eight consecutive epochs, refining the convergence process. A batch size of 2 was selected to balance memory requirements for processing 1024 × 1024 resolution images while capturing urban thermal patterns effectively. This combination of extended training, dynamic learning rate adjustments, and optimal batch size supported the model in learning complex spatial dependencies, ensuring robust performance across various urban environments.

2.4.2 Visual Analysis of Predicted Heat Maps

A comparison of satellite images, ground truth heat maps, and U-Net model's predicted heat maps for four distinct urban areas (A, B, C, and D) is presented in (Figure 2A.4). Starting with Area A, the model successfully identified high-temperature zones within densely built-up regions, while cooler temperatures were accurately represented in nearby green spaces. This differentiation between urban infrastructure and vegetated areas closely aligned with the ground truth, highlighting the model's sensitivity to variations in land cover.

In Area B, which features an open green space surrounded by residential buildings, the model predictions showed lower temperatures within the vegetative area and higher temperatures in the adjacent built environment. This result reflects the model's capability to capture thermal gradients influenced by both natural cooling effects and nearby urban structures. In densely populated Area C, elevated temperatures appeared primarily along major roads and rooftops, where surface materials are known to absorb and retain heat. The predicted map closely mirrored the ground truth, demonstrating the efficacy of the model in identifying the typical thermal patterns associated with high-density urban zones.

Finally, Area D, with its complex layout of curved streets and varied building densities, further illustrates the model's ability to generalise across intricate urban environments. Temperature gradients were accurately represented, with higher temperatures in dense building clusters and cooler values in vegetated zones. This consistency across diverse urban settings underscores the robustness and adaptability of the model in predicting thermal patterns across structured and complex landscapes.

Although the model performed consistently well, it exhibited minor inaccuracies in specific urban environments, particularly in areas with high variability in building density, surface materials, and environmental context. In these heterogeneous settings, the model showed slight deviations from the ground truth heat maps, particularly in regions where mixed land cover introduced complex thermal patterns.

In densely built urban centres with minimal vegetation, the model sometimes underestimated the intensity of UHIs. In contrast, in highly vegetated zones, the model occasionally predicted marginally higher temperatures than observed. These discrepancies were most prominent in urban areas with a combination of concrete and vegetative surfaces, where thermal variation was complex.

2.4.3 Quantitative Evaluation Metrics

The final model performance was assessed using MSE, metrics well-suited for continuous temperature prediction. On the validation set, the model achieved an MSE of 0.00383, reflecting its ability to produce pixel-level predictions that closely matched the ground truth heat intensities. MSE provided insight into the model's effectiveness in minimising larger errors.

These metrics were selected to assess the accuracy of model predictions in capturing real urban heat distributions. Due to the continuous nature of the output, segmentation metrics such as Intersection over Union (IoU) and pixel accuracy were not applicable, as they are intended for categorical segmentation tasks. Nonetheless, MSE directly assessed the capacity of the model to predict fine-scale temperature gradients, affirming its suitability for urban heat mapping applications. In addition to its predictive accuracy, the model demonstrated efficient processing, with each image processed in under 30 seconds. This rapid processing time is advantageous for large-scale urban heat mapping.

To further evaluate the visual and structural quality of the generated heat maps, we also calculated PSNR and SSIM. A PSNR of approximately 18.7 dB indicates that the predicted heat maps are visually consistent with the ground truth references, offering sufficient detail for spatial interpretation. Additionally, the SSIM value of up to 0.378 demonstrates the model's ability to preserve spatial thermal patterns across different urban morphologies. Together, these complementary metrics confirm the model's robustness in both numerical prediction and visual fidelity—attributes that are critical for practical urban analysis and planning applications.

2.5 Findings and Discussion

This study demonstrates the efficacy of the U-Net model, an advanced Urban AI tool, in generating high-resolution urban heat maps, providing precise assessments of thermal distribution patterns across Adelaide, South Australia. The model's high predictive accuracy, validated by low MSE value, underscores its ability to capture nuanced thermal variations within diverse urban landscapes. Its rapid processing time of under 30 seconds per image highlights its computational efficiency, making it an essential resource for large-scale urban heat mapping applications as well as emergency response during extreme heat events. These findings represent significant contributions to the broader field of urban heat assessment, addressing both accuracy and scalability—key elements for effective climate resilience planning.

2.5.1 Model Performance and Implications

As an Urban AI solution, the U-Net model delivers high-resolution outputs that are particularly valuable for identifying microscale temperature variations in urban environments. This granular spatial detail enables urban planners to pinpoint heat-prone zones within dense infrastructure, facilitating targeted interventions such as green infrastructure expansion and the use of reflective materials, both of which are proven to reduce UHI intensity (Bowler et al., 2010; Gago et al., 2013; Shaamala et al., 2024). Furthermore, the model's rapid processing capability aligns with operational requirements for extreme weather events, offering real-time decision-making support for mitigating heat risks (Chen et al., 2006; Johnson et al., 2012).

The U-Net model's robust performance across varied urban morphologies highlights its adaptability. It accurately identified high-temperature zones in densely built areas, such as roads and rooftops, where impervious surfaces retain heat (Rizwan et al., 2008; Voogt & Oke, 2003). In suburban areas, it effectively captured lower temperatures associated with vegetation, reflecting greenery's cooling effects (Bowler et al., 2010; Weng, 2009). However, discrepancies arose in mixed land cover areas, where complex interactions between concrete and vegetation create challenging thermal patterns. These findings emphasise the importance of integrating contextual environmental variables, such as shading and material reflectivity, to further enhance accuracy in heterogeneous urban environments (Gago et al., 2013; Middel et al., 2014).

Compared to traditional methods, the U-Net model exemplifies the transformative potential of Urban AI. Traditional techniques often rely on moderate-resolution satellite data (e.g., MODIS and Landsat) with extensive preprocessing, limiting their effectiveness for precise UHI analysis (Shi et al., 2021). In contrast, the U-Net's encoder–decoder architecture, combined with skip connections, preserves critical pixel-level spatial features, enabling highly accurate identification of small-scale thermal gradients (Briegel et al., 2023; Ronneberger et al., 2015). This capability is indispensable for managing heterogeneous urban environments, where targeted interventions are vital for mitigating localised heat challenges (Diem et al., 2024; Elmarakby & Elkadi, 2024).

The U-Net model's operational efficiency further cements its role as a foundational Urban AI agent. Its ability to rapidly assess thermal patterns supports timely decision-making during extreme heat events, addressing climate resilience needs effectively. Unlike traditional LST approaches that face significant delays (Tomlinson et al., 2011; Zhou et al., 2018), the U-Net model ensures prompt, data-driven urban planning and climate-adaptive design (Almashhour et al., 2024; Kumar et al., 2024).

Future advancements could include integrating additional environmental variables, such as humidity, wind patterns, and solar radiation, to better contextualise temperature patterns and enhance sensitivity to complex urban environments (Ghorbany et al., 2024; Liu & Morawska, 2020; Tehrani et al., 2024). Expanding the training data to encompass cities with diverse climates and urban morphologies would strengthen the model's adaptability and robustness, ensuring its usability in varied geographic contexts. Further exploration of hybrid models combining U-Net with temporal data analysis, such as recurrent neural networks, could refine its predictive accuracy for continuous seasonal assessments (Demiss & Elsaigh, 2024; Nazir et al., 2024).

Evaluating the U-Net model across multiple cities would validate its scalability and establish a framework for Urban AI-enabled heat mapping. Such a framework could provide tailored insights into diverse thermal profiles, enabling planners to design interventions for location-specific UHI challenges (Li et al., 2017; Zhou et al., 2011). By aligning with global urban resilience goals, this Urban AI-driven approach fosters sustainable urban development, improves liveability, and enhances climate adaptability (Degirmenci et al., 2021; Ramyar et al., 2021).

2.5.2 Urban Heat Resilience and Vulnerability in Adelaide

This study highlights significant variations in urban heat resilience across metropolitan Adelaide, shaped by land use, urban morphology, and vegetation cover. High-resolution mapping using the U-Net model reveals that densely built areas, including the CBD, industrial precincts, and transport corridors, record the highest LSTs due to impervious surfaces and heat-retaining materials (Figure 2A.5).

Conversely, suburban neighbourhoods with green spaces exhibit moderate resilience, benefiting from tree cover and open spaces that help regulate temperature extremes. However, new high-density residential developments with limited vegetation contribute to localised warming, underscoring the need for climate-responsive urban design.

Coastal zones, parklands, and peri-urban areas demonstrate the greatest heat resilience, with lower LSTs due to the cooling effects of vegetation, water bodies, and open landscapes. These findings are in line with Morgan et al. (2024) and reinforce the crucial role of green and blue infrastructure in mitigating urban heat stress and enhancing urban climate resilience.

2.5.3 Practical Implications for Urban Planning

Urban planners and local governments currently rely on a combination of coarse-resolution thermal satellite imagery (e.g., MODIS or Landsat), empirical models, and in situ measurements to inform their urban heat mitigation strategies. These include tree canopy targets, zoning incentives for reflective materials, and emergency heatwave plans. However, existing approaches often lack the spatial granularity and real-time responsiveness required for street-level interventions.

The proposed U-Net model addresses these limitations by generating high-resolution, near real-time heat maps from widely available RGB imagery, requiring less than 30 seconds per image. This enables time-sensitive applications, such as identifying high-risk zones during extreme heat events and supporting proactive planning decisions. For instance, city councils could integrate these maps into GIS platforms to prioritise tree planting on heat-vulnerable streets, evaluate zoning compliance, or assess the thermal impact of new developments.

Moreover, prior studies have validated the effectiveness of RGB-based LST estimation, confirming that models trained on morphological and spatial patterns can reliably predict thermal behaviour across different urban contexts—even in the absence of thermal infrared data (Abunnasr & Mhawej, 2023; Zhang et al., 2013; Zhao et al., 2023). This makes RGB-based approaches especially valuable for cities with limited access to thermal sensors or commercial satellite products.

The model also complements existing strategies by offering a scalable, low-cost solution that supports a continuous assessment of urban thermal dynamics. Nonetheless, its current reliance on LST restricts its ability to capture air temperature or thermal comfort directly. Therefore, it should be seen as a complementary decision support tool rather than a replacement for physical sensor networks. Future integration with real-time sensing and urban analytics platforms could further enhance its value and planning impact.

2.5.4 Limitations and Future Research

While the U-Net model demonstrated strong overall performance, certain limitations emerged in areas with extreme thermal dynamics, presenting unique challenges. For example, in highly urbanised zones with minimal vegetation, the model occasionally underestimated peak temperatures. This underestimation likely stems from the limited representation of high-density UHI conditions in the training dataset. Incorporating additional data that captures extreme urban heat scenarios could improve the model's sensitivity, ensuring more accurate predictions in such contexts. Expanding the dataset to include nuanced environmental variables would enhance the model's capacity to perform reliably in diverse urban environments.

Another limitation lies in the model's exclusive reliance on data from Adelaide, South Australia, which restricts its generalisability to regions with varying climates, urban layouts, and vegetation types. Although the model achieved commendable results within Adelaide, its application to other geographic areas may necessitate retraining or fine-tuning to account for unique environmental characteristics. Broadening the training dataset to include a wider range of urban forms, climates, and land use types would significantly enhance the model's robustness and adaptability, enabling it to deliver accurate results across diverse global landscapes.

Additionally, the static nature of the data used in this study constrains the model's ability to account for seasonal variations in urban heat dynamics. Changes in solar radiation, atmospheric conditions, and vegetation states significantly influence surface temperature patterns, which are not captured in the current model. Incorporating seasonal variability into the dataset could provide a more dynamic perspective on urban thermal patterns, strengthening the model's predictive power for year-round applications. Such enhancements would allow for more comprehensive analyses of heat distribution under varying environmental conditions.

Furthermore, while the model estimates LST, it does not capture air temperature or thermal comfort directly. This distinction is important, and future work could incorporate sensor-based measurements or predictive thermal comfort indices to enhance the relevance for human-centred heat risk assessment.

Addressing these limitations through data expansion, the integration of additional variables, temporal modelling, and computational optimisation would considerably improve the U-Net model's adaptability, accuracy, and scalability. These advancements would enable the model to perform effectively across a broader range of urban settings, enhancing its role as an invaluable tool for sustainable urban planning and climate resilience. By refining its capabilities, the model could support more precise and actionable insights, empowering urban planners and policymakers to implement targeted interventions that mitigate UHI effects and promote climate-adaptive urban development.

2.6 Conclusion

This study demonstrates the U-Net DL model as a transformative tool for high-resolution urban heat mapping, offering an innovative approach to predicting LST from RGB satellite imagery. By eliminating reliance on direct thermal data, this method provides a cost-effective, scalable, and accessible solution for urban heat monitoring, particularly in cities where thermal imagery is limited. The encoder–decoder architecture with skip connections enables pixel-level precision, effectively capturing fine-scale thermal variations across Adelaide's diverse urban landscape. This capability is particularly valuable for identifying localised heat-prone areas, allowing planners to implement targeted heat mitigation strategies, such as expanding green infrastructure or incorporating reflective materials. The model's robust performance, as reflected in low MSE value, confirms its suitability for real-world urban climate analysis.

One of the U-Net model's standout contributions is its ability to process high-resolution heat maps in under 30 seconds per image, a marked improvement over traditional thermal mapping methods, which often require extensive preprocessing and offer less spatial precision. This computational efficiency makes the model highly practical for large-scale and near real-time applications, particularly during heat emergencies when rapid decision-making is crucial for public health interventions. By bridging the gap between advanced geospatial modelling and practical application, this study supports dynamic urban planning, providing data-driven insights that policymakers can use to promote sustainable urban development and climate resilience.

While the U-Net model has demonstrated strong potential, opportunities exist for further refinement. Expanding the training dataset to include diverse urban morphologies, climates, and land use types would improve generalisability and accuracy. Additionally, integrating environmental variables such as wind patterns, humidity, and solar radiation could enhance predictive capability by offering a comprehensive understanding of urban microclimates. Future research could explore hybrid models that combine U-Net with temporal analysis techniques, improving the ability to predict seasonal variations and adapt to extreme climate conditions.

In conclusion, this study highlights the potential of RGB-based Urban AI models in high-resolution urban heat mapping, providing an innovative solution to a critical urban challenge. This framework empowers planners and policymakers to mitigate UHI effects, optimise green infrastructure, and enhance public health strategies by delivering precise, rapid, and actionable insights. Its scalability and adaptability position it as a valuable tool for climate-responsive urban planning, supporting the transition towards sustainable, liveable, and thermally resilient cities.

Acknowledgements

This chapter, with permission from the copyright holder, is a reproduced version of the following journal article: Shaamala, A., Tilly, N., & Yigitcanlar, T., (2025). Leveraging Urban AI for high-resolution urban heat mapping towards climate-resilient cities. *Environment and Planning B: Urban Analytics and City Science, 52*(9), 2251–2266.

Appendix

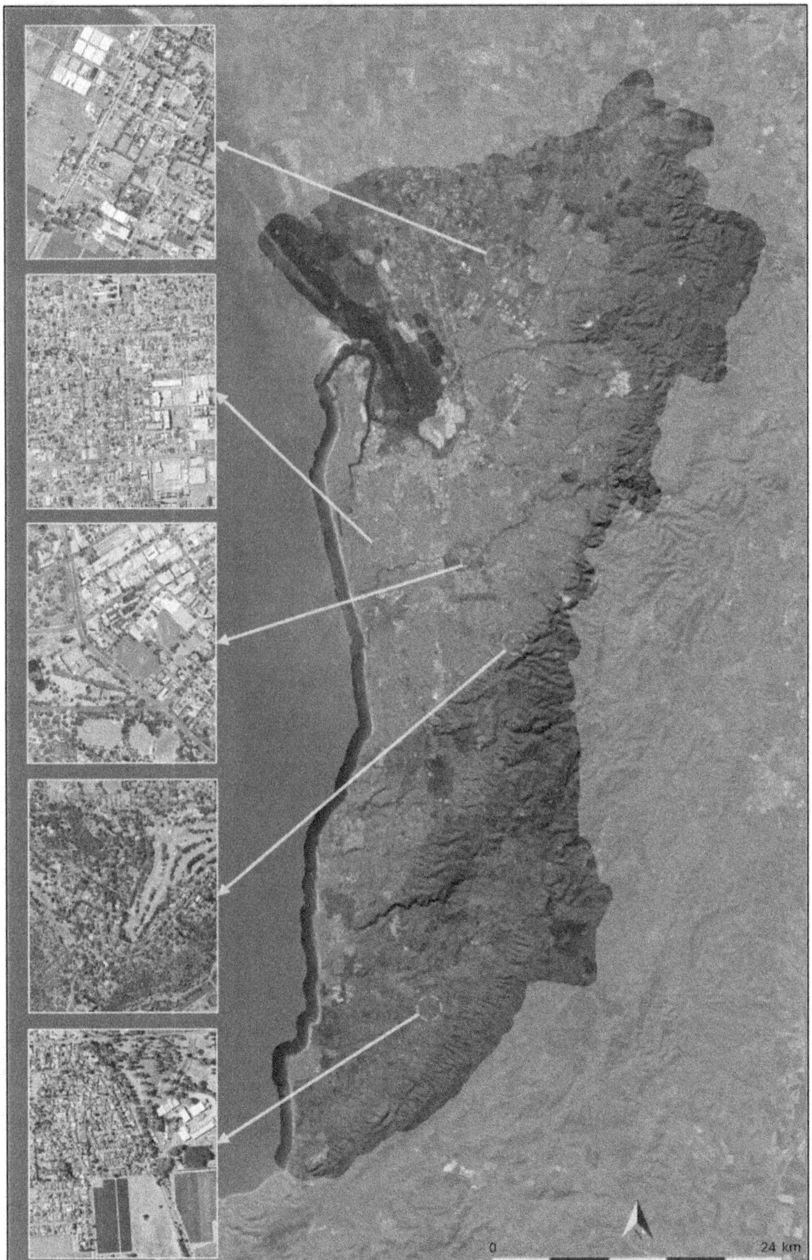

FIGURE 2A.1
Map of metropolitan Adelaide with selected locations showing various urban morphologies.

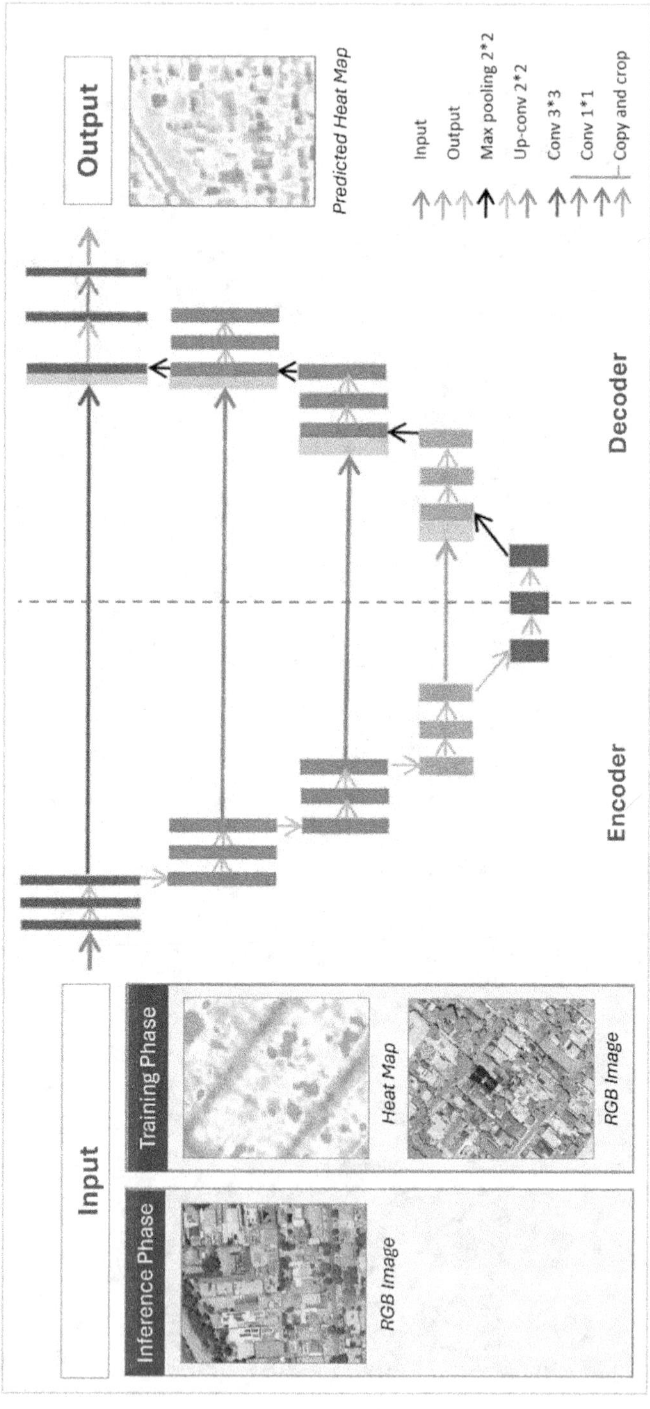

FIGURE 2A.2
U-Net architecture used in the study for urban heat mapping.

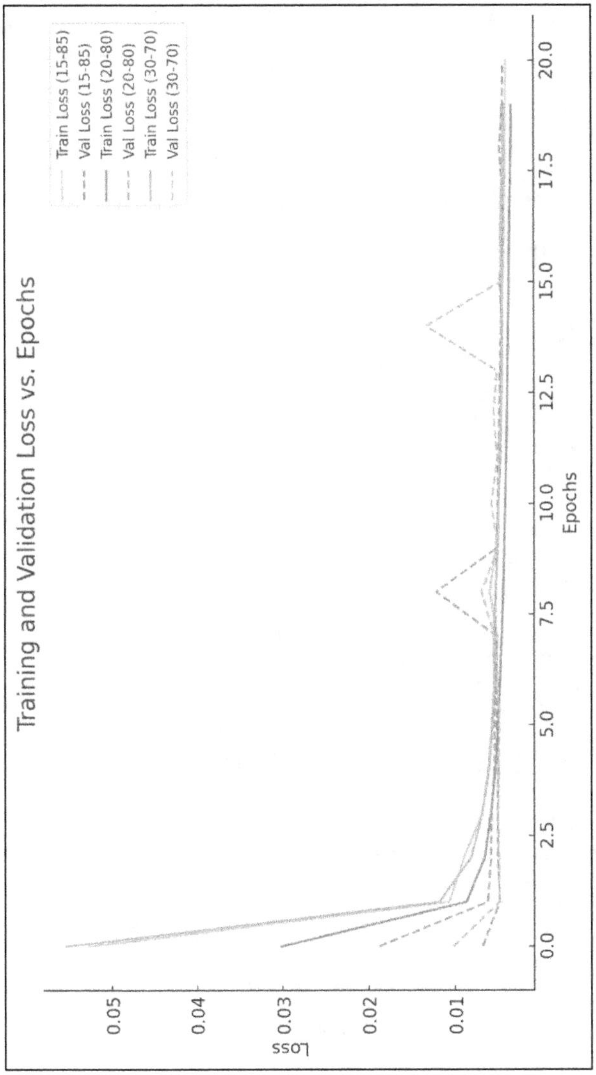

FIGURE 2A.3
Loss curves for different data splits.

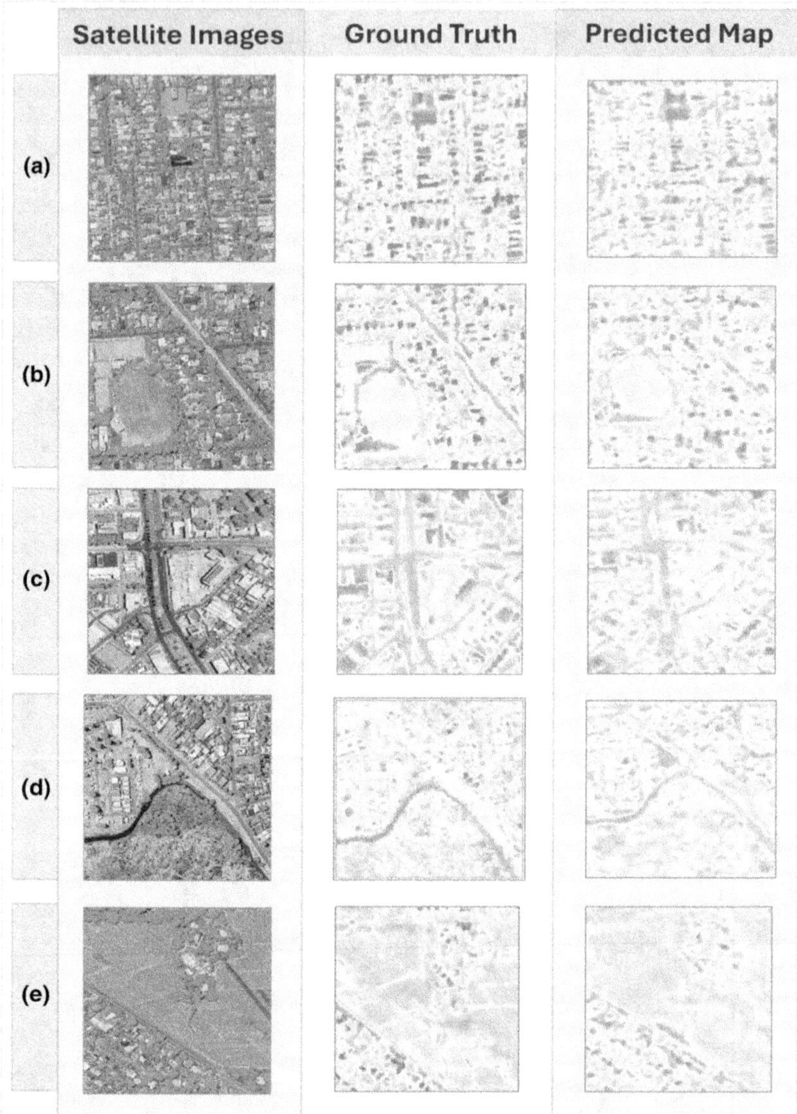

FIGURE 2A.4
Comparison of satellite images, ground truth, and predicted heat maps for urban areas.

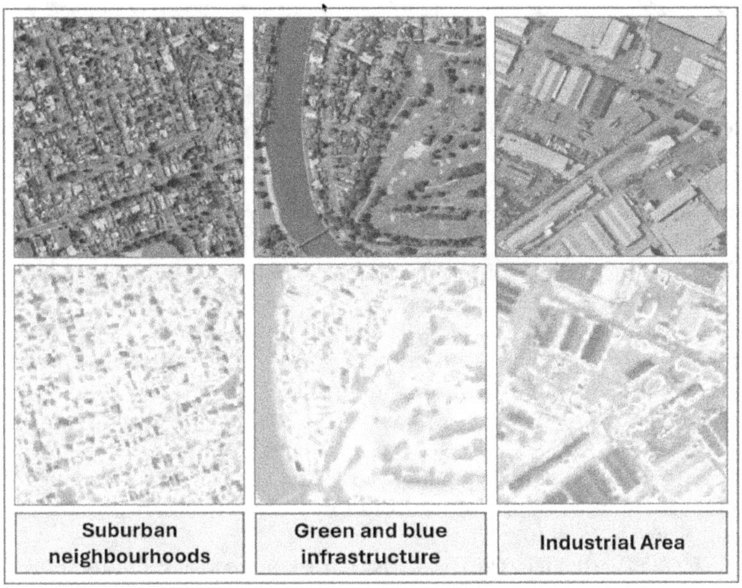

FIGURE 2A.5
Satellite imagery and heat maps showing urban heat variations across Adelaid City.

References

Abunnasr, Y., & Mhawej, M. (2023). Fully automated land surface temperature downscaling based on RGB very high spatial resolution images. *City and Environment Interactions, 19,* 100110.

Afshari, A., Vogel, J., & Chockalingam, G. (2023). Statistical downscaling of SEVIRI land surface temperature to WRF near-surface air temperature using a deep learning model. *Remote Sensing, 15,* 4447.

Akbari, H., Cartalis, C., Kolokotsa, D., …, & Zinzi, M. (2016). Local climate change and urban heat island mitigation techniques—The state of the art. *Journal of Civil Engineering and Management, 22*(1), 1–16.

Almashhour, R., Kolo, J., & Beheiry, S. (2024). Critical reflections on strategies for mitigating and adapting to urban heat islands. *International Journal of Urban Sustainable Development, 16*(1), 144–162.

Bhakare, S., Dal Gesso, S., Venturini, M., …, & Petitta, M. (2024). Intercomparison of machine learning models for spatial downscaling of daily mean temperature in complex terrain. *Atmosphere, 15*(9), 1085.

Bowler, D., Buyung-Ali, L., Knight, T., & Pullin, A. (2010). Urban greening to cool towns and cities. *Landscape and Urban Planning, 97*(3), 147–155.

Briegel, F., Makansi, O., Brox, T., Matzarakis, A., & Christen, A. (2023). Modelling long-term thermal comfort conditions in urban environments using a deep convolutional encoder-decoder as a computational shortcut. *Urban Climate, 47,* 101359.

Chen, X., Zhao, H., Li, P., & Yin, Z. (2006). Remote sensing image-based analysis of the relationship between urban heat island and land use/cover changes. *Remote Sensing of Environment, 104*(2), 133–146.

Cheval, S., Amihăesei, V., Chitu, Z., …, & Tudose, N. (2024). A systematic review of urban heat island and heat waves research (1991–2022). *Climate Risk Management, 44*, 100603.

Coutts, A., Harris, R., Phan, T., Livesley, S., Williams, N., & Tapper, N. (2016). Thermal infrared remote sensing of urban heat. *Remote Sensing of Environment, 186*, 637–651.

Degirmenci, K., Desouza, K., Fieuw, W., Watson, R., & Yigitcanlar, T. (2021). Understanding policy and technology responses in mitigating urban heat islands. *Sustainable Cities and Society, 70*, 102873.

Demiss, B., & Elsaigh, W. (2024). Application of novel hybrid deep learning architectures combining convolutional neural networks and recurrent neural networks. *Engineering Research Express, 6*(3), 032102.

Diem, P., Nguyen, C., Diem, N., …, & Phan, T. (2024). Remote sensing for urban heat island research. *Remote Sensing Applications, 33*, 101081.

Dimitrov, S., Iliev, M., Borisova, B., Semerdzhieva, L., & Petrov, S. (2024). A methodological framework for high-resolution surface urban heat island mapping. *Remote Sensing, 16*, 4007.

Elmarakby, E., & Elkadi, H. (2024). Prioritising urban heat island mitigation interventions. *Science of the Total Environment, 948*, 174927.

Elmes, A., Rogan, J., Williams, C., …, & Martin, D. (2017). Effects of urban tree canopy loss on land surface temperature magnitude and timing. *ISPRS Journal of Photogrammetry and Remote Sensing, 128*, 338–353.

Fabrizi, R., Bonafoni, S., & Biondi, R. (2010). Satellite and ground-based sensors for the urban heat island analysis in the City of Rome. *Remote Sensing, 2*(5), 1400–1415.

Faisal, A., Yigitcanlar, T., Kamruzzaman, M., & Paz, A. (2021). Mapping two decades of autonomous vehicle research. *Journal of Urban Technology, 28*(3–4), 45–74.

Gago, E., Roldan, J., Pacheco-Torres, R., & Ordóñez, J. (2013). The city and urban heat islands. *Renewable and Sustainable Energy Reviews, 25*, 749–758.

Ge, J., Wang, Y., Zhou, D., Gu, Z., & Meng, X. (2024). Effects of urban vegetation on microclimate and building energy demand in winter. *Sustainable Cities and Society, 102*, 105199.

Ghorbany, S., Hu, M., Yao, S., & Wang, C. (2024). Towards a sustainable urban future. *Sustainability, 16*(11), 4609.

Guo, R., Yang, B., Guo, Y., …, & Wang, F. (2024). Machine learning-based prediction of outdoor thermal comfort. *Building and Environment, 254*, 111301.

Hung, T., Uchihama, D., Ochi, S., & Yasuoka, Y. (2006). Assessment with satellite data of the urban heat island effects in Asian mega cities. *The International Journal of Applied Earth Observation and Geoinformation, 8*, 34–48.

Irfeey, A., Chau, H., Sumaiya, M., Wai, C., Muttil, N., & Jamei, E. (2023). Sustainable mitigation strategies for urban heat island effects in urban areas. *Sustainability, 15*(14), 10767.

Johnson, D., Stanforth, A., Lulla, V., & Luber, G. (2012). Developing an applied extreme heat vulnerability index utilizing socioeconomic and environmental data. *Applied Geography, 35*(1–2), 23–31.

Kamruzzaman, M., Deilami, K., & Yigitcanlar, T. (2018). Investigating the urban heat island effect of transit-oriented development in Brisbane. *Journal of Transport Geography, 66*, 116–124.

Kumar, P., Debele, S., Khalili, S., Halios, C., Sahani, J., …, & Jones, L. (2024). Urban heat mitigation by green and blue infrastructure. *Innovation, 5*(2), 100588.

Li, F., Yigitcanlar, T., Nepal, M., Nguyen, K., & Dur, F. (2023). Machine learning and remote sensing integration for leveraging urban sustainability. *Sustainable Cities and Society, 96*, 104653.

Li, J., Mao, Y., Ouyang, J., & Zheng, S. (2022). A review of urban microclimate research based on CiteSpace and VOSviewer analysis. *International Journal of Environmental Research and Public Health, 19*(8), 4741.

Li, J., Yan, Y., & Stouffs, R. (2024). Simulation and prediction of daytime surface urban heat island intensity under multiple scenarios via fully connected neural network. *Sustainable Cities and Society, 116*, 105922.

Li, X., Zhou, Y., Asrar, G., Imhoff, M., & Li, X. (2017). The surface urban heat island response to urban expansion. *Science of The Total Environment, 605–606*, 426–435.

Liu, N., & Morawska, L. (2020). Modeling the urban heat island mitigation effect of cool coatings in realistic urban morphology. *Journal of Cleaner Production, 264*, 121560.

Mathew, A., Sreekumar, S., Khandelwal, S., & Kumar, R. (2019). Prediction of land surface temperatures for surface urban heat island assessment over Chandigarh city using support vector regression model. *Solar Energy, 186*, 404–415.

McCarthy, M., Best, M., & Betts, R. (2010). Climate change in cities due to global warming and urban effects. *Geophysical Research Letters, 37*(9), L09705.

Middel, A., Häb, K., Brazel, A., Martin, C., & Guhathakurta, S. (2014). Impact of urban form and design on mid-afternoon microclimate in Phoenix Local Climate Zones. *Landscape and Urban Planning, 122*, 16–28.

Mohan, M., & Kandya, A. (2015). Impact of urbanization and land-use/land-cover change on diurnal temperature range. *Science of the Total Environment, 506*, 453–465.

Morgan, G., Zlotnick, D., North, L., Smith, C., & Stevenson, L. (2024). Deep learning for urban tree canopy coverage analysis. *Geomatics, 4*(1), 412–432.

Nazir, A., He, J., Zhu, N., …, & Pathan, M. (2024). A deep learning-based novel hybrid CNN-LSTM architecture for efficient detection of threats in the IoT ecosystem. *Ain Shams Engineering Journal, 15*(7), 102777.

Nguyen, B., Tian, G., Vo, M., …, & Granero-Belinchon, C. (2022). Convolutional Neural Network Modelling for MODIS Land Surface Temperature Super-Resolution. European Signal Processing Conference, 1806–1810.

Nichol, J., Fung, W., Lam, K., & Wong, M. (2009). Urban heat island diagnosis using ASTER satellite images and 'in situ' air temperature. *Atmospheric Research, 94*(2), 276–284.

Ramani, V., Arjunan, P., Poolla, K., & Miller, C. (2024). Semantic segmentation of longitudinal thermal images for identification of hot and cool spots in urban areas. *Building and Environment, 249*, 111112.

Rampal, N., Gibson, P., Sood, A., …, & Meyers, T. (2022). High-resolution downscaling with interpretable deep learning. *Weather and Climate Extremes, 38*, 100525.

Ramyar, R., Ackerman, A., & Johnston, D. (2021). Adapting cities for climate change through urban green infrastructure planning. *Cities, 117*, 103316.

Rizwan, A., Dennis, L., & Liu, C. (2008). A review on the generation, determination and mitigation of Urban Heat Island. *Journal of Environmental Sciences, 20*(1), 120–128.

Ronneberger, O., Fischer, P., & Brox, T. (2015). U-Net: Convolutional networks for biomedical image segmentation. *Lecture Notes in Computer Science, 9351*, 234–241.

SA Dataset. (2024). South Australia Dataset. Retrieved November 13, 2024, from https://data.sa.gov.au/data/dataset.

Sarrat, C., Lemonsu, A., Masson, V., & Guedalia, D. (2006). Impact of urban heat island on regional atmospheric pollution. *Atmospheric Environment, 40*, 1743–1758.

Shaamala, A., Yigitcanlar, T., Nili, A., & Nyandega, D. (2024). Strategic tree placement for urban cooling. *Urban Climate, 56*, 102084.

Shi, H., Xian, G., Auch, R., Gallo, K., & Zhou, Q. (2021). Urban heat island and its regional impacts using remotely sensed thermal data. *Land, 10*(8), 867.

Shivanna, K. R. (2022). Climate change and its impact on biodiversity and human welfare. *Proceedings of the Indian National Science Academy, 88*(2), 160–171.

Sidiqui, P., Roös, P., & Herron, M. et al (2022). Urban Heat Island vulnerability mapping using advanced GIS data and tools. *Journal of Earth System Science, 131*, 266.

Supritha, N., & Murthy, N. (2023). A comparative study of the CNN based models used for remote sensing image classification. *International Journal of Electrical and Electronics Research, 11*(3), 646–651.

Tehrani, A., Veisi, O., Kia, K., …, & Mehan, A. (2024). Predicting urban heat island in European cities. *Urban Climate, 56*, 102061.

Tomlinson, C., Chapman, L., Thornes, J., & Baker, C. (2011). Remote sensing land surface temperature for meteorology and climatology. *Meteorological Applications, 18*(3), 296–306.

Vinayak, B., Lee, H., Gedam, S., & Latha, R. (2022). Impacts of future urbanization on urban microclimate and thermal comfort over the Mumbai metropolitan region. *Sustainable Cities and Society, 79*, 103703.

Voogt, J., & Oke, T. (2003). Thermal remote sensing of urban climates. *Remote Sensing of Environment, 86*(3), 370–384.

Wang, H., Yang, J., Chen, G., Ren, C., & Zhang, J. (2023). Machine learning applications on air temperature prediction in the urban canopy layer. *Urban Climate, 49*, 101499.

Wang, Y., Liang, Z., Ding, J., Shen, J., Wei, F., & Li, S. (2022). Prediction of urban thermal environment based on multi-dimensional nature and urban form factors. *Atmosphere, 13*(9), 1493.

Weng, Q. (2009). Thermal infrared remote sensing for urban climate and environmental studies. *ISPRS Journal of Photogrammetry and Remote Sensing, 64*(4), 335–344.

Wu, J., Xie, C., Zhang, Z., & Zhu, Y. (2022). A deeply supervised attentive high-resolution network for change detection in remote sensing images. *Remote Sensing, 15*(1), 45.

Ye, X., Yigitcanlar, T., Goodchild, M., …, & Newman, G. (2025). Artificial intelligence in urban science. Annals of GIS, https://doi.org/10.1080/19475683.2025.2469110.

Yigitcanlar, T. (2006). Australian local governments' practice and prospects with online planning. *URISA Journal, 18*(2), 7–17.

Yigitcanlar, T. (2024a). *Urban artificial intelligence: A guidebook for understanding concepts and technologies*. CRC Press.

Yigitcanlar, T. (2024b). *Urban artificial intelligence: A guidebook for understanding perceptions and ethics*. CRC Press.

Yigitcanlar, T., Agdas, D., & Degirmenci, K. (2023a). Artificial intelligence in local governments. *AI & Society, 38*(3), 1135–1150.

Yigitcanlar, T., Li, R., Beeramoole, P., & Paz, A. (2023b). Artificial intelligence in local government services. *Government Information Quarterly, 40*(3), 101833.

Yigitcanlar, T., David, A., Li, W., Fookes, C., Bibri, S., & Ye, X. (2024c). Unlocking artificial intelligence adoption in local governments. *Smart Cities, 7*(4), 1576–1625.

Yigitcanlar, T., Senadheera, S., Marasinghe, R., Bibri, S., Sanchez, T., Cugurullo, F., & Sieber, R. (2024d). Artificial intelligence and the local government. *Cities, 152,* 105151.

Yin, Z., Liu, Z., Liu, X., Zheng, W., & Yin, L. (2023). Urban heat islands and their effects on thermal comfort in the US. *Ecological Indicators, 154,* 110765.

Zhang, G., Cai, M., & Hu, A. (2013). Energy consumption and the unexplained winter warming over northern Asia and North America. *Nature Climate Change, 3*(5), 466–470.

Zhao, L., Zhang, L., Huang, X., & Wang, J. (2023). Estimating land surface temperature using visible bands and deep learning. *Remote Sensing Applications, 29,* 100946.

Zhou, D., Xiao, J., Bonafoni, S., …, & Sobrino, J. A. (2018). Satellite remote sensing of surface urban heat islands. *Remote Sensing, 11*(1), 48.

Zhou, W., Huang, G., & Cadenasso, M. (2011). Does spatial configuration matter? *Landscape and Urban Planning, 102*(1), 54–63.

3

Microclimate Risk Monitoring through Real-Time Platforms

3.1 Introduction and Background

Urban environments are increasingly vulnerable to the impacts of microclimatic variations, driven by phenomena such as urban heat islands, localised air pollution, and changes in land cover and vegetation patterns (Elliott et al., 2020; Kong et al., 2021; Li et al., 2024a). These factors significantly influence urban weather patterns, resulting in variations of temperatures and rainfall in different parts within a city (Dizdaroglu & Yigitcanlar, 2014; Liu et al., 2024). Such disparities can have profound impacts on residents, especially vulnerable populations such as children, the elderly, individuals with chronic illnesses, or those with disabilities (Kousis et al., 2022; Li et al., 2024b; Yigitcanlar, 2010). To mitigate these challenges, existing weather monitoring systems and alert platforms in cities employ various approaches to enhance forecast accuracy and provide timely warnings.

Many of these systems leverage centralised meteorological data, integrating historical climate records with real-time observations from centralised weather stations to generate city-wide forecasts (Murray, 2021; Oswald et al., 2024). Advancements in artificial intelligence (AI) have enabled the development of models that can predict weather in microclimates (Han et al., 2024). Many cities deploy extensive sensor networks across various urban areas to monitor environmental conditions in real time (Frustaci et al., 2022; Maclean et al., 2021). These networks provide detailed data on environmental conditions, capturing variations in temperature, humidity, air quality, and other parameters crucial for understanding urban microclimates. Mobile microclimate stations have emerged as an innovative solution to urban climate monitoring (Shi et al., 2021; Tsin et al., 2016). These stations offer the flexibility to monitor key variables that regulate urban environments across different locations, offering a complementary data source to satellite imagery and fixed weather stations.

Weather monitoring agencies utilise data from centralised and city-wide sensor networks to predict weather conditions and issue alerts, primarily at the city-wide scale (Ostrometzky & Messer, 2024; Rashid & Rehmani, 2016). To effectively communicate these warnings, local governments and city authorities disseminate notifications to residents through text-based alerts (via SMS), official websites, and dedicated mobile applications (Ascione et al.,

 DOI: 10.1201/9781003737056-4

2024; Johnson et al., 2022). Mobile weather applications also play a crucial role in delivering real-time alerts to users, often sourcing predictions directly from government websites via their application programming interfaces (APIs) (Kaginalkar et al., 2021; Zabini, 2016). While existing weather monitoring systems provide high accuracy in weather forecasts, and alert platforms notify residents before weather events, they exhibit several critical limitations (Yassaghi et al., 2019).

Firstly, weather alerts are usually disseminated uniformly across the city, without considering the specific locations of residents, their ages, health conditions, or disabilities (Acosta-Coll et al., 2018; Casanueva et al., 2019). Vulnerable groups, such as children, the elderly, and individuals with health conditions or disabilities, lack access to tailored information that could help them make informed decisions about outdoor activities and necessary preparations (Connon & Hall, 2021).

Secondly, existing systems typically treat the city as a homogenous entity, sending alerts to all residents irrespective of the specific weather conditions in their local area. This one-size-fits-all approach overlooks the microclimatic differences within a city, where weather conditions can vary significantly over short distances (Shaamala et al., 2024). For instance, a slight increase in temperature in cooler parts of a city might have a greater impact on residents there than in consistently warmer areas (Wong et al., 2021).

Finally, local authorities, including municipal governments, lack effective tools to accurately monitor and understand the number of individuals "at-risk" across various urban locations in real time or in anticipation of upcoming weather events, hindering effective resource allocation and proactive emergency response planning (Chitwatkulsiri & Miyamoto, 2023).

To address these challenges, this study proposes a novel platform architecture rooted in platform urbanism and driven by the Artificial-Intelligence-of-Things (AIoT) concept. The platform captures real-time weather data from sensors located in various parts of a city and integrates it with the population's demographic and health information for tailored analysis. By analysing this combined data, the system automatically generates personalised alerts for residents and equips local authorities with insights through a dashboard to monitor and respond to populations at risk. This study uses Brisbane, Australia, as a case study to underscore the significance of such a platform and to demonstrate its effectiveness in addressing localised weather-related challenges.

The integration of demographic and health data of Brisbane into the platform was designed with a strong emphasis on data validity, privacy, and ethical considerations. Demographic and health statistics were sourced from publicly available datasets, primarily from the Australian Bureau of Statistics (ABS), ensuring accuracy and reliability. These datasets undergo rigorous validation processes before publication, ensuring their suitability for research and policy applications (Fair et al., 2019). It is important to note that this study did not utilise actual personal information, such as individuals' "real" age, location, or specific health conditions. Instead, we employed aggregated population statistics—such as the total number of residents within different age groups and those with specific health conditions in each suburb. This approach enabled a realistic simulation of the platform's capabilities while adhering to ethical research guidelines.

This study begins by examining the variation in temperature and rainfall across different suburbs in Brisbane, emphasising the shortcomings of centralised weather alert systems that overlook these microclimatic differences. Subsequently, the study introduces the theoretical framework underpinning the platform, grounded in the concepts of platform urbanism and AIoT, followed by the architecture of the platform, implementation, and its demonstration.

3.1.1 Localised Weather Variations in the Case Study Area of Brisbane

Brisbane, the capital of the state of Queensland, is situated on the eastern coast of Australia. Brisbane has a subtropical climate characterised by warm, humid summers and mild, dry winters (Yang et al., 2023). Spanning an area of over 15,800 km^2, Brisbane accommodates a population exceeding 2.5 million, distributed over 190 suburbs across a wide range of urban and peri-urban regions (Hou et al., 2023). The city's unique topography, encompassing a river system, coastal influences, and varying elevations, contributes to diverse microclimatic conditions across its metropolitan area (Javanroodi et al., 2023).

The Queensland Government maintains a network of over 50 sensors across the state, including devices to monitor air quality, organic matter, hydrogen, metals, meteorological conditions, and smoke (Li et al., 2024a; QLD, 2024). Within the Brisbane City, nine meteorological sensors are positioned in various suburbs, as illustrated in Figure 3.1. The spatial distribution of these sensors reveals both clustered and isolated patterns. To explore localised temperature and rainfall variations in the vicinity of these sensors, this study utilised the past 10 years' meteorological data (2014–2023) from all stations, except for Deagon. The Deagon sensor, being a recent addition to the network, has only recorded data since 2021. The data was downloaded from the Queensland Government's Open Data Portal.

The average daytime (6:00 AM to 6:00 PM) and nighttime (6:00 PM to 12:00 AM) temperatures across various seasons—Autumn (March, April, May), Spring (September, October, November), Summer (December, January, February), and Winter (June, July, August)—reveal notable spatial and temporal variations in Brisbane's weather conditions, as depicted in Figure 3.2. In Summer, the differences in daytime average temperatures between weather stations are the most pronounced, ranging from 28.7°C in South Brisbane to 25.8°C in Wynnum North. Similarly, nighttime temperatures also vary, though less dramatically, between 23.6°C in South Brisbane and 22.1°C in Wynnum North. Conversely, during Winter, temperature anomalies between stations are reduced, with differences narrowing to around 1°C. These findings indicate that temperature is not uniformly distributed across the city,

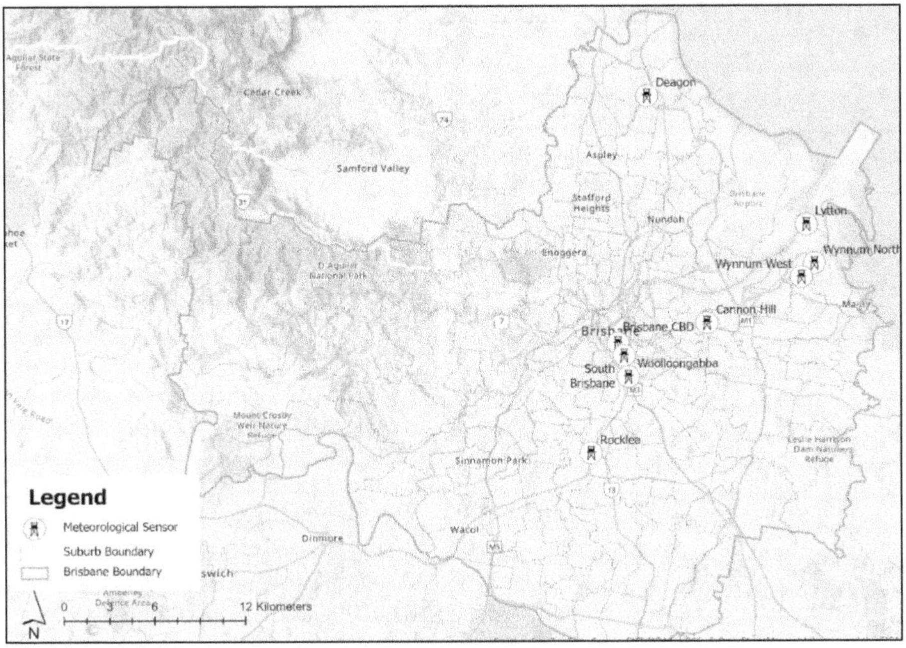

FIGURE 3.1
Locations of meteorological sensors in Brisbane operated by the Queensland State Government.

even within a relatively compact area, due to microclimatic factors such as heat islands, vegetation cover, land use patterns, and proximity to water bodies (Kong et al., 2021; Li et al., 2024a).

To further highlight the significance of localised temperature variations, an interpolation approach was applied using the inverse distance weighting (IDW) method to spatially represent temperature distributions across Brisbane suburbs (Masoudi, 2021). The focus was on surrounding suburbs of existing stations only due to the uneven distribution of weather sensors, which limits the comprehensive city-wide coverage. Figure 3.3(a–d) illustrates the interpolated average daytime temperatures across all seasons, and Figure 3.3(e–h) presents the interpolated average nighttime temperatures across all seasons in Brisbane. These figures provide insights into how temperatures vary across different parts of the city, emphasising the higher temperatures in core urban areas, including the Brisbane City, South Brisbane, and Woolloongabba, which consistently exhibit higher average temperatures across all seasons. On the contrary, suburbs located in the outskirts demonstrate comparatively cooler conditions, highlighting localised temperature gradients influenced by factors like urban heat islands, vegetation cover, and proximity to water bodies (Li et al., 2024a; Li et al., 2024b).

Another important matrix for understanding microclimatic patterns, particularly temperature variations, is the analysis of warm days, warm

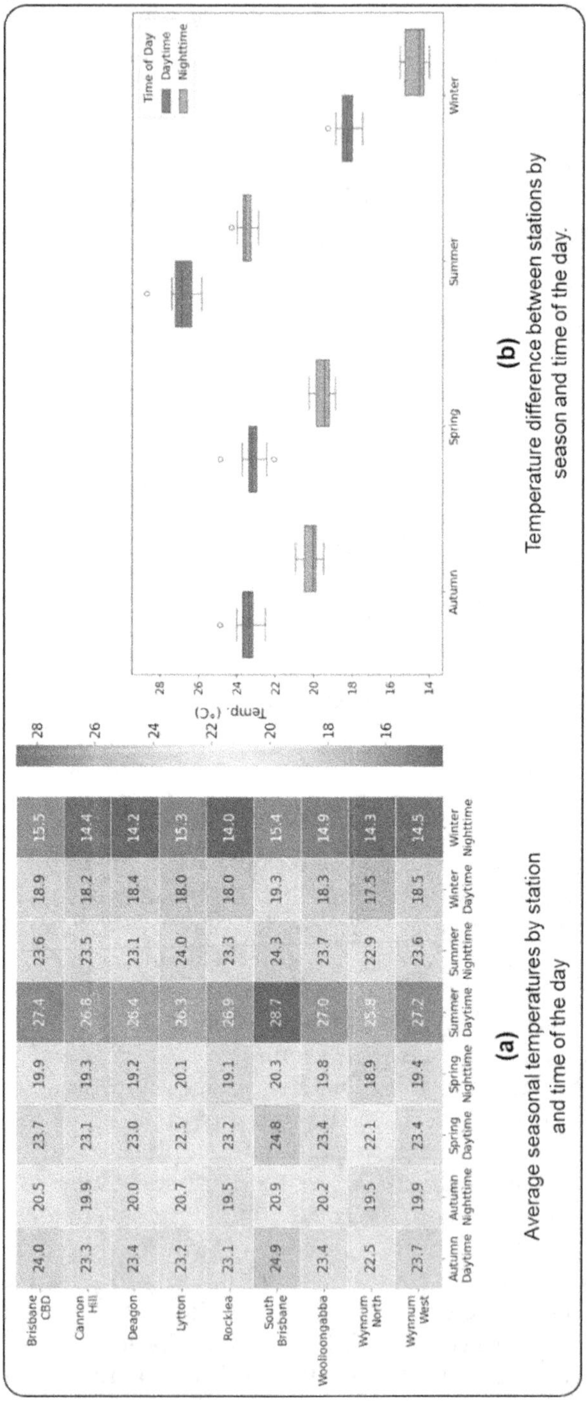

(a) Average seasonal temperatures by station and time of the day

(b) Temperature difference between stations by season and time of the day.

FIGURE 3.2
(a, b) Seasonal daytime and nighttime temperature and variations across Brisbane based on daily data from 2014 to 2023.

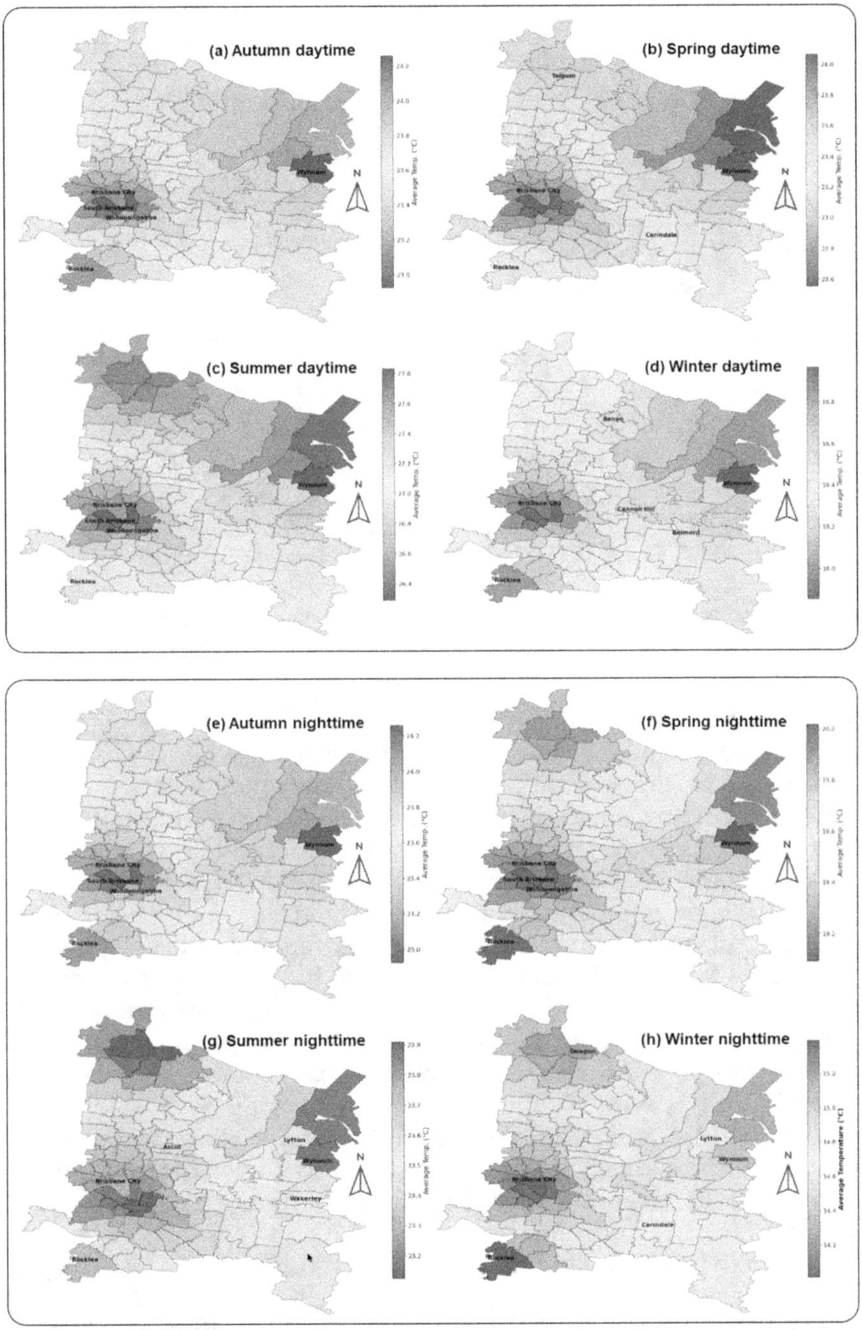

FIGURE 3.3
(a–d) average daytime temperature across Brisbane suburbs for all seasons based on daily data from 2014 to 2023. (e–h) average nighttime temperature across Brisbane suburbs for all seasons based on daily data from 2014 to 2023.

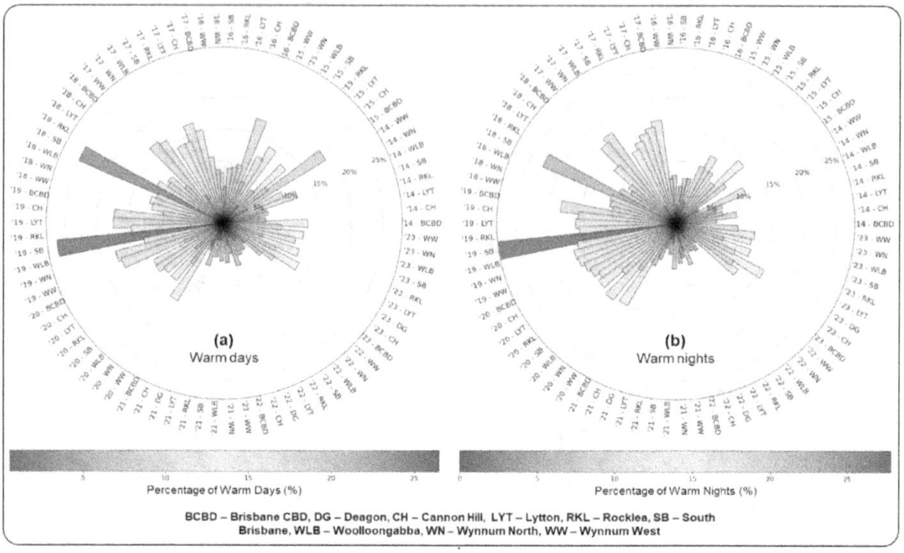

FIGURE 3.4
Percentage of (a) warm days and (b) warm nights recorded across Brisbane weather sensors (2014–2023).

nights, cool days, and cool nights. The Bureau of Meteorology (BoM) of the Australian Government defined a warm day as a day when maximum temperature exceeds the 90th percentile of baseline values and a warm night as when minimum temperature surpasses the 90th percentile (BoM, 2024; Yassaghi et al., 2019). Conversely, a cool day and cool night occur when maximum and minimum temperatures fall below the 10th percentile, respectively (BoM, 2024). Figures 3.4 and 3.5 present the percentage of warm and cool days and nights over the past decade across different weather stations. Core urban areas like South Brisbane recorded the highest proportion of warm days (up to 28.7%) and warm nights, reflecting reduced nocturnal cooling in densely built environments (Li et al., 2024). In contrast, outer suburbs like Wynnum North and Rocklea exhibited higher percentages of cool days and nights (Figure 3.5), with Wynnum North experiencing nearly 17.5% of cool nights annually, highlighting the moderating influence of vegetation and coastal proximity. These temperature variations underscore the potential impacts on health, sleep, and outdoor activities of residents (Kam et al., 2023).

Extreme daytime temperatures can increase the risk of heat exhaustion and dehydration, while elevated nighttime temperatures can disrupt sleep patterns, leading to long-term health consequences (Buguet et al., 2023; Chevance et al., 2024). The proposed platform directly addresses these challenges by providing real-time, location-specific weather alerts tailored to the needs of different population groups. By integrating demographic and health data, the platform identifies individuals most susceptible to temperature-related

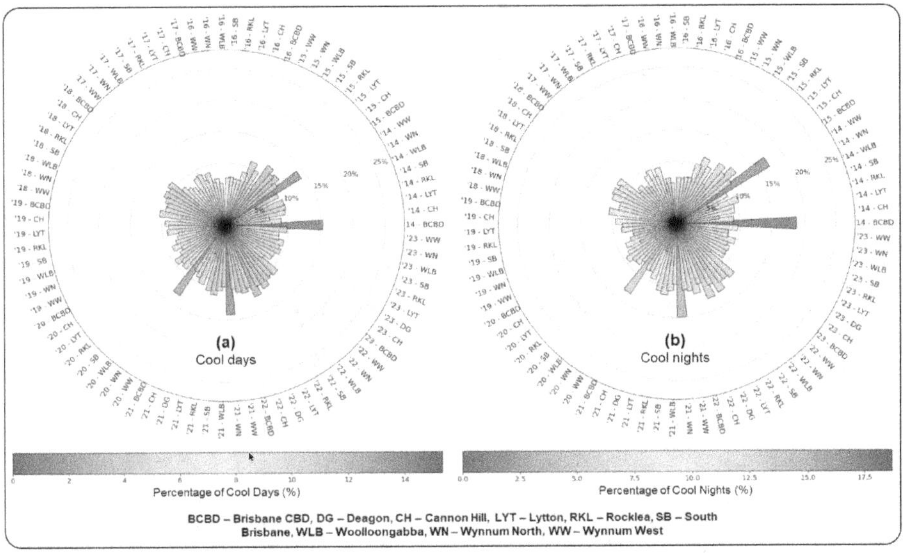

FIGURE 3.5
Percentage of (a) cool days and (b) cool nights recorded across Brisbane weather sensors (2014–2023).

risks and ensures that alerts are sent to those in need, rather than issuing broad, non-targeted warnings. This ensures that individuals in suburbs with historically cooler climates receive timely alerts when temperatures rise beyond their usual range, helping them prepare for unexpected heat stress. Conversely, in consistently warmer areas, alerts can be adjusted to reflect the higher threshold of temperature tolerance among residents while still prioritising the most vulnerable.

The platform equips local authorities with a dashboard that visualises at-risk populations across different suburbs, enabling targeted, localised, and personalised planning, mitigation, and intervention. During prolonged high temperatures or heavy rainfall events, authorities can identify vulnerable areas and implement adaptive strategies such as opening cooling centres, improving drainage systems, distributing tailored weather advisories, or deploying emergency response teams where they are most needed. By providing real-time, location-specific insights, the system facilitates proactive and precise risk management, mitigating climate-related health impacts and ensuring that no vulnerable groups are overlooked.

To measure and classify the intensity of precipitation, the BoM defines wet days as those with daily precipitation equal to or exceeding 1 mm, heavy precipitation days as those with daily rainfall equal to or exceeding 10 mm, and very heavy precipitation days as those with daily rainfall equal to or exceeding 30 mm (BoM, 2024). The spatial variability in wet days across Brisbane is evident in Figure 3.6, with Wynnum North recording the highest annual

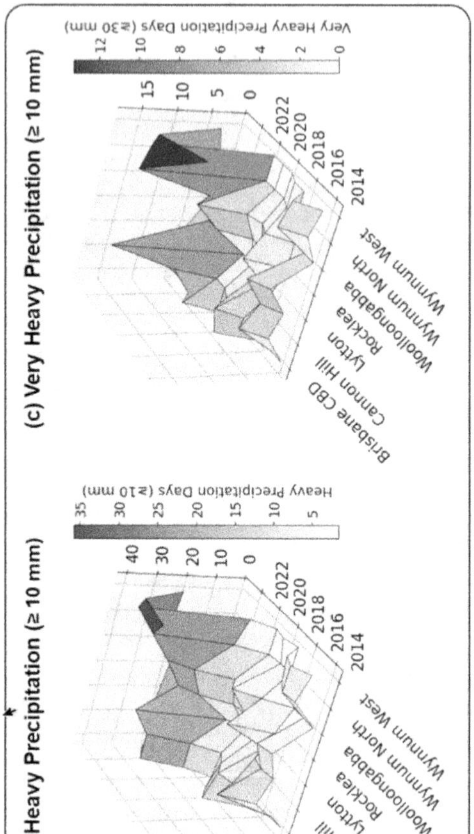

FIGURE 3.6
(a–c) Yearly counts of wet, heavy, and very heavy precipitation days in Brisbane from 2014 to 2023.

count of wet days in 2022, reaching 107 days, while Wynnum West closely follows with 95 wet days. Stations such as Lytton and Cannon Hill also experienced significantly wet years, with 93 wet days in Lytton during 2022.

On the other hand, some stations recorded notably fewer wet days in specific years; for instance, Rocklea only had 24 wet days in 2016. Heavy precipitation days (≥10 mm) are less frequent than wet days, but their occurrence significantly varies by station and year. Wynnum West experienced 42 heavy precipitation days in 2022, marking it as a hotspot for such events, followed by Wynnum North, which recorded 35 heavy precipitation days in 2021. In contrast, Rocklea and Brisbane central business district (CBD) have relatively fewer heavy precipitation days, with Rocklea experiencing 9 days in 2019 and Brisbane CBD seeing just 16 in 2020. Very heavy precipitation days (≥30 mm) are rare but highly impactful. Coastal areas such as Wynnum North recorded a peak of 17 days in 2021, while Wynnum West experienced 16 days in the same year, highlighting these areas as hotspots for extreme rainfall. Conversely, stations like Brisbane CBD had fewer instances, with its maximum being only seven days in 2020.

3.1.2 Local Weather Impacts on Residents and the Need for Microclimate-Based Monitoring Systems

The uneven distribution of thermal stress across Brisbane has significant implications for public health, liveability, and regular activities (Li et al., 2024b). Areas experiencing more frequent warm days and nights require increased public alerts due to heightened health risks. Nonetheless, individuals in these regions may be more accustomed to such microclimates, potentially exhibiting greater tolerance to incremental increases in thermal stress (Elliott et al., 2020). Conversely, in suburbs with fewer warm days and nights, even slight increases in temperature or a rise in warmer days can significantly disrupt residents' comfort and routines, as they are less acclimated to these changes (Wong et al., 2021).

While the definitions are location-specific, traditional systems often assume uniformity in the application of these metrics, focusing on centralised locations like the Brisbane City (Murray, 2021). However, such an approach disregards microclimatic differences across suburbs, leading to potential inequities in alerts. For example, for Brisbane City and Wynnum, the warm days might be based on a maximum temperature of around 33°C and 30°C, respectively, reflecting the city's historical data. Despite experiencing more frequent warm days compared to Brisbane City, Wynnum residents, including 2523 elderly individuals, and 1461 people with respiratory issues, may not receive sufficient alerts under a centralised system. For Rocklea and nearby suburbs, where temperatures are usually cooler, modest increase in temperature or number of warm days or warm nights, could lead to discomfort for about 264 children, and 224 elderly, along with 265 residents with various health conditions such as asthma, heart condition, and respiratory issues. Similarly, significant variability in number of wet days, heavy precipitation days, very heavy precipitation days calls localised alert systems, particularly in areas prone to higher rainfall intensities. For example, Wynnum (both north and west) experience high intensity of precipitation, whereas suburbs like Rocklea, with fewer heavy precipitation days, may face challenges adapting to sudden increases in heavy rainfall.

Localised weather monitoring and alert systems, thus, should be implemented to provide real-time, community-specific warnings (Clegg et al., 2022). These systems must account for microclimatic conditions and prioritise vulnerable populations by offering proactive notifications (Ascione et al., 2024). By integrating demographic and health data, such systems can enhance urban resilience and public safety (Haraguchi et al., 2022). Recognising these localised disparities in weather impacts and the limitations of traditional weather monitoring and alert systems, this research poses the following question:

- How can a localised weather monitoring and alert platform be designed to account for microclimatic variations and address the needs of vulnerable populations?

To address this, we developed a platform architecture that leverages real-time weather data, demographic profiles, and microclimatic thresholds to provide targeted alerts for each community. The subsequent sections of this chapter delve into the shortcoming of traditional weather monitoring and alerts systems, theoretical concept of the platform, discuss the proposed platform's architecture, a demonstration using the Brisbane City's data, platform's usability and scalability, limitations, and future research avenues.

3.2 Theoretical Framework: Platform Urbanism and AIoT

The rapid growth of urbanisation and the increasing complexity of urban challenges have led to the emergence of platform urbanism as a critical concept for understanding and addressing city-wide issues (D'Amico et al., 2020; Repette et al., 2021). Platform urbanism refers to the increasing reliance on digital platforms to reshape how cities collect, analyse, and respond to data (Bibri & Jagatheesaperumal, 2023). The proposed platform in this study provides flexibility to integrate various weather prediction models—whether they are global, regional, or city-specific—making it adaptable and addressing the gap between advanced meteorological forecasting and its practical, real-time application in microclimate level. To achieve this adaptability, the platform incorporates the concept of AIoT.

The term AIoT may initially suggest a focus on physical devices such as devices equipped with technologies like NVIDIA Jetson modules or Google's Edge TPU that can perform real-time AI computations at the edge, close to the data

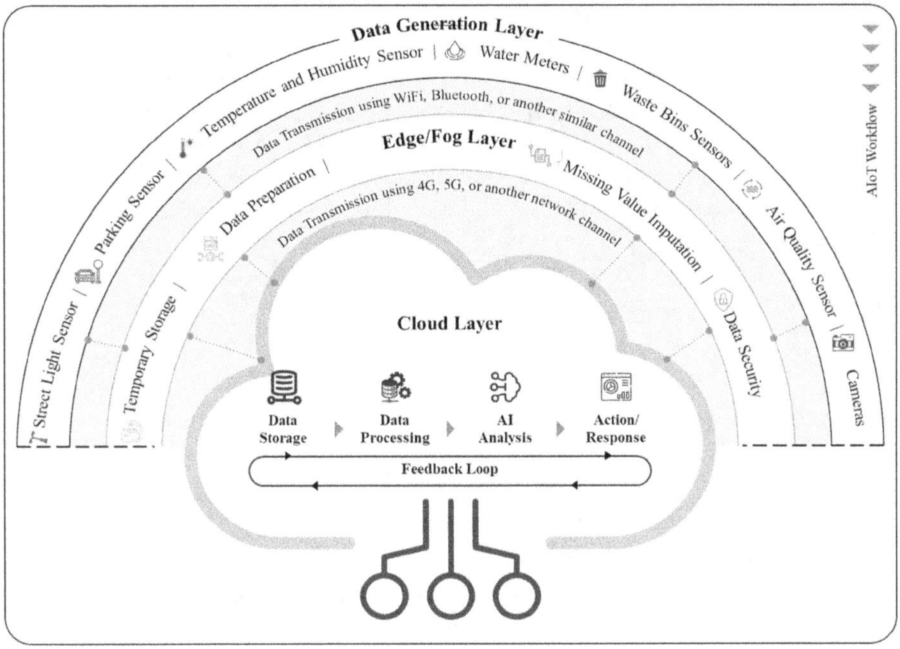

FIGURE 3.7
AIoT-based platform architecture, derived from Bibri et al. (2024), Jagatheesaperumal et al. (2024), and Muhammed et al. (2024).

source, it extends far beyond physical devices in a smart city context (Kuguoglu et al., 2021; Pise et al., 2022). Researchers and practitioners in smart cities conceptualise AIoT mostly as a cloud-based platform (Figure 3.7) or an entire ecosystem that seamlessly integrates Internet-of-Things (IoT) data with AI-driven analytics and actions (Forkan et al., 2024; Pise et al., 2022). The AIoT system captures data through various IoT devices such as sensors or cameras (Anthony, 2024). This data is then transmitted to the edge/fog layer using connectivity technologies like Wi-Fi, Bluetooth, Zigbee, LoRaWAN, and 5G (Muhammed et al., 2024). The edge/fog layer is a crucial processing tier between data sources and cloud infrastructure, ensuring data quality through preprocessing tasks like filtering and anomaly detection (Song et al., 2021). It supports edge computing capabilities to perform real-time data analytics close to the data source, thereby reducing latency and bandwidth and enhancing the responsiveness of the smart city applications (Almalki et al., 2023).

After initial processing in the edge/fog layer, data is transmitted to the cloud layer through high-speed communication channels such as 4G, 5G, or equivalent technologies (Toh, 2020). The cloud layer serves as a central hub for comprehensive data management and advanced analytics. In this layer, data undergoes storage, further processing, and AI analysis using advanced machine learning (ML) algorithms to extract insights and make predictions. Data undergoes further processing, involving transformations, integration, and enhancement to support advanced analytics (Forkan et al., 2024). AI-driven analysis is then applied, enabling insights, predictions, and decision-making for applications such as resource allocation, environmental monitoring, and public safety (Bibri & Jagatheesaperumal, 2023). Based on these insights, automated actions like alert notifications or system optimisations are triggered (Kuguoglu et al., 2021; Pise et al., 2022). The cloud layer incorporates a feedback loop, where system responses are re-evaluated and fed back to improve AI models, ensuring adaptability and ongoing optimisation in response to changing conditions (Belli et al., 2020).

3.3 Existing Weather Monitoring and Alert Systems and Limitations

Existing weather and health alert systems have made significant strides in protecting public health during extreme weather events. Country-specific systems such as BoM in Australia, the National Weather Service (NWS) Alert System in the USA, and the Met Office Weather Warnings in the UK are widely regarded for their ability to disseminate weather alerts at a city or regional level (Casanueva et al., 2019; Neußner, 2021). Global platforms such as AccuWeather and Weather.com provide weather updates and alerts across diverse locations, leveraging sophisticated models and extensive datasets (Atencio et al., 2024). Similarly, state- or city-specific platforms such as VicEmergency for the State of Victoria in Australia, Emergency WA in the State of Western Australia, NYC Emergency Management Notify NYC in New York, Chicago OEMC Alert System, Singapore's MyENV, Tokyo's Disaster Preparedness Tokyo, or London Resilience Alerts offer localised emergency alerts. Despite their utility, most of these tools exhibit notable limitations when addressing localised microclimatic variability and the diverse vulnerabilities of urban populations (Li et al., 2024c). Despite their accuracy for city-specific warnings or alerts, these systems or platforms often overlook the microclimatic variations that can occur within the urban city (Kousis et al., 2022).

In Brisbane, weather-related alerts are primarily disseminated through the State of Queensland Government's Queensland Alert platform and the Brisbane City Council's alert system. Both systems rely on warnings issued by the BoM, which uses data from a limited number of centralised weather stations to issue alerts for broader regions or cities. While accurate at a macro level, this approach often fails to account for microclimatic variations across suburbs, as discussed in the introduction section (Localised Weather Variations in Brisbane) above. Reliance on a single weather station or generalised dataset for an entire city is a common characteristic in weather monitoring and alert systems around the world (Hahn et al., 2022).

Another limitation is the lack of personalised alerts. These existing systems fail to account for crucial variables such as a person's specific location, age, health conditions, or disabilities—factors that significantly influence an individual's vulnerability to extreme weather events. For instance, during heavy rainfall, notifications sent to all areas may lack specificity, failing to focus on flood-prone neighbourhoods. In Wynnum North, which experiences more frequent and intense precipitation, residents in low-lying flood-prone areas would benefit from targeted alerts, while similar alerts may be unnecessary for elevated areas like Cannon Hill.

Moreover, many platforms are subscription-based, requiring residents to register to receive alerts. The subscription-based alert system risks excluding vulnerable populations who may lack awareness, access, or resources to subscribe. Marginalised groups—often those most affected by weather extremes—may remain uninformed about imminent risks (Dodd et al., 2024). Furthermore, existing systems typically lack the capability to incorporate microclimate-specific models or demographic data (Barriopedro et al., 2023). For example, warm days and warm nights are defined based on percentile thresholds that vary across suburbs (Barriopedro et al., 2023).

The Brisbane City might define a warm day as a maximum temperature exceeding 33°C, whereas Wynnum North might set this threshold at 30°C. Generalised thresholds often fail to accommodate these differences, leaving many residents unprepared for significant weather events. While existing systems provide essential services, they often lack the granularity needed to address the complex interplay between microclimates, individual health status, and location-specific risks. Our research emphasises the need to integrate microclimate data, resident-specific information, and location-based risk assessments for more effective and targeted alerts.

To address these challenges, we developed a platform using ArcGIS GeoEvent, which enables real-time data processing. During the literature review, no evidence was found of GeoEvent being used for a platform specifically designed to generate weather alerts that consider microclimate variability. Nevertheless, several Environmental Systems Research Institute (ESRI) reports and publications have highlighted the potential of GeoEvent for such applications, indicating its capacity to support microclimate-based alerting systems (ESRI, 2021; McLaughlin, 2017; Milazzo et al., 2009).

The ArcGIS ecosystem was selected for this study due to its widespread use among planners and local government authorities, making it a familiar and trusted platform for urban management. Its comprehensive suite of spatial analysis tools, coupled with the real-time data integration and analytical capabilities of GeoEvent, enables seamless integration of geospatial data from diverse sources. This functionality supports the generation of location-specific insights essential for microclimate monitoring, aligning with the aim of this research. The following section provides an overview of the software and its various elements.

3.4 ArcGIS GeoEvent

GeoEvent Server, developed by ESRI, is a server-based application that facilitates real-time geospatial data management, transforming static geographic information system (GIS) applications into dynamic decision support systems (Zhang et al., 2018). Its primary function is to integrate real-time, event-based data streams into the ArcGIS Enterprise ecosystem, allowing for the seamless ingestion, filtering, processing, and analysis of incoming data (McLaughlin, 2017). GeoEvent supports automated notifications when predefined conditions are met, enabling rapid response to critical events (ESRI, 2021). For example, transportation authorities can utilise GeoEvent Server to analyse real-time traffic data and dynamically adjust speed limits based on current road conditions and weather patterns, optimising traffic flow and enhancing road safety.

Similarly, water authorities can leverage ArcGIS GeoEvent Server to enhance environmental monitoring by ingesting real-time sensor data from water stations, tracking critical parameters such as pH levels, dissolved oxygen content, and pollutant concentrations, and automatically alerting relevant authorities when ecological thresholds are exceeded, enabling swift and targeted responses to potential environmental threats. The system's core functionality (Figure 3.8) is driven by GeoEvent Services, which facilitate workflows connecting Input Connectors (ICs), Filters, Processors, and Output Connectors (OCs), enabling efficient data transformation and delivery of actionable insights. This architecture allows for the continuous processing of event-driven data, supporting timely alerts and decision-making across diverse operational contexts.

3.4.1 Real-Time Data Ingestion

GeoEvent revolutionises data ingestion into the GIS ecosystem and enables real-time data integration from various sources through ICs (Johnson et al., 2022).

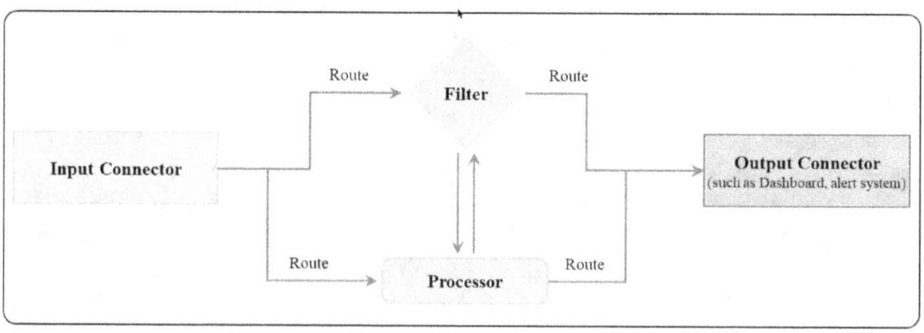

FIGURE 3.8
GeoEvent server elements.

This functionality allows organisations to add new data into their daily operations and significantly enhance their situational awareness on the current desired event, enabling faster and more informed decision-making. The ICs within GeoEvent extract data or attribute values from each event, constructing a GeoEvent that can be routed through optional Filters and Processors before reaching an OC. GeoEvent Services require at least one IC, though multiple Connectors can be utilised for complex data sources. GeoEvent offers a wide range of ICs or sources to ingest data as shown in Figure 3.9.

3.4.2 Real-Time Analytics

The real-time analytics capability of GeoEvent represents a significant advancement in geospatial data processing and analysis (Yao & Wang, 2020). By employing a variety of Filters and Processors, users can define specific criteria for event detection, spatial relationships, and attribute-based conditions. This functionality facilitates the identification of patterns, anomalies, and trends within streaming data, enabling rapid response to evolving situations. Filters are configurable elements used to remove event-data that do not meet specified criteria. These Filters can be based on attributes, spatial conditions, or a combination of both. Filters within GeoEvent include Attribute Filters, Spatial Filters, Property Filters, Tag-based Filters, Regular Expression Filters, and Custom Filters.

Processors on the other hand perform specific actions on event data, including data aggregation, analysis, calculation, or transformation. GeoEvent includes following Processors: Add XYZ Values, Bearing Calculator, Buffer Creator, Convex Hull Creator, Difference Creator, Envelope Creator, Event Joiner, Event Volume Controller, Feature to Point, Field Calculator, Field Calculator (Regular Expression), Field Enricher (Feature Service), Field Enricher (File), Field Mapper, Field Reducer, GeoTagger, Incident Detector, Intersector, No Operation, Projector, Range Fan Calculator, Simplifier, Symmetric Difference Creator, Track Gap Detector, and Union Creator. Users can also develop custom Processors using the GoeEvent Server SKD.

3.4.3 Real-Time Data Visualisation and Notification

GeoEvent's ability to visualise assets on a web map, either as discrete features or aggregated data, offers a significant operational advantage by providing a clear understanding of what is happening and where it is occurring in real time (Trilles et al., 2017; Yao & Wang, 2020). Along with visualisation, GeoEvent has the capability to alert and notify key stakeholders about events as they occur or shortly thereafter ensures that critical information is delivered promptly, enabling informed and timely decision-making. The visualisation and notification functionalities within GeoEvent provide a comprehensive spatial context for ongoing events and asset statuses, facilitating enhanced situational awareness and decision-making processes.

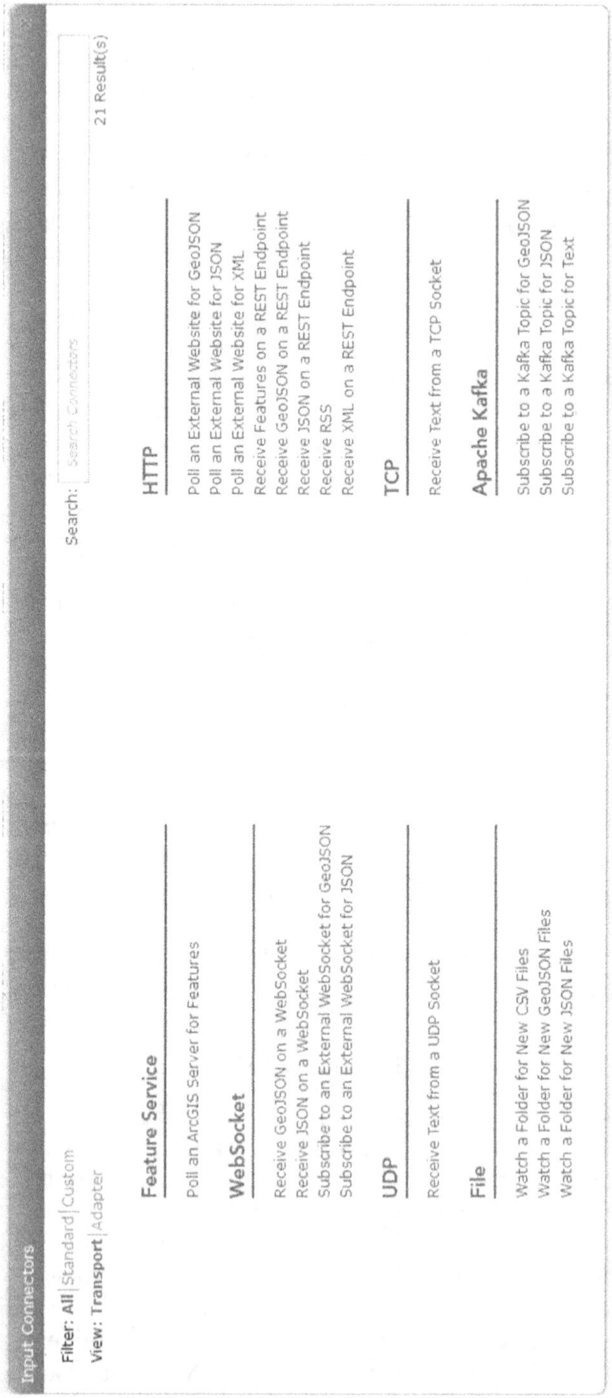

FIGURE 3.9
Available ICs in ArcGIS GeoEvent, screenshot of ICs from the GeoEvent application, November 2024.

The OCs within GeoEvent are used to visualise and notify event data to a designated customer in an appropriate format. A GeoEvent Service must have an Output to operate. GeoEvent Server also includes the following, but not limited to, OCs: Add a Feature, Update a Feature, Send an Email, Write to a CSV File, Write to a GeoJSON File, Write to a JSON File, Send an Instant Message, Write GeoJSON to a Kafka Topic, Write JSON to a Kafka Topic, Publish Text to a UDP or External TCP Socket, Add/Update a Feature to a Spatiotemporal Big Data Store, and Send Features to a Stream Service. GeoEvent also offers creating customised OCs using GeoEvent Manager or the SDK, allowing precise control over data delivery.

3.5 A Novel Platform Architecture

The platform's architecture is designed to efficiently process and integrate real-time weather data with spatial information. The platform comprises three ICs, three OCs, three GeoEvent Services, and two ArcGIS Notebook Schedules, details of ICs and OCs are presented in Table 3.1 followed by the platform's conceptual block diagram in Figure 3.10 and detailed architecture in Figure 3.11.

Figure 3.10 illustrates the conceptual block diagram of the platform, highlighting its key components and workflows. The process begins with the ingestion of meteorological data from weather sensors via API into ArcGIS GeoEvent. Filters and Processors ensure data quality by checking for missing values before updating meteorological parameters in the weather station Feature Layer. Two ArcGIS Notebooks then activate sequentially. The first Notebook employs an AI-/ML-based weather prediction model to forecast weather conditions for the next 12 hours. The second Notebook assesses predicted weather data to identify predefined conditions, such as warm

TABLE 3.1

Input and OCs Required in the Proposed Platform

Input Connector (IC)	Output Connector (OC)
IC1 utilises the "Poll an External Website for JSON" feature to capture sensor data via API at 60-minute intervals.	**OC1** uses "Update a Feature" to modify the weather station Feature Layer.
IC2 employs "Poll an ArcGIS Server for Features" to ingest weather station point Feature Layer from ArcGIS web map.	**OC2** employs "Update a Feature" to update residents' information Feature Layer.
IC3 uses "Poll an ArcGIS Server for Features" to ingest residents' location as point Feature Layer from ArcGIS web map.	**OC3** utilises "Send a Text Message" and/or "Send an Email" feature for notifications.

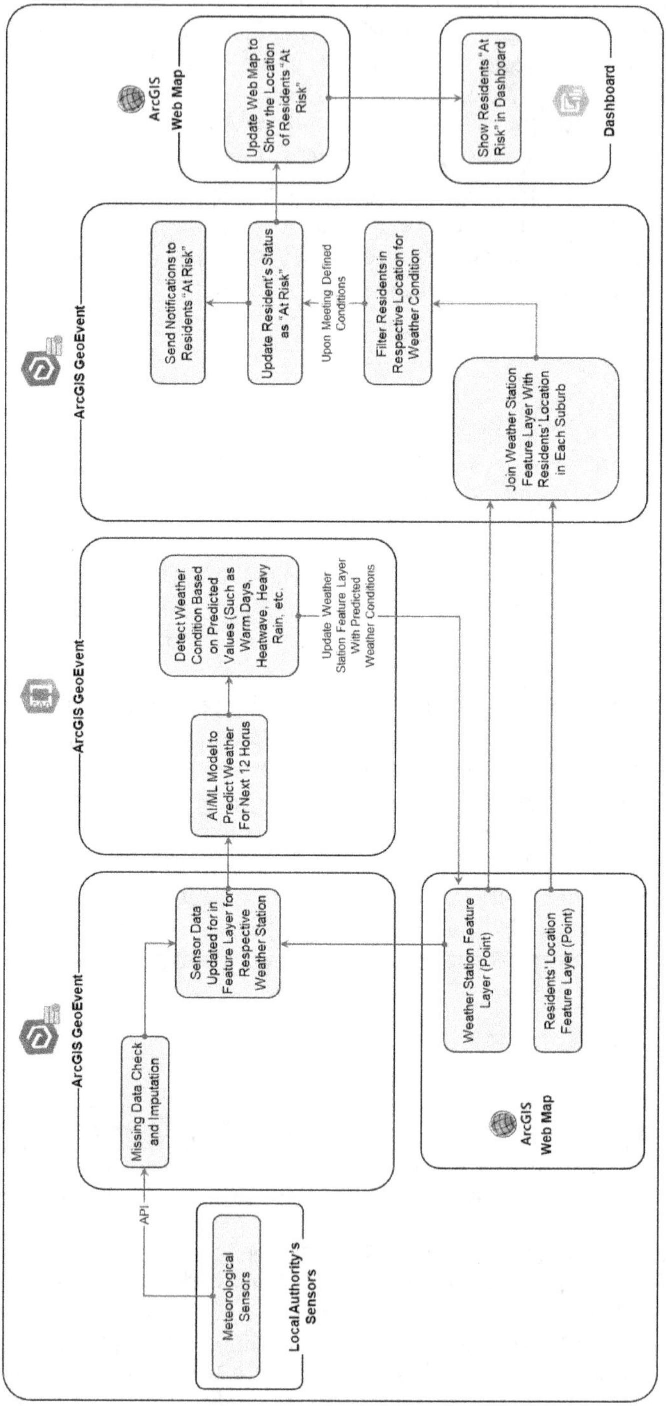

FIGURE 3.10
Conceptual block diagram of the platform.

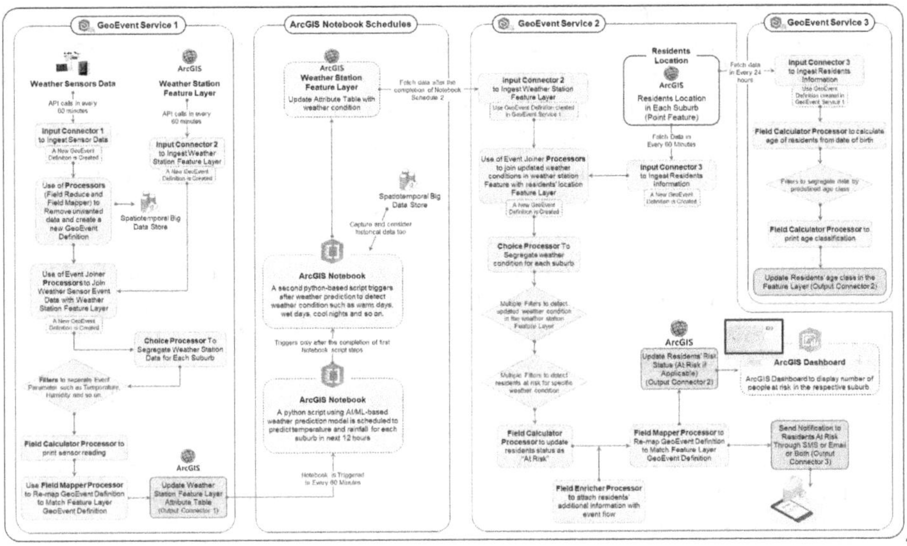

FIGURE 3.11
Architecture of the real-time and predictive risk monitoring and alerts system.

days, cool days, wet days, and extreme events, updating the weather station Feature Layer with respective conditions for each suburb.

To identify residents at risk, the platform integrates weather data with the Resident Feature Layer using GeoEvent's Event Joiner Processor. This combined dataset is processed using Filters to detect weather-based risks in each suburb. If a risk is identified, the status of residents in the affected suburb is updated to "At-Risk" within the residents' Feature Layer. This information is displayed in an ArcGIS Dashboard, enabling local government officials to visualise the number and location of at-risk residents. Additionally, OCs trigger notifications via SMS or email to inform residents, ensuring timely, location-specific alerts. The detailed architecture of the platform is depicted in Figure 3.11, accompanied by a description of each step and process.

3.5.1 GeoEvent Service 1: Integrating and Processing Real-Time Weather Data

Service 1 is dedicated to processing real-time sensor data and integrating it with the weather station Feature Layer. IC1 ingests sensor data from a data source using their API and creates a GeoEvent Definition upon the initial ingestion, dynamically updating the definition if the schema changes (e.g., new parameters are added). A GeoEvent Definition serves as the data schema or structure that defines the attributes, field names, and data types for incoming event data. It ensures that data from various sources, such as weather sensors or resident datasets, can be properly interpreted, processed, and merged within the system. IC2 ingests the weather station Feature Layer data hosted in ArcGIS web map, creating a new Definition on the first call. Due to the static nature of the Feature Layer's attribute structure, subsequent calls do not require additional Definition creation.

The sensor data ingested by IC1 passes through a Field Reducer Processor, which filters out extraneous parameters to reduce event data volume and optimise processing speed. The streamlined data is forwarded to a Field Mapper, ensuring compatibility with the Definition of weather station Feature Layer. The Event Joiner Processor then merges the filtered sensor data with the weather station Feature Layer data based on a common field, creating a unified Definition that includes both real-time and static attributes. This combined dataset is then processed through a Choice Processor, which segregates data based on suburbs. The segregated data is then routed through a series of Filters to isolate specific weather parameters such as air temperature, humidity, precipitation, and wind speed. The filtered data is processed by corresponding Field Calculators to compute and assign values to the relevant attributes in the weather station Feature Layer. Finally, the data passes through a second Field Mapper to re-map the Definition to match the structure of the Feature Layer. The updated data is sent to OC1, which updates the weather parameters in the ArcGIS web map in real time.

3.5.2 ArcGIS Notebook Schedules: Predictive Modelling and Weather Condition Analysis

The platform architecture integrates two Python-based ArcGIS Notebook Schedules designed to enhance predictive capabilities and automate weather-related alert systems. These Notebooks are strategically scheduled to run in every 60 minutes only after the weather station Feature Layer has been updated by GeoEvent Service 1. Notebook Schedule 1 is designed to execute ML-based weather prediction models, which can be flexibly adapted to integrate various AI/ML models, depending on the context and requirements. This adaptability is a key feature of the platform, allowing the integration of existing weather prediction models that utilise weather station data alone or combine it with spatial factors such as elevation and land cover. The platform's design within the ArcGIS environment further facilitates spatially driven analyses, a method frequently employed by researchers. By leveraging the platform's capability to work seamlessly with spatial data and scheduling Python scripts, the execution of such models can be fully automated, providing an efficient and scalable solution for predicting weather conditions at a microclimatic level.

The second Notebook builds upon the outputs of the first, using predictive results alongside historical data stored in the ArcGIS Spatiotemporal Big Data Store to analyse and detect predefined weather conditions. These can include but are not limited to detecting heatwave, heavy precipitation, cool nights, or warm days. The Notebook processes these results to update weather conditions for the next 12 hours and systematically records these statuses in the weather station Feature Layer.

3.5.3 GeoEvent Service 2: Dynamic Risk Detection and Resident Notification

Service 2 incorporates critical event flows to leverage predictive analytics with residents' risk levels and disseminate risk information through a synchronised dashboards and multi-channel notifications. This service leverages data from two ICs: IC2 and IC3. IC2 utilises the existing Definition from GeoEvent Service 1, maintaining consistency without requiring re-mapping. IC3, upon initial data ingestion, generates a Definition that remains static. An Event Joiner Processor combines these two data streams into a unified Definition, which then cascades through the workflow. Firstly, a Choice Processor segregates data based on suburbs. From here, multiple Filters detect updated weather conditions, as determined by ArcGIS Notebook Schedule 2, for each suburb. Upon identifying a specific weather condition for a suburb, another series of Filters assess the resident data to determine individuals at risk for that condition. A Field Calculator is then employed to label these residents as "At-Risk," updating their status in the event workflow.

The Field Enricher Processor plays a pivotal role in this workflow by augmenting the GeoEvent with additional resident-specific information stored in an external, secure data repository. For privacy and security reasons, the resident data Feature Layer includes only anonymised identifiers and essential fields such as location and age. The Field Enricher Processor accesses supplementary data, such as contact information, from an encrypted local repository using a shared identifier between the GeoEvent and the external dataset. This approach ensures compliance with data privacy standards and safeguards sensitive information, mitigating risks associated with unauthorised access or cyber threats. The enriched data is then processed through a Field Mapper to re-align the Definition with the structure of the residents' Feature Layer. This refined dataset is forwarded to OC2, which update the ArcGIS web map. An ArcGIS Dashboard is linked to this web map, configured synchronous refreshing at predefined intervals. This setup provides real-time visualisation of residents at risk. Concurrently, OC3 facilitates the transmission of personalised notifications to affected residents via SMS, email, or both, ensuring timely and direct communication.

3.5.4 GeoEvent Service 3: Updating and Managing Resident Demographic Profiles

Service 3 is responsible for dynamically updating residents' age categories within the Feature Layer, ensuring demographic data remains accurate and relevant for risk assessment workflows. This service uses the same Definition created in GeoEvent Service 2 for IC3, which ingests the residents' Feature Layer data in every 24 hours. Once the data is ingested, it flows directly into a Field Calculator Processor, where the age of each resident is recalculated based on their date of birth. This recalculated data is routed through a series of Filters that classify residents into predefined age categories such as children, adults, and elderly. Each category is assigned through dedicated Field Calculators, which label the respective residents accordingly. The updated data from the Field Calculator is then sent directly to OC2, which updates the ArcGIS web map.

3.6 Case Study: Demonstrating the Platform's Architecture and Practical Application in Brisbane

To demonstrate the platform's architecture in a real-world scenario, weather station data from Brisbane CBD and Wynnum North were utilised, along with demographic data from these two suburbs. The selection criteria focused on suburbs with existing weather stations to ensure access to accurate, localised data. Brisbane CBD and Wynnum North were chosen due to their contrasting geographic and microclimatic contexts (Figure 3.1)—Brisbane CBD representing the urban core and Wynnum North situated in a coastal area. This contrast offers a compelling case for evaluating the platform's capabilities in diverse climatic conditions. Table 3.2 presents the demographic and health profiles of these two suburbs based on data from the ABS.

The platform's architecture, demonstrated in the Brisbane context, is presented across multiple segments. It follows the conceptual framework illustrated in Figure 3.11, with detailed breakdowns provided in Figures 3.13, 3.16, and 3.19 for greater clarity. The demonstration platform's architecture adheres to the same structure outlined in the "Proposed Platform Architecture" section, maintaining a consistent number and functionality of Connectors, Processors, Filters, GeoEvent Services, and Notebook Schedules. Figure 3.12 provides a detailed snapshot of the ArcGIS GeoEvent

TABLE 3.2

Demographic and Health Profiles of Suburbs with Weather Sensors

	Population				Health Record				
Suburb	**Total**	**Children**	**Adults**	**Elderly**	**Asthma**	**Diabetes**	**Heart**	**Lung**	**Disability**
Brisbane CBD	12,579	604	10,937	1038	234	226	788	917	216
Wynnum (North)	14,034	2441	9070	2523	516	657	1483	1461	769

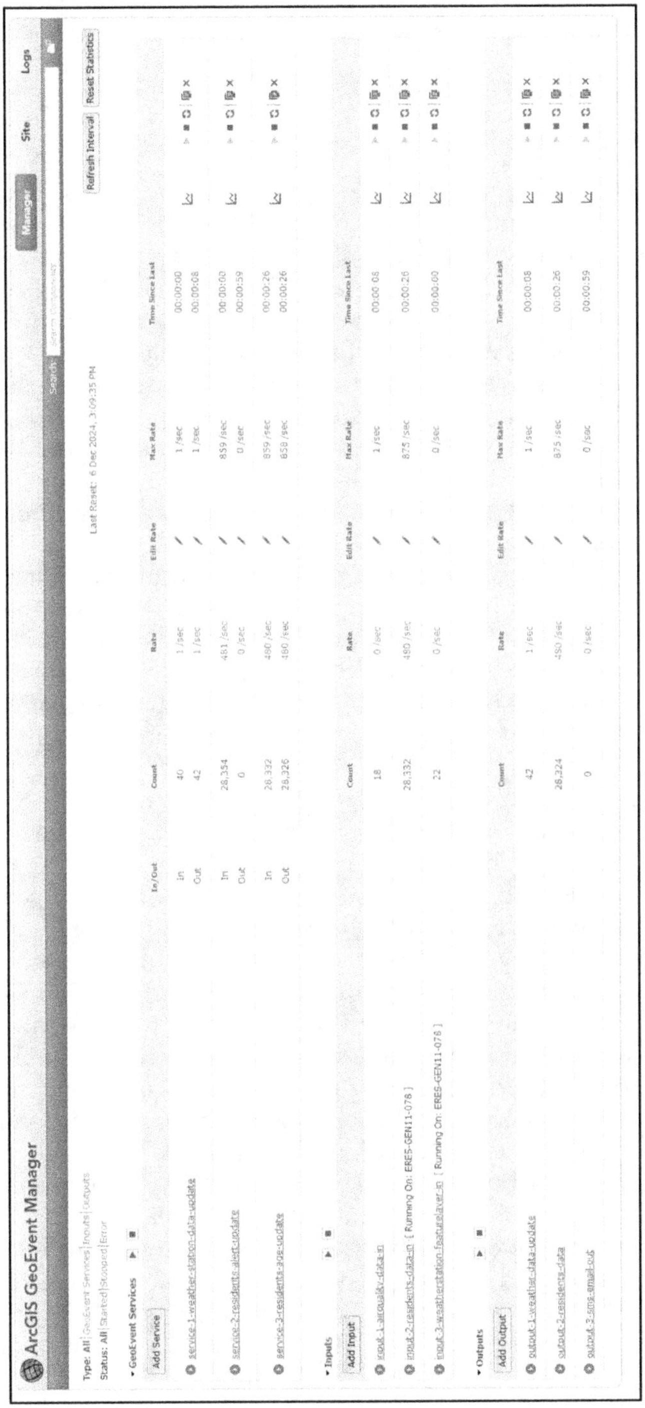

FIGURE 3.12
ArcGIS GeoEvent manager displaying ICs, OCs, and GeoEvent services with Brisbane event data flow.

Manager interface, illustrating the configuration and operation of the platform's ICs, OCs, and GeoEvent Services, along with their respective data flow rates.

3.6.1 GeoEvent Service 1

Figure 3.13 illustrates the architecture and workflow of GeoEvent Service 1, which integrates real-time weather data into the platform. Two ICs were set up for this service following the configuration in Figure 3.14. IC1 ingests meteorological Data from the Queensland Government Open Data Portal via their Air Quality Monitoring Data API (https://airquality.des.qld.gov.au/v1). The API provides a comprehensive dataset that includes meteorological parameters as well as air quality metrics, such as particulate matter and other pollutants. The data fetch frequency is configured to 3600 seconds, corresponding to an interval of 60 minutes. The Default Spatial Reference 4326 represents the WGS 84 (World Geodetic System 1984), a widely used geographic coordinate system based on latitude and longitude. The ingested data undergoes processing through a Field Reducer, which selectively retains essential meteorological parameters (temperature, humidity, rainfall, wind speed, wind direction, and barometric pressure) while discarding extraneous information. From the Field Reducer Processor, the data flows to a Field Mapper Processor. Initially, the ingested data adheres to a default definition that includes a broader set of parameters. The Field Mapper reconfigures the data to a refined definition (as shown in Figure 3.15), ensuring compatibility with downstream processing stages.

Concurrently, IC2 retrieves weather station Feature Layer, hosted in the ArcGIS Online web map. An Event Joiner Processor then merges this spatial data with the sensor data, combining two distinct Definitions using the common field "suburb," as illustrated in Figure 3.15. The Event Joiner Processor is set with a cache size of 1000 and the "Clear After Join" option, ensuring efficient memory management by maintaining a limited number of records in the cache, thereby optimising performance. Following the Event Joiner, the data enters a Choice Processor, which filters information based on the associated suburbs. While this demonstration involves only two suburbs, the Choice Processor is particularly advantageous for larger datasets, such as Brisbane's 190+ suburbs, as it consolidates decision-making and reduces the complexity of workflows. Separate data goes through six Filters and verify the existence of valid weather station data for each suburb Post-filtering, the data undergoes through Field Calculators to update with existing values or in case of missing values, previously predicted data for the current hour is printed (discussed further under Notebook Schedule 1). From Field Calculators, data goes to a Field Mapper Processor to align with the target Definition before updating the weather station Feature Layer in ArcGIS web map.

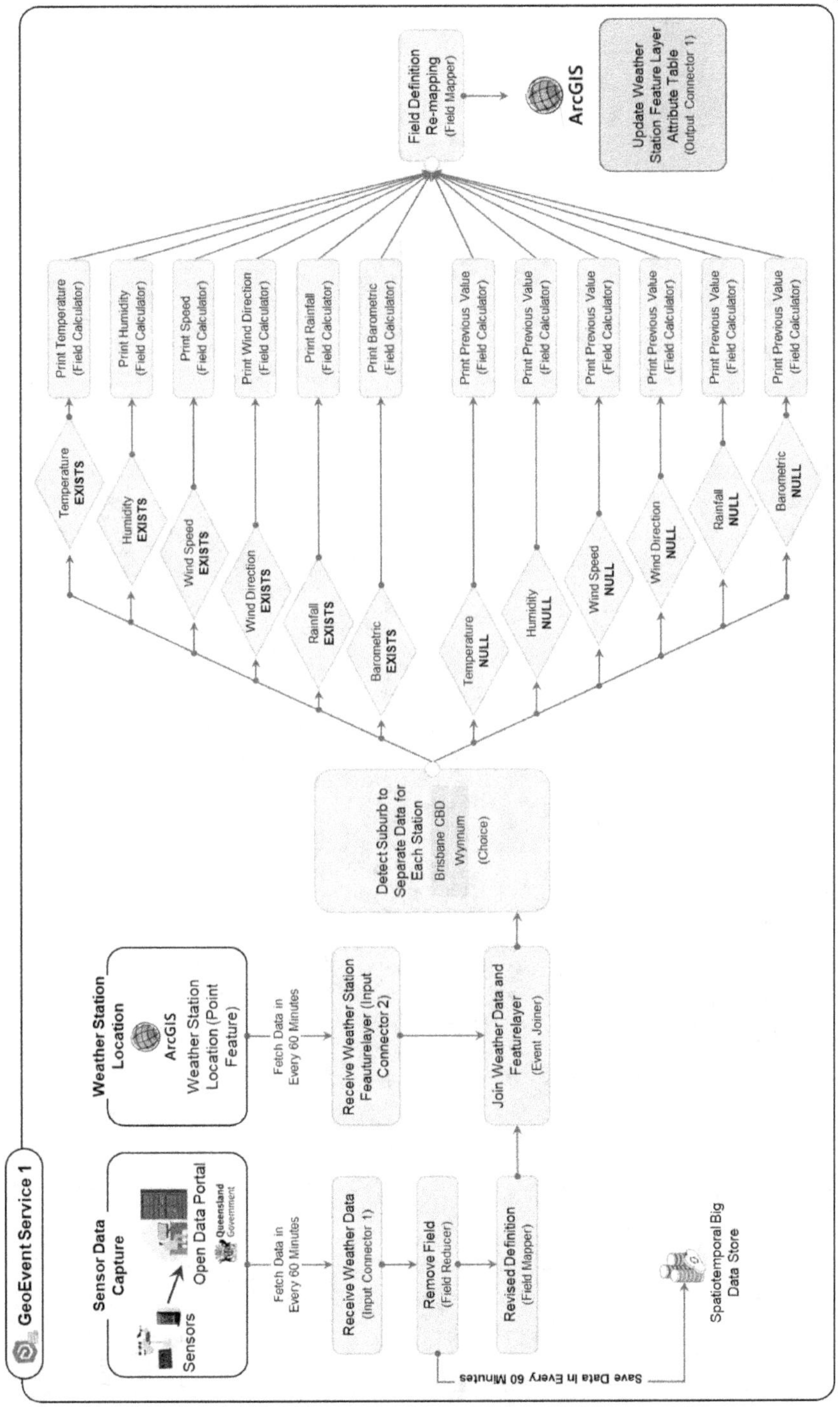

FIGURE 3.13
GeoEvent Service 1 architecture.

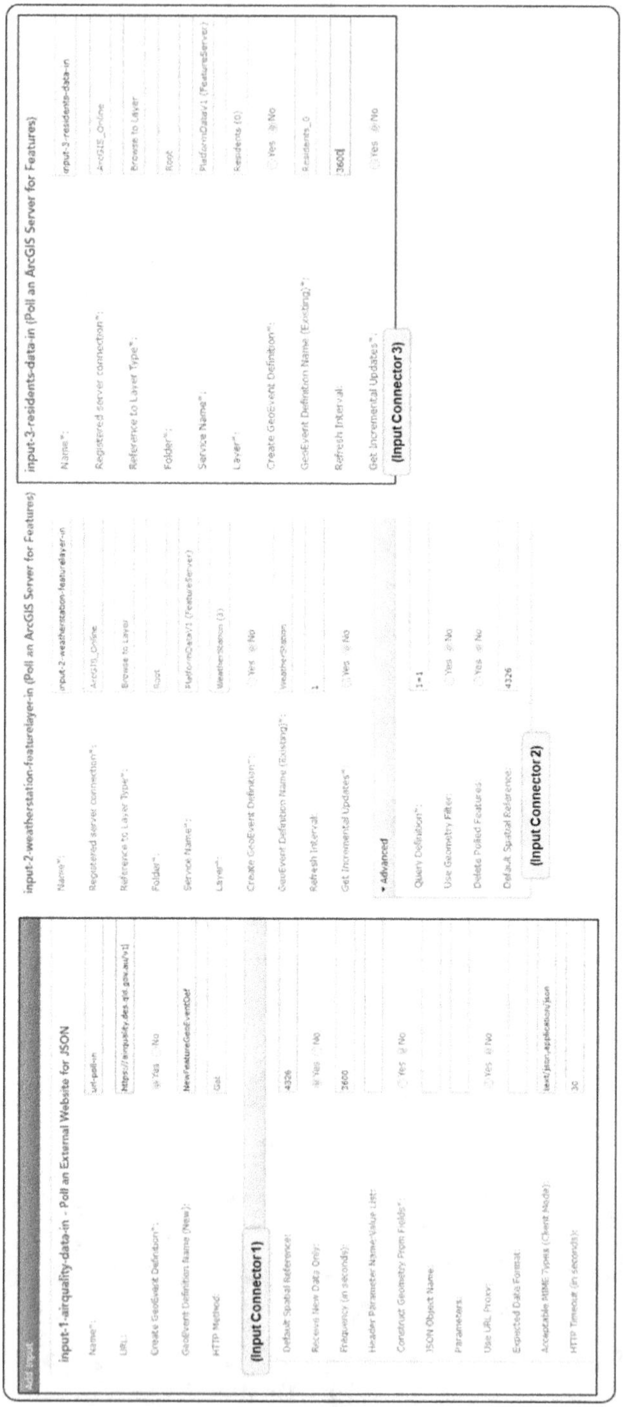

FIGURE 3.14
Configuration of IC1, IC2, and IC3 utilised for demonstration within the ArcGIS GeoEvent environment.

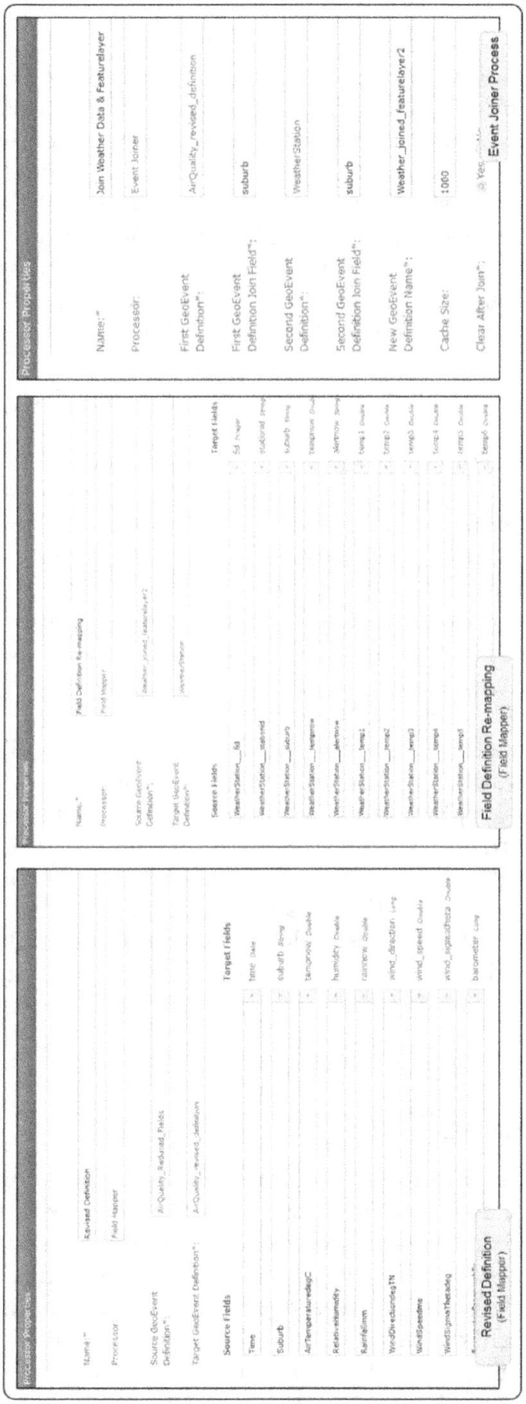

FIGURE 3.15
Configuration of Field Mappers and Event Joiner Processors used in GeoEvent Service 1.

3.6.2 Notebook Schedules 1 and 2

ArcGIS Notebooks are scheduled to execute every 60 minutes as integral components of the platform. The first script implements a ML-based weather prediction model, drawing inspiration from studies by Ren et al. (2021) and Scher and Messori (2018). For this demonstration, we employed a random forest regression model, which has shown promising results in short-term weather forecasting (He et al., 2022; Massaoudi et al., 2021). While more sophisticated or geographically tailored weather prediction models could enhance accuracy, the focus of this research was to showcase the platform's capability to integrate AI and ML methodologies seamlessly, rather than developing new predictive models. As highlighted in the introduction, considerable academic research has already been dedicated to the development of complex weather prediction models (He et al., 2022; Li et al., 2024; Massaoudi et al., 2021).

Our research aims to bridge the gap by creating a platform capable of integrating such models into actionable workflows. It is worth noting that the development of more sophisticated, location-specific models presents an avenue for future research, as discussed in our future research directions section. The prediction model utilises weather data from the weather station Feature Layer, updated by GeoEvent Service 1, to forecast the next 12 hours of weather conditions on an hourly basis. These predictions are then used to update the respective fields in the attribute table for each station.

The second ArcGIS Notebook script is triggered subsequently, leveraging historical data stored within the ArcGIS Spatiotemporal Big Data Store. This script analyses various weather conditions, including warm days, warm nights, cool days, cool nights, wet days, heavy precipitation, and very heavy precipitation, adhering to the definitions established by the Queensland Government (as outlined in the introduction). For this demonstration, the focus was limited to the conditions listed above. However, the platform is inherently flexible, allowing for the inclusion of additional conditions to meet the specific needs of different regions or cities. This adaptability reinforces the platform's potential as a scalable tool for localised weather monitoring and alert systems.

3.6.3 GeoEvent Service 2

GeoEvent Service 2 is a critical component of the architecture, designed to detect weather conditions and identify residents at risk in specific locations. As illustrated in Figure 3.16, two critical datasets underpin this service: the weather station Feature Layer, updated by ArcGIS Notebook Schedule 2 to provide real-time weather condition data, and the Resident Feature Layer, which contains spatial and demographic information. Both datasets are ingested at 60-minute intervals through ICs. Since these datasets have distinct schemas, an Event Joiner Processor merges their

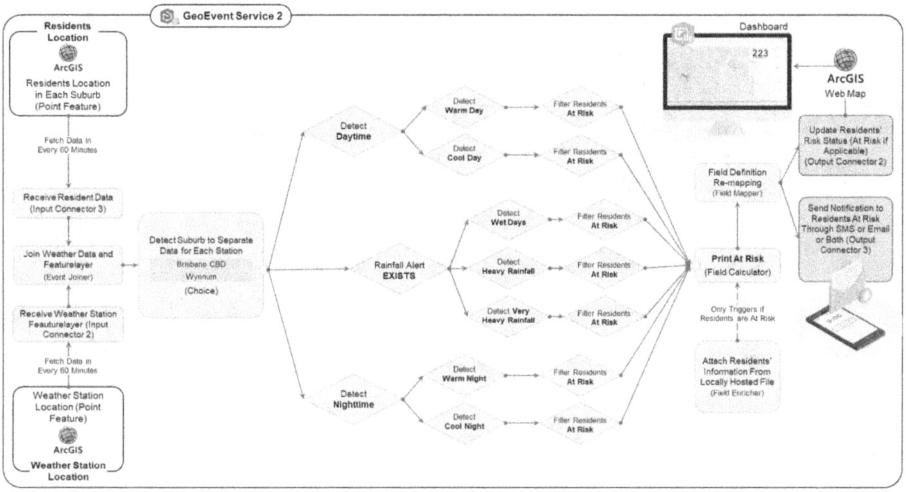

FIGURE 3.16
GeoEvent Service 2 architecture.

definitions into a unified schema referred to as the Secondary GeoEvent Definition, ensuring seamless integration of spatial and non-spatial attributes for analysis.

The combined dataset is then routed through a Choice Processor, which segregates data based on suburb names. For each suburb, the data undergoes a series of Filters designed to detect specific weather conditions. Initially, three primary Filters identify daytime periods, nighttime periods, and precipitation alerts. Data from the daytime Filter is further analysed to detect warm days and cool days, while the nighttime filter output is examined for warm nights and cool nights. Additionally, the precipitation alert Filter detects precipitation conditions such as wet days, heavy precipitation, and very heavy precipitation through three distinct Filters.

When any of the specified conditions are detected for a particular suburb, the subsequent set of Filters, each linked to the corresponding seven condition-detection Filters, updates the status of residents in that suburb based on the risk matrix illustrated in Table 3.3. This matrix, developed for demonstration purposes and informed by methodologies from Sharma and Bhaskaran (2024) and Chesnais et al. (2019), assigns risk levels to residents based on the detected conditions. While the matrix is tailored to this demonstration, it can be adapted to meet the needs of specific regions.

The Field Enricher (File) Processor accesses securely stored .txt or .csv files containing sensitive personal information. It links this data to the Resident Feature Layer using a common ID, ensuring personal information remains separate from the main Feature Layer to uphold data security, as outlined in the platform's architecture. Two OCs (OC2 and OC3) are configured with GeoEvent Service 2. The OC2 is configured to print risk status of the residents

TABLE 3.3

Risk Matrix for Weather Stressor and Vulnerable Groups Considered for Demonstration Purpose Only

	Warm Days	Warm Nights	Cool Days	Cool Nights	Wet Days	Heavy Precipitation	Very Heavy Precipitation
Children	Risk	Risk	Risk	Risk	Risk	Risk	Risk
Adults	–	–	–	–	–	–	–
Elderly	Risk	Risk	Risk	Risk	Risk	Risk	Risk
Asthma patients	Risk	Risk	Risk	Risk	–	–	–
Diabetic patients	Risk	Risk	–	–	–	–	–
Heart diseases	Risk	Risk	–	–	–	–	–
Lung diseases	Risk	Risk	Risk	Risk	–	–	–
Disability	Risk	–	–	–	Risk	Risk	Risk

and update the Feature Layer in the web map. This web map within ArcGIS Online is connected to a dashboard that shows the number of people at risks and their location as illustrated in Figure 3.17. OC3 is configured as shown in Figure 3.18 to send messages and/or emails to residents "at risk." The function ${FIELD-NAME} is used here to dynamically calls for the respective information of residents from the. csv file through reading by the Field Enricher Processor.

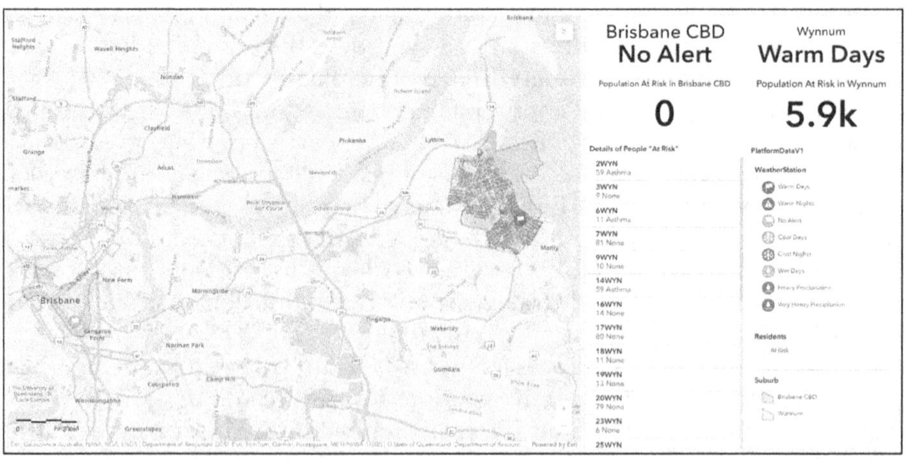

FIGURE 3.17
Dashboard displaying residents "at-risk" to authorised officers.

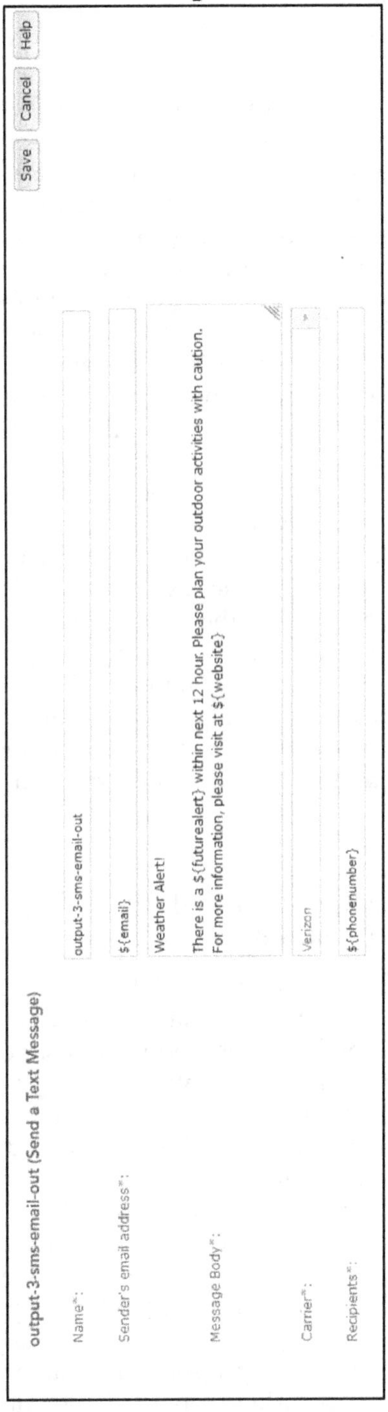

FIGURE 3.18
OC3 to send alert to residents.

3.6.4 GeoEvent Service 3

Service 3, as illustrated in Figure 3.19, is designed to calculate and update residents' age data. Recognising the sensitivity of personal data, we adhered to strict ethical guidelines and did not access or attempt to use exact resident locations or personal details. Instead, for demonstration purposes, we utilised publicly available demographic data from the ABS 2021 census. This dataset provided an overview of the population figures, age distributions, and the prevalence of various health conditions and disabilities within Brisbane suburbs. Dummy point feature data were then randomly generated within the boundaries of the respective suburbs in ArcGIS web map to simulate a resident database. This approach ensured that the platform's functionalities could be tested without compromising privacy.

IC3 ingests the residents' Feature Layer every 24 hours at a rate of 448 events per second. The data is processed through a Field Calculator to compute and update residents' ages in the attribute table. Three Filters then categorise residents into Children (0–14 years), Adults (15–64 years), and Elderly (65+ years), with each group assigned its respective classification via dedicated Field Calculators. OC2 updates the residents' age categories in the Feature Layer every 24 hours. While the demonstration did not incorporate real-time location updates due to the unavailability of precise locational data, this capability remains a key future enhancement. Local governments could utilise housing authority data through an API to periodically update residents' residential addresses, ensuring accurate and current information. Additionally, GeoEvent's GeoTag feature could enable real-time tracking of vulnerable populations, providing critical support during extreme weather events or emergencies.

3.6.5 Customised Notifications: Potential Technical and Social Challenges

In the case study, the platform demonstrated its ability to personalise notifications by integrating real-time weather data with demographic and health profiles to identify and prioritise at-risk populations. For example, during warm nights in Wynnum North, notifications were prioritised for elderly residents and individuals with respiratory conditions, given that prolonged nighttime heat can exacerbate cardiovascular and respiratory issues. Similarly, during heavy rainfall events, alerts were directed to residents in flood-prone areas, particularly those with mobility impairments, to provide timely warnings and support proactive decision-making. The ability to dynamically adjust notification priorities based on weather severity and demographic vulnerability demonstrates the platform's potential for context-sensitive risk communication.

Since this was a demonstration project, no technical challenges were encountered. However, in real-world deployment, potential technical issues

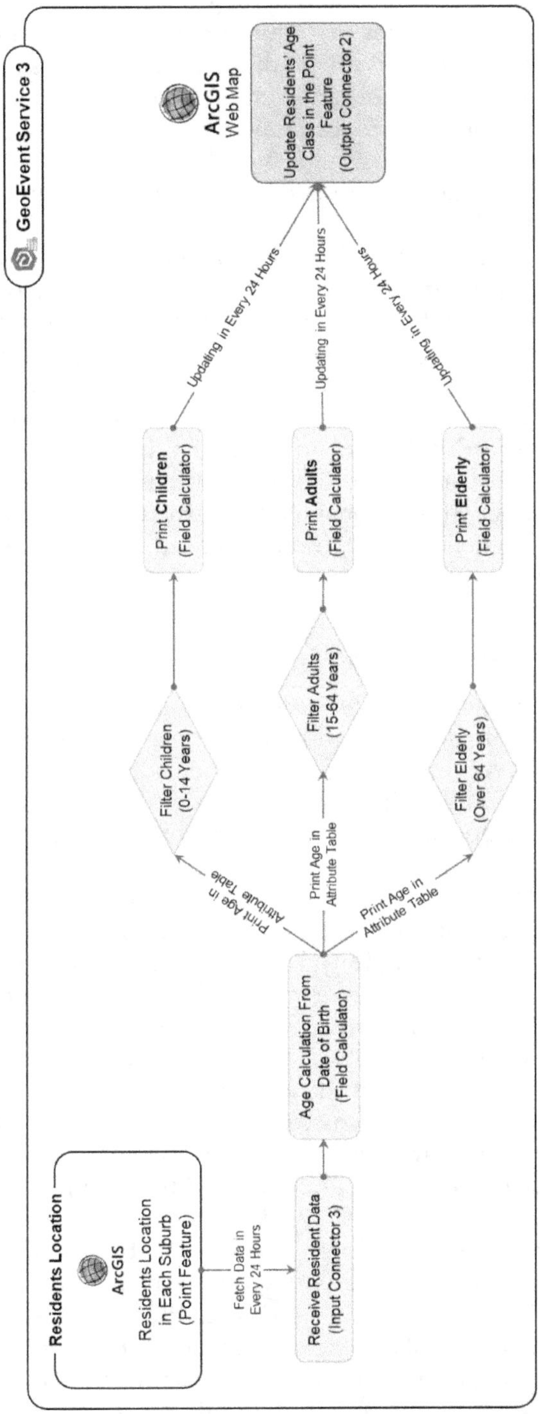

FIGURE 3.19
GeoEvent Service 3 architecture.

could include incomplete or outdated demographic datasets, requiring continuous updates and validation to maintain accuracy. Additionally, integrating real-time sensor data with predictive models across diverse urban environments could present calibration challenges, particularly when adapting the platform to cities with different data infrastructures or weather monitoring capabilities. Addressing these challenges would require interoperability between datasets, automated validation processes, and adaptive calibration techniques.

Beyond technical considerations, social acceptance and public engagement are crucial factors influencing the effectiveness of personalised alerts. Challenges such as alert fatigue—where frequent notifications may lead residents to ignore warnings—would need to be addressed through careful optimisation of notification thresholds and customisation options. To mitigate this, future iterations of the platform could incorporate user preference settings and feedback mechanisms, allowing residents to customise alert thresholds while ensuring high-risk individuals remain adequately informed. By addressing these considerations, the platform has the potential to enhance urban resilience through personalised weather risk communication tailored to diverse urban contexts.

3.6.6 Privacy and Cybersecurity

The proposed platform incorporates various forms of sensitive data, including residents' demographic details, health conditions, and location-based information. As such, privacy and cybersecurity considerations are crucial to its design, development, and deployment. Ensuring the confidentiality, integrity, and security of this information is essential for maintaining public trust and ensuring compliance with legal and regulatory standards, such as Australia's Privacy Act 1988 (Yuvaraj, 2018). Organisations should ensure informed consent when collecting data and maintain transparency regarding its usage. Robust security measures, including access controls and encryption protocols, are critical for safeguarding sensitive information from unauthorised access.

Adhering to a privacy-by-design approach, the platform incorporates measures to protect the confidentiality, integrity, and security of sensitive data throughout its operations. One of the key features supporting privacy is the Field Enricher (File) Processor, which isolates personally identifiable information (PII) from the primary Resident Feature Layer. Instead of storing sensitive details like phone numbers and email addresses in the Feature Layer, this information is securely stored in encrypted. txt or. csv files within an access-controlled environment. These files are linked to the Resident Feature Layer via a shared unique identifier, allowing the platform to access essential personal information only when an "at-risk" alert is triggered.

To further enhance security, it is recommended to implement role-based access control for local government officials, ensuring that access to the dashboard and the platform's backend configuration is restricted to authorised personnel only. Additionally, regular security audits and assessments are advised to identify and mitigate potential vulnerabilities within the GeoEvent Services, web maps, and dashboards. These measures can collectively strengthen the platform's overall security, promoting data privacy, integrity, and confidentiality in alignment with data protection laws and ethical standards.

3.6.7 Platform Prospects, Limitations, and Integration into Local Government System

Although this study demonstrated the platform's capabilities in Brisbane CBD and Wynnum North, its modular architecture and customisable components enable flexible adaptation to diverse urban contexts beyond Brisbane. Its flexibility stems from three key design features: API-driven interoperability, adaptability of AI-based weather prediction model, and configurable demographic integration. By modifying API configurations, cities can integrate local weather monitoring networks and tailor the platform to their specific meteorological data sources.

A key feature of the platform is its ability to integrate diverse AI and ML-based weather prediction models, allowing for region-specific forecasting through Python scripts or spatial analytical models. This capability addresses the critical need identified in AI weather prediction systems for location-sensitive modelling, particularly when adapting to cities with distinct topographic features or unique microclimate dynamics. Beyond meteorological data integration, the platform's modular architecture allows for the seamless incorporation of city-specific demographic and health datasets within ArcGIS. This capability enables local governments and researchers to input customised population distributions, health vulnerabilities, and risk thresholds, ensuring that risk assessments reflect the specific socio-environmental dynamics of a given region. Such adaptability is particularly crucial for cities with varied population densities, infrastructure profiles, and environmental risk factors, where the effects of extreme weather events can differ significantly. Future research can involve testing the platform in cities with diverse topographies, climatic conditions, and population densities to more comprehensively evaluate its generalisability and effectiveness across varied urban settings.

The platform's modular design allows for dynamic adjustments to suit varying climatic conditions, geographic regions, and user-specific requirements. Customisation options include region-specific definitions for weather thresholds and conditions, ensuring that alerts are contextually relevant. Its architecture supports the inclusion of additional Connectors and Processors to incorporate new data streams, such as real-time energy consumption or social media feeds for disaster response, thereby extending its utility beyond weather monitoring to a broader spectrum of urban resilience applications (McLaughlin, 2017; Yao & Wang, 2020).

While the platform offers a robust solution for localised weather monitoring and alerting, its reliance on API-based sensor data introduces several challenges. API disruptions, such as server outages, connectivity issues, or high-frequency request limitations, can result in data gaps that compromise the platform's ability to deliver timely and accurate weather insights. Additionally, delays in API-configured data updates reduce the platform's capacity to respond to rapidly changing weather conditions, limiting real-time alerting efficiency. Addressing these limitations requires future iterations of the platform to incorporate strategies for handling missing or delayed data, such as interpolative modelling or the integration of alternative data sources to ensure continuous system functionality.

Local governments aiming to deploy this platform require an ArcGIS Enterprise subscription, which many local government authorities already maintain for urban planning and management purposes. This subscription provides access to ArcGIS GeoEvent and Portal for ArcGIS, enabling the creation of web maps and dashboards to support real-time data visualisation and monitoring. Alternatively, web maps and dashboards can be developed in ArcGIS Online and integrated with GeoEvent by configuring the ArcGIS Online server as a registered server within the GeoEvent environment. Another option for local governments is ArcGIS Velocity, a fully cloud-based solution that offers similar functionality but does not require manual installation or local server configuration. Unlike GeoEvent, which necessitates manual installation and the establishment of connections between ArcGIS platforms, Velocity operates as a complete Software as a Service (SaaS) solution. However, Velocity requires a separate subscription, which may result in additional costs for local governments. Given these considerations, GeoEvent was selected for this research, as it is included within the ArcGIS Enterprise subscription and is already widely utilised by local governments, particularly in Australia. The use of GeoEvent avoids the need for additional subscription costs while still supporting the platform's functional requirements for real-time data integration, processing, and visualisation.

3.7 Conclusion

This research presents a novel platform for real-time weather monitoring and personalised alert generation that addresses the need for localised weather monitoring and risk-based alert systems. Leveraging platform urbanism and AIoT concepts, the platform enables dynamic data integration to detect weather conditions and issue targeted alerts for residents considering their location, age, health conditions, and disabilities. The proposed platform offers distinct advantages over existing weather monitoring and alert systems in terms of cost-effectiveness, scalability, and ease of implementation.

The platform's scalable architecture allows for the integration of multiple environmental hazard assessments, positioning it as a cost-effective and adaptable solution for climate risk management. By integrating additional data streams and analytical models, the system can evolve into a comprehensive multi-hazard monitoring platform capable of addressing threats such as floods, bushfires, heatwaves, and air pollution events. Consolidating hazard assessments within a single interface simplifies operations for local governments, eliminating the need to manage separate systems for heatwave, flood, and air quality monitoring. Additionally, as the platform is built within the widely adopted ArcGIS ecosystem, local governments and urban planners can deploy it without requiring entirely new infrastructure, minimising implementation costs. The use of automated data ingestion, processing, and notification dissemination reduces manual intervention, ensuring operational efficiency.

We identified several future research areas to enhance its capabilities and impact. One promising direction involves the development of more accurate microclimate-based weather prediction models tailored to integrate seamlessly into the platform. Along with the station data, these models can incorporate other microclimate variables such as elevation, land cover, proximity to water bodies, and urban heat island effects. Future research directions include enhancing cybersecurity and data security, particularly as the platform manages sensitive resident information. While current measures involve data anonymisation and encrypted storage, addressing potential vulnerabilities in data transmission and system integration remains a critical focus. Advancements in encryption protocols, secure data transmission methods, and access control systems are key areas for exploration.

Another direction involves refining the platform's risk classification system from a binary "at-risk" model to a more nuanced approach with high, moderate, and low-risk categories, requiring interdisciplinary collaboration between climatologists, public health experts, and data scientists. Expanding the platform to incorporate a broader range of risk events, such as floods, bushfires, and air quality issues, presents an opportunity to streamline emergency response and strengthen community resilience. Additionally, exploring the integration of citizen science and crowdsourced data through mobile applications could enhance data granularity, supporting more accurate risk analysis and prediction.

Additionally, the strategic placement of weather sensors in urban areas requires further research, particularly in cities (like in our case study context of Brisbane) where weather station distribution is uneven. With nine weather stations in Brisbane, some clustered together, an estimated 34%–44% of the city lacks coverage, assuming a 3–5 km radius per station. Addressing these gaps could enhance the capture of microclimatic variations across diverse urban environments. Prospective research could explore methods to optimise sensor placement for comprehensive city-wide coverage, thereby improving data accuracy and inclusivity. Advancing these areas would contribute to a more robust platform for urban weather monitoring and risk management, ultimately supporting urban resilience and public safety.

Acknowledgements

This chapter, with permission from the copyright holder, is a reproduced version of the following journal article: Hossain, T., Yigitcanlar, T., Nguyen, K., & Xu, Y. (2025). Platform urbanism for resident safety: a real-time predictive microclimate risk monitoring and alert system. Urban Climate, 61, 102445.

References

Acosta-Coll, M., Ballester-Merelo, F., Martinez-Peiró, M., & De la Hoz-Franco, E., 2018. Real-time early warning system design for pluvial flash floods—A review. *Sensors, 18*(7), 2255. https://doi.org/10.3390/s18072255

Almalki, F. A., Alsamhi, S. H., Sahal, R., Hassan, J., Hawbani, A., Rajput, N. S., Saif, A., Morgan, J., & Breslin, J., 2023. Green IoT for eco-friendly and sustainable smart cities: Future directions and opportunities. *Mobile Networks and Applications, 28*(1), 178–202. https://doi.org/10.1007/s11036-021-01790-w

Anthony, B., 2024. Decentralized AIoT based intelligence for sustainable energy prosumption in local energy communities: A citizen-centric prosumer approach. *Cities, 152*, 105198. https://doi.org/10.1016/j.cities.2024.105198

Ascione, F., de Rossi, F., Iovane, T., Manniti, G., & Mastellone, M., 2024. Energy demand and air quality in social housing buildings: A novel critical review. *Energy and Buildings, 319*, 114542. https://doi.org/10.1016/j.enbuild.2024.114542

Atencio, E., Lozano, F., Alfaro, I., Lozano-Galant, J. A., & Muñoz-La Rivera, F., 2024. Integrating web-based weather data into building information modeling models through robot process automation. *Applied Sciences, 14*(19), 9109. https://doi.org/10.3390/app14199109

Barriopedro, D., García-Herrera, R., Ordóñez, C., Miralles, D. G., & Salcedo-Sanz, S., 2023. Heat waves: Physical understanding and scientific challenges. *Reviews of Geophysics, 61*(2), e2022RG000780. https://doi.org/10.1029/2022RG000780

Belli, L., Cilfone, A., Davoli, L., Ferrari, G., Adorni, P., Di Nocera, F., Dall'Olio, A., Pellegrini, C., Mordacci, M., & Bertolotti, E., 2020. IoT-enabled smart sustainable cities: Challenges and approaches. *Smart Cities, 3*(3), 1039–1071. https://doi.org/10.3390/smartcities3030052

Bibri, S. E., Huang, J., Jagatheesaperumal, S. K., & Krogstie, J., 2024. The synergistic interplay of artificial intelligence and digital twin in environmentally planning sustainable smart cities: A comprehensive systematic review. *Environmental Science and Ecotechnology, 20*, 100433. https://doi.org/10.1016/j.ese.2024.100433

Bibri, S. E., & Jagatheesaperumal, S. K., 2023. Harnessing the potential of the metaverse and artificial intelligence for the internet of city things: Cost-effective XReality and synergistic AIoT technologies. *Smart Cities, 6*(5), 2397–2429. https://www.mdpi.com/2624-6511/6/5/109

BoM. (2024). *About the climate extremes analyses. Bureau of Meteorology.* Retrieved 30 November 2024 from http://www.bom.gov.au/climate/change/about/extremes.shtml.

Buguet, A., Radomski, M. W., Reis, J., & Spencer, P. S., 2023. Heatwaves and human sleep: Stress response versus adaptation. *Journal of the Neurological Sciences, 454,* 120862. https://doi.org/10.1016/j.jns.2023.120862

Casanueva, A., Burgstall, A., Kotlarski, S., Messeri, A., Morabito, M., Flouris, A. D., Nybo, L., Spirig, C., & Schwierz, C., 2019. Overview of existing heat-health warning systems in Europe. *International Journal of Environmental Research and Public Health, 16*(15), 2657. https://doi.org/10.3390/ijerph16152657

Chesnais, M., Green, A., Phillips, B., Aitken, P., Dyson, J., Trancoso, R., Rajan, J., & Dunbar, C. (2019). *Queensland – State heatwave assessment 2019.* The State of Queensland (Queensland Fire and Emergency Services). https://www.disaster.qld.gov.au/__data/assets/pdf_file/0026/339308/QFES-Heatwave-Risk-Assessment.pdf

Chevance, G., Minor, K., Vielma, C., Campi, E., O'Callaghan-Gordo, C., Basagaña, X., Ballester, J., & Bernard, P., 2024. A systematic review of ambient heat and sleep in a warming climate. *Sleep Medicine Reviews, 75,* 101915. https://doi.org/10.1016/j.smrv.2024.101915

Chitwatkulsiri, D., & Miyamoto, H., 2023. Real-time urban flood forecasting systems for southeast Asia—A review of present modelling and its future prospects. *Water, 15*(1), 178. https://doi.org/10.3390/w15010178

Clegg, G., Haigh, R., & Amaratunga, D., 2022. Towards an improved understanding of participation in natural hazard early warning systems. *International Journal of Disaster Resilience in the Built Environment, 13*(5), 615–631. https://doi.org/10.1108/IJDRBE-11-2020-0120

Connon, I. L. C., & Hall, E., 2021. 'It's not about having a back-up plan; it's always being in back-up mode': Rethinking the relationship between disability and vulnerability to extreme weather. *Geoforum, 126,* 277–289. https://doi.org/10.1016/j.geoforum.2021.08.008

D'Amico, G., L'Abbate, P., Liao, W., Yigitcanlar, T., & Ioppolo, G., 2020. Understanding sensor cities: Insights from technology giant company driven smart urbanism practices. *Sensors, 20*(16), 4391. https://www.mdpi.com/1424-8220/20/16/4391

Dizdaroglu, D., & Yigitcanlar, T., 2014. A parcel-scale assessment tool to measure sustainability through urban ecosystem components: The MUSIX model. *Ecological Indicators, 41,* 115–130. https://doi.org/10.1016/j.ecolind.2014.01.037

Dodd, S., Kragh-Furbo, M., Davies, J., Butterfield, S., Morris, A., & Brown, H., 2024. Health impacts of climate change in the UK: A qualitative synthesis detailing the conjuncture of social structure, extreme weather, and mental health. *SSM – Qualitative Research in Health, 6,* 100475. https://doi.org/10.1016/j.ssmqr.2024.100475

Elliott, H., Eon, C., & Breadsell, J. K., 2020. Improving city vitality through urban heat reduction with green infrastructure and design solutions: A systematic literature review. *Buildings, 10*(12), 219. https://doi.org/10.3390/buildings10120219

ESRI. (2021). *ArcGIS GeoEvent Server quick start guide.*

Fair, K. M., Zachreson, C., & Prokopenko, M., 2019. Creating a surrogate commuter network from Australian Bureau of Statistics census data. *Scientific Data, 6*(1), 150. https://doi.org/10.1038/s41597-019-0137-z

Forkan, A. R. M., Kang, Y.-B., Marti, F., Banerjee, A., McCarthy, C., Ghaderi, H., Costa, B., Dawod, A., Georgakopolous, D., & Jayaraman, P. P., 2024. AIoT-citysense: AI and IoT-driven city-scale sensing for roadside infrastructure maintenance. *Data Science and Engineering, 9*(1), 26–40. https://doi.org/10.1007/s41019-023-00236-5

Frustaci, G., Pilati, S., Lavecchia, C., & Montoli, E. M., 2022. High-resolution gridded air temperature data for the urban environment: The Milan data set. *Forecasting, 4*(1), 238–261. https://www.mdpi.com/2571-9394/4/1/14

Hahn, C., Garcia-Marti, I., Sugier, J., Emsley, F., Beaulant, A.-L., Oram, L., Strandberg, E., Lindgren, E., Sunter, M., & Ziska, F., 2022. Observations from personal weather stations—EUMETNET interests and experience. *Climate, 10*(12), 192. https://doi.org/10.3390/cli10120192

Han, J., Chong, A., Lim, J., Ramasamy, S., Wong, N. H., & Biljecki, F., 2024. Microclimate spatio-temporal prediction using deep learning and land use data. *Building and Environment, 253*, 111358. https://doi.org/10.1016/j.buildenv.2024.111358

Haraguchi, M., Nishino, A., Kodaka, A., Allaire, M., Lall, U., Kuei-Hsien, L., Onda, K., Tsubouchi, K., & Kohtake, N., 2022. Human mobility data and analysis for urban resilience: A systematic review. *Environment and Planning B: Urban Analytics and City Science, 49*(5), 1507–1535. https://doi.org/10.1177/23998083221075634

He, Y., Chen, C., Li, B., & Zhang, Z., 2022. Prediction of near-surface air temperature in glacier regions using ERA5 data and the random forest regression method. *Remote Sensing Applications: Society and Environment, 28*, 100824. https://doi.org/10.1016/j.rsase.2022.100824

Hou, K. M., Diao, X., Shi, H., Ding, H., Zhou, H., & de Vaulx, C., 2023. Trends and challenges in AIoT/IIoT/IoT implementation. *Sensors, 23*(11), 5074. https://www.mdpi.com/1424-8220/23/11/5074

Jagatheesaperumal, S. K., Bibri, S. E., Huang, J., Rajapandian, J., & Parthiban, B., 2024. Artificial intelligence of things for smart cities: Advanced solutions for enhancing transportation safety. *Computational Urban Science, 4*(1), 10. https://doi.org/10.1007/s43762-024-00120-6

Javanroodi, K., Perera, A. T. D., Hong, T., & Nik, V. M., 2023. Designing climate resilient energy systems in complex urban areas considering urban morphology: A technical review. *Advances in Applied Energy, 12*, 100155. https://doi.org/10.1016/j.adapen.2023.100155

Johnson, C., Schmidt, S., Taylor, J., & deLaChapelle, J. (2022). *Geospatial database development: Supporting geohazard risk assessments through real-time data and geospatial analytics* [2022 14th International Pipeline Conference].

Kaginalkar, A., Kumar, S., Gargava, P., & Niyogi, D., 2021. Review of urban computing in air quality management as smart city service: An integrated IoT, AI, and cloud technology perspective. *Urban Climate, 39*, 100972. https://doi.org/10.1016/j.uclim.2021.100972

Kam, S., Hwang, B. J., & Parker, E. R., 2023. The impact of climate change on atopic dermatitis and mental health comorbidities: A review of the literature and examination of intersectionality. *International Journal of Dermatology, 62*(4), 449–458. https://doi.org/10.1111/ijd.16557

Kong, J., Zhao, Y., Carmeliet, J., & Lei, C., 2021. Urban heat island and its interaction with heatwaves: A review of studies on mesoscale. *Sustainability, 13*(19), 10923. https://doi.org/10.3390/su131910923

Kousis, I., Manni, M., & Pisello, A. L., 2022. Environmental mobile monitoring of urban microclimates: A review. *Renewable and Sustainable Energy Reviews, 169*, 112847. https://doi.org/10.1016/j.rser.2022.112847

Kuguoglu, B. K., van der Voort, H., & Janssen, M., 2021. The giant leap for smart cities: Scaling up smart city artificial intelligence of things (AIoT) initiatives. *Sustainability, 13*(21), 12295. https://doi.org/10.3390/su132112295

Li, F., Yigitcanlar, T., Li, W., Nepal, M., Nguyen, K., & Dur, F., 2024a. Understanding urban heat vulnerability: Scientometric analysis of five decades of research. *Urban Climate, 56*, 102035. https://doi.org/10.1016/j.uclim.2024.102035

Li, F., Yigitcanlar, T., Nepal, M., Nguyen, K., Dur, F., & Li, W., 2024b. Assessing heat vulnerability and multidimensional inequity: Lessons from indexing the performance of Australian capital cities. *Sustainable Cities and Society, 115*, 105875. https://doi.org/10.1016/j.scs.2024.105875

Li, F., Yigitcanlar, T., Nepal, M., Thanh, K. N., & Dur, F., 2024c. A novel urban heat vulnerability analysis: Integrating machine learning and remote sensing for enhanced insights. *Remote Sensing, 16*(16), 3032. https://doi.org/10.3390/rs16163032

Li, H., Zhao, Y., Wang, C., Ürge-Vorsatz, D., Carmeliet, J., & Bardhan, R., 2024d. Harnessing cooling from urban trees: Interconnecting background climates, urban morphology, and tree traits. *EGUsphere, 2024*, 1–50. https://doi.org/10.5194/egusphere-2024-234

Li, M., Shao, Q., Dabrowski, J. J., Rahman, A., Powell, A., Henderson, B., Hussain, Z., & Steinle, P., 2024e. Developing a statistical approach of evaluating daily maximum and minimum temperature observations from third-party automatic weather stations in Australia. *Quarterly Journal of the Royal Meteorological Society, 150*(760), 1624–1642. https://doi.org/10.1002/qj.4662

Liu, Y., Guaralda, M., Yigitcanlar, T., Limb, M., & Garcia-Hansen, V., 2024. Navigating urban climate design implementation challenges: insights from Brisbane's built environment experts. *Journal of Urban Design*, 1–20. https://doi.org/10.1080/13574809.2024.2376680

Maclean, I. M. D., Duffy, J. P., Haesen, S., Govaert, S., De Frenne, P., Vanneste, T., Lenoir, J., Lembrechts, J. J., Rhodes, M. W., & Van Meerbeek, K., 2021. On the measurement of microclimate. *Methods in Ecology and Evolution, 12*(8), 1397–1410. https://doi.org/10.1111/2041-210X.13627

Masoudi, M., 2021. Estimation of the spatial climate comfort distribution using tourism climate index (TCI) and inverse distance weighting (IDW) (case study: Fars Province, Iran). *Arabian Journal of Geosciences, 14*(5), 363. https://doi.org/10.1007/s12517-021-06605-6

Massaoudi, M., Chihi, I., Sidhom, L., Trabelsi, M., Refaat, S. S., & Oueslati, F. S., 2021. Enhanced random forest model for robust short-term photovoltaic power forecasting using weather measurements. *Energies, 14*(13), 3992. https://doi.org/10.3390/en14133992

McLaughlin, C. L., 2017. Improving research methods for the study of geography and mental health: Utilization of social networking data and the ESRI GeoEvent Processor. *School Psychology International, 38*(4), 398–407. https://doi.org/10.1177/0143034317714617

Milazzo, M. F., Ancione, G., Lisi, R., Vianello, C., & Maschio, G., 2009. Risk management of terrorist attacks in the transport of hazardous materials using dynamic geoevents. *Journal of Loss Prevention in the Process Industries, 22*(5), 625–633. https://doi.org/10.1016/j.jlp.2009.02.014

Muhammed, D., Ahvar, E., Ahvar, S., Trocan, M., Montpetit, M.-J., & Ehsani, R., 2024. Artificial intelligence of things (AIoT) for smart agriculture: A review of architectures, technologies and solutions. *Journal of Network and Computer Applications, 228*, 103905. https://doi.org/10.1016/j.jnca.2024.103905

Murray, A. T., 2021. Contemporary optimization application through geographic information systems. *Omega, 99*, 102176. https://doi.org/10.1016/j.omega.2019.102176

Neußner, O., 2021. Early warning alerts for extreme natural hazard events: A review of worldwide practices. *International Journal of Disaster Risk Reduction, 60,* 102295. https://doi.org/10.1016/j.ijdrr.2021.102295

Ostrometzky, J., & Messer, H., 2024. Opportunistic weather sensing by smart city wireless communication networks. *Sensors, 24*(24), 7901. https://doi.org/10.3390/s24247901

Oswald, S. M., Schneider, S., Hahn, C., Žuvela-Aloise, M., Schmederer, P., Wastl, C., & Hollosi, B., 2024. High-resolution air temperature forecasts in urban areas: A meteorological perspective on their added value. *Atmosphere, 15*(12), 1544. https://doi.org/10.3390/atmos15121544

Pise, A. A., Almuzaini, K. K., Ahanger, T. A., Farouk, A., pant, K., Pareek, P. K., & Nuagah, S. J., 2022. Enabling artificial intelligence of things (AIoT) healthcare architectures and listing security issues. *Computational Intelligence and Neuroscience, 2022*(1), 8421434. https://doi.org/10.1155/2022/8421434

QLD. (2024). *Live air data. Queensland Government.* Retrieved 7 December 2024 from https://apps.des.qld.gov.au/air-quality/

Rashid, B., & Rehmani, M. H., 2016. Applications of wireless sensor networks for urban areas: A survey. *Journal of Network and Computer Applications, 60,* 192–219. https://doi.org/10.1016/j.jnca.2015.09.008

Ren, X., Li, X., Ren, K., Song, J., Xu, Z., Deng, K., & Wang, X., 2021. Deep learning-based weather prediction: A survey. *Big Data Research, 23,* 100178. https://doi.org/10.1016/j.bdr.2020.100178

Repette, P., Sabatini-Marques, J., Yigitcanlar, T., Sell, D., & Costa, E., 2021. The evolution of city-as-a-platform: smart urban development governance with collective knowledge-based platform urbanism. *Land, 10*(1), 33. https://doi.org/10.3390/land10010033

Scher, S., & Messori, G., 2018. Predicting weather forecast uncertainty with machine learning. *Quarterly Journal of the Royal Meteorological Society, 144*(717), 2830–2841. https://doi.org/10.1002/qj.3410

Shaamala, A., Yigitcanlar, T., Nili, A., & Nyandega, D., 2024. Strategic tree placement for urban cooling: A novel optimisation approach for desired microclimate outcomes. *Urban Climate, 56,* 102084. https://doi.org/10.1016/j.uclim.2024.102084

Sharma, A. R., & Bhaskaran, S., 2024. Deriving community vulnerability indices by analyzing multi-resolution space-borne data and demographic data for extreme weather events in global cities. *Remote Sensing Applications: Society and Environment, 33,* 101086. https://doi.org/10.1016/j.rsase.2023.101086

Shi, R., Hobbs, B. F., Zaitchik, B. F., Waugh, D. W., Scott, A. A., & Zhang, Y., 2021. Monitoring intra-urban temperature with dense sensor networks: Fixed or mobile? An empirical study in Baltimore, MD. *Urban Climate, 39,* 100979. https://doi.org/10.1016/j.uclim.2021.100979

Song, T., Cai, J., Chahine, T., & Li, L., 2021. Towards smart cities by internet of things (IoT)—a silent revolution in China. *Journal of the Knowledge Economy, 12*(2), 1–17. https://doi.org/10.1007/s13132-017-0493-x

Toh, C. K., 2020. Security for smart cities. *IET Smart Cities, 2*(2), 95–104. https://doi.org/10.1049/iet-smc.2020.0001

Trilles, S., Calia, A., Belmonte, Ó., Torres-Sospedra, J., Montoliu, R., & Huerta, J., 2017. Deployment of an open sensorized platform in a smart city context. *Future Generation Computer Systems, 76,* 221–233. https://doi.org/10.1016/j.future.2016.11.005

Tsin, P. K., Knudby, A., Krayenhoff, E. S., Ho, H. C., Brauer, M., & Henderson, S. B., 2016. Microscale mobile monitoring of urban air temperature. *Urban Climate, 18,* 58–72. https://doi.org/10.1016/j.uclim.2016.10.001

Wong, N. H., Tan, C. L., Kolokotsa, D. D., & Takebayashi, H., 2021. Greenery as a mitigation and adaptation strategy to urban heat. *Nature Reviews Earth & Environment,* 2(3), 166–181. https://doi.org/10.1038/s43017-020-00129-5

Yang, T., Ding, Y., Li, B., & Athienitis, A. K., 2023. A review of climate adaptation of phase change material incorporated in building envelopes for passive energy conservation. *Building and Environment, 244,* 110711. https://doi.org/10.1016/j.buildenv.2023.110711

Yao, F., & Wang, Y., 2020. Towards resilient and smart cities: A real-time urban analytical and geo-visual system for social media streaming data. *Sustainable Cities and Society, 63,* 102448. https://doi.org/10.1016/j.scs.2020.102448

Yassaghi, H., Mostafavi, N., & Hoque, S., 2019. Evaluation of current and future hourly weather data intended for building designs: A Philadelphia case study. *Energy and Buildings, 199,* 491–511. https://doi.org/10.1016/j.enbuild.2019.07.016

Yigitcanlar, T., 2010. *Rethinking sustainable development: Urban management, engineering, and design.* IGI Global.

Yuvaraj, J., 2018. How about me? The scope of personal information under the Australian Privacy Act 1988. *Computer Law & Security Review,* 34(1), 47–66. https://doi.org/10.1016/j.clsr.2017.05.019

Zabini, F., 2016. Mobile weather apps or the illusion of certainty. *Meteorological Applications,* 23(4), 663–670. https://doi.org/10.1002/met.1589

Zhang, X., Chen, N., Chen, Z., Wu, L., Li, X., Zhang, L., Di, L., Gong, J., & Li, D., 2018. Geospatial sensor web: A cyber-physical infrastructure for geoscience research and application. *Earth-Science Reviews, 185,* 684–703. https://doi.org/10.1016/j.earscirev.2018.07.006

4

Integrated Planning for Climate-Resilient Cities

4.1 Introduction

Urban areas face unprecedented vulnerability to climate change, acting as the epicentres of both risk and opportunity. Climate extremes such as intense heatwaves, powerful storms, and destructive floods are increasingly common, intensifying risks for urban centres where population density, infrastructure, and economic concentration exacerbate the impacts of climate events. Globally, climate-related disasters impose economic losses estimated at $200 million per day, according to the World Meteorological Organisation, a figure that continues to grow with the escalating frequency and severity of extreme weather. For cities, this growing threat is compounded by social inequities and infrastructural limitations, underscoring the need for cities to play a proactive role in addressing the climate crisis.

To support this transition, the IPCC's *Special Report on Climate Change and Cities* is organised around six critical focus areas, each addressing core aspects of urban resilience and adaptation. These focus areas—Biophysical and Environmental Factors, Impacts and Risks, Sectoral Development and Adaptation, Energy and Emissions, Governance and Policy, and Social Cohesion and Equity—collectively provide a comprehensive roadmap for cities to address the interlinked challenges of climate resilience (IPCC, 2024). The report emphasises that cities must transcend traditional growth-oriented planning models and embrace resilience-based frameworks that integrate science, community needs, and long-term sustainability goals.

Taken together, these challenges and guiding frameworks underscore the need for a more integrated and evidence-informed understanding of how cities can navigate escalating climate risks. Building on the IPCC's thematic foundations, the remainder of this chapter examines how urban planning, climate science, governance, and equity considerations can converge to shape more resilient and inclusive urban futures. By outlining the study's research design, presenting the key analytical insights, and discussing their implications for contemporary urban climate practice, this chapter aims to contribute a coherent perspective on how cities can operationalise climate resilience in ways that are scientifically grounded, socially just, and adaptable to rapidly changing conditions.

 DOI: 10.1201/9781003737056-5

4.2 Research Design

This chapter adopts a structured qualitative research design supported by targeted documentary analysis and thematic synthesis. The purpose of the research is to distil how the IPCC's *Special Report on Climate Change and Cities* can inform an integrated urban resilience framework, with particular emphasis on convergence across urban planning, climate science, governance, and socio-ecological equity.

The research proceeded in three stages. Firstly, a comprehensive review of the report's six focus areas was undertaken, complemented by a broader scan of peer-reviewed literature on climate adaptation, resilience planning, and more-than-human urbanism. This enabled the identification of conceptual linkages, thematic patterns, and gaps regarding the role of cities as integrators of resilience policy and practice.

Secondly, an interpretive synthesis approach was applied to analyse the interactions between the report's focus areas—Biophysical and Environmental Factors, Impacts and Risks, Sectoral Development and Adaptation, Energy and Emissions, Governance and Policy, and Social Cohesion and Equity. This involved mapping cross-cutting themes related to convergence, urban adaptive capacity, and equitable resilience, which together formed the analytical backbone of the study.

Finally, the study critically examined how these themes align with contemporary urban planning discourses, including nature-based solutions, governance reform, community participation, and more-than-human perspectives. This triangulated analysis enabled the development of an integrated conceptual framework that positions cities as key agents of multi-scalar climate resilience.

Through this design, the study provides a robust, evidence-informed synthesis that advances theoretical understanding of convergence-based urban resilience while offering practical insights for policy, planning, and local governance.

4.3 Analysis and Results

4.3.1 The Imperative of Convergent Resilience Frameworks

Considering compounding climate risks, the need for a convergence framework is paramount (Ye & Niyogi, 2022). This approach integrates urban planning, climate science, public health, environmental policy, and social equity to construct a holistic and adaptive strategy. Cities must navigate the complex interactions between environmental systems, human health, and infrastructure resilience, all of which are impacted by climate change. The IPCC's *Special Report on Climate Change and Cities* underscores this in its *Biophysical and Environmental Factors* focus area, which examines how urban systems influence and are influenced by natural processes, from hydrological cycles to urban heat islands. By understanding these biophysical dynamics, cities can implement measures that mitigate environmental stresses, improve air quality, and manage stormwater, reducing overall vulnerability to climate impacts.

To operationalise these insights, cities must adopt governance structures that promote cross-departmental collaboration, support resilient decision-making, and address urban vulnerabilities from multiple angles. The report's *Impacts and Risks* focus area delves into these compounded climate risks, highlighting the vulnerabilities of critical urban infrastructure, such as energy and water systems, and examining the cascading effects on the food–energy–water–health nexus. Recognising these compounded risks allows cities to prioritise adaptation strategies, ensuring that infrastructure is resilient not only to today's climate but also to future extremes. Through this convergence, cities can foster resilience that addresses both immediate and structural vulnerabilities, supporting health, economic stability, and environmental integrity.

4.3.2 Cities as Integrators: Mobilising Science, Policy, and Practice

The role of cities as integrators is essential to creating cohesive responses to climate adaptation. Cities bring together a diverse network of policymakers, scientists, business leaders, and residents, each with unique insights and resources, to co-create climate solutions that are both innovative and scalable. By integrating multiple sectors, cities can serve as models for urban resilience, where collaborative efforts support adaptive infrastructure projects and nature-based solutions that respond to both environmental and social needs. This is underscored by the IPCC's focus area on *Sectoral Development and Adaptation*, which encourages cities to adopt nature-based solutions like urban greening, green roofs, and wetland restoration to enhance resilience while creating liveable spaces.

Additionally, by promoting cross-sector collaboration and leveraging public–private partnerships, cities can address the *Energy and Emissions* focus area of the IPCC report, which emphasises that cities are key contributors to global greenhouse gas emissions and must lead in energy transition efforts (Solecki et al., 2024). Cities can reduce their carbon footprint by transitioning to renewable energy sources, improving urban energy management, and establishing carbon accounting mechanisms that enable transparent reporting and accountability. As hubs of innovation and adaptability, cities can bridge knowledge and policy across sectors, becoming catalysts for resilience and sustainability that other urban areas can emulate.

4.3.3 Modernising Building Codes and Governance for Urban Resilience

Building codes and governance structures form the backbone of urban resilience, yet many cities operate under codes that fail to reflect current climate risks. Updated building codes can mandate climate-resilient materials, energy-efficient designs, and features that support sustainable urban environments, such as flood-resistant structures and permeable surfaces. The IPCC's *Governance and Policy* focus area stresses that effective governance is essential for cities to achieve climate resilience, outlining the importance of climate finance, supportive regulatory frameworks, and alignment with the Sustainable Development Goals (SDGs). By facilitating public–private partnerships, attracting investments, and creating incentives for sustainable practices, cities can foster an environment where resilience is embedded within urban systems.

Governance reforms should also ensure transparency and community engagement, allowing residents to participate directly in resilience planning and decisions. Zoning laws, infrastructure investments, and emergency response protocols can be reoriented to support compact and sustainable growth patterns, which not only reduce sprawl but also limit exposure to environmental hazards. When aligned with updated building codes, these governance reforms create urban landscapes that are prepared for the challenges of a rapidly changing climate, protecting both people and infrastructure while promoting inclusive and adaptive growth.

4.3.4 Bridging Socio-Economic Disparities for Equitable Resilience

The unequal distribution of climate impacts within cities underscores the need for an equitable approach to resilience. Vulnerable communities—such as low-income families, the elderly, and marginalised groups—often bear the brunt of climate risks due to systemic inequities in housing, infrastructure, and access to resources. Recognising this, the IPCC's focus on *Social Cohesion and Equity* stresses that climate adaptation efforts must prioritise social equity, ensuring that resilience strategies are inclusive and benefit all urban residents. By addressing these socio-economic disparities, cities can reduce climate vulnerability and foster stronger, more cohesive communities.

Equitable urban planning should prioritise resilient housing, green infrastructure, and accessible public transit in underserved areas, ensuring that all residents have access to safe, sustainable environments. Cities should also implement inclusive planning processes that engage marginalised communities, incorporating their voices into resilience strategies that directly impact their lives. In doing so, cities can bridge the "resilience divide", making climate adaptation a universal right rather than a privilege. Through equitable resilience, cities can build cohesive urban systems where social justice and environmental health reinforce one another.

4.3.5 Convergence of Urban Planning and Climate Science: A Blueprint for Future Resilience

The intersection of urban planning and climate science provides cities with a powerful framework to tackle the complex challenges of climate change.

By leveraging geospatial data, predictive climate models, and community-focused design principles, cities can anticipate climate risks and implement adaptive strategies. The IPCC's *Sectoral Development and Adaptation* focus area calls for cities to adopt adaptive planning that incorporates these insights, using climate models to inform zoning, urban design, and resource allocation in ways that reduce exposure to risks like flooding, extreme heat, and storms.

Aligning urban planning with scientific insights allows cities to implement resilience strategies that are both proactive and responsive, addressing the immediate impacts of climate change while preparing for future shifts. This convergence ensures that cities do not merely react to climate events but actively build capacity for long-term resilience. It enables cities to make informed, data-driven decisions that align local resilience goals with global climate targets, setting a standard for integrated, science-informed urban planning.

4.3.6 Engaging Stakeholders and Enhancing Public Knowledge

Resilience is a community effort, and effective urban resilience strategies require transparent, inclusive engagement among all stakeholders. Cities can create knowledge hubs that provide accessible climate data, educational resources, and real-time information on environmental risks, empowering residents to make informed decisions and participate actively in resilience planning. The IPCC's focus on *Social Cohesion and Equity* highlights the importance of including diverse community voices in the climate adaptation process, ensuring that resilience strategies reflect the needs and priorities of all residents, especially those from underserved populations.

Through public forums, partnerships, and open data initiatives, cities can foster a culture of resilience that extends beyond government actions, building a community-driven climate movement. Stakeholder engagement should prioritise input from marginalised communities to ensure that resilience strategies are equitable and inclusive, strengthening social bonds and creating a shared responsibility for climate adaptation. By fostering public knowledge and collaboration, cities can establish a resilient urban culture that values sustainability, preparedness, and collective action.

4.3.7 More-Than-Human Cities: Envisioning a Shared Future for All

Beyond the IPCC's focal areas, envisioning a truly sustainable urban future calls for the concept of *"More-Than-Human Cities,"* which expands urban planning to encompass the needs and agency of both human and non-human entities (Yigitcanlar, 2018). This approach recognises cities as shared environments where various species coexist, shaping ecological and social dynamics alike. Adopting a more-than-human perspective encourages planners to design spaces that balance human infrastructure with habitats essential for diverse life forms to thrive, enriching the urban landscape.

Incorporating this more-than-human perspective addresses significant challenges such as biodiversity loss, climate change, and environmental degradation (Yigitcanlar et al., 2019). Nature-based solutions like urban forests, green roofs, and wildlife corridors not only contribute to urban resilience against climate extremes but also provide critical habitats within the cityscape. Moreover, water-sensitive urban design elements, such as rain gardens and permeable pavements, support natural hydrological cycles and enhance the urban ecosystem's resilience. These solutions improve urban conditions by cooling cities, managing stormwater, and enhancing air quality, creating environments where both human and non-human entities benefit (Steele et al., 2019).

Recognising the rights and needs of non-human entities can foster a more inclusive urban resilience. Governance models that incorporate non-human interests—through biodiversity action plans, habitat protections, or environmental justice frameworks—help ensure that development aligns with ecological preservation and community resilience. The more-than-human approach requires rethinking governance structures, with urban policies explicitly accounting for and prioritising ecological health as part of the city's resilience strategy (Maller, 2021).

In sum, the concept of more-than-human cities proposes a vision where urban areas are designed to foster ecological interdependence and resilience. This shared urban future supports a vibrant ecosystem, where human and non-human communities alike contribute to and benefit from a balanced, thriving environment. By embracing a more-than-human approach, cities can become models of sustainability and inclusivity, fostering an urban resilience that respects and nurtures the broader web of life (Maller, 2018).

4.4 Findings and Discussion

The analysis demonstrates that effective climate resilience in cities is fundamentally a multi-scalar governance challenge, requiring coordinated action across global, national, metropolitan, and local levels. Rather than emerging from isolated interventions, urban resilience is produced through the interaction of institutional frameworks, sectoral systems, and community capacities operating across scales. Integrating the IPCC's six focus areas through a multi-scalar lens reveals how governance alignment (or misalignment) shapes the effectiveness, equity, and durability of urban climate responses.

4.4.1 Global and National Frameworks as Enabling Conditions

At the upper governance scales, global and national frameworks (such as the IPCC assessments, the Paris Agreement, and the SDGs) provide the normative, scientific, and policy foundations for urban climate action. The findings indicate that these frameworks play a critical role in standardising climate knowledge, defining risk narratives, and legitimising adaptation and mitigation priorities. Nonetheless, their influence on urban resilience remains largely indirect unless translated into actionable planning instruments at the city level.

The analysis highlights a recurring implementation gap between high-level climate commitments and local planning practice. While national climate strategies often articulate ambitious resilience goals, they frequently lack mechanisms to support cities with data access, regulatory flexibility, and long-term financing. This disconnect limits cities' ability to operationalise climate science in land use planning, infrastructure investment, and social policy, reinforcing the need for stronger vertical integration across governance scales.

Figure 4.1 presents a multi-scalar governance framework that synthesises the IPCC's focus areas by illustrating how global and national climate frameworks, city-level governance, local systems, and community and ecological processes interact to shape urban climate resilience.

4.4.2 City and Metropolitan Governance as the Integrative Core

Cities and metropolitan governments emerge from the analysis as the primary integrators of climate resilience, positioned between global frameworks and local systems. This scale is where climate science, urban planning, and policy converge most directly. The findings show that cities with stronger cross-departmental coordination (linking planning, environment, infrastructure, public health, and social services) are better able to manage compounded climate risks.

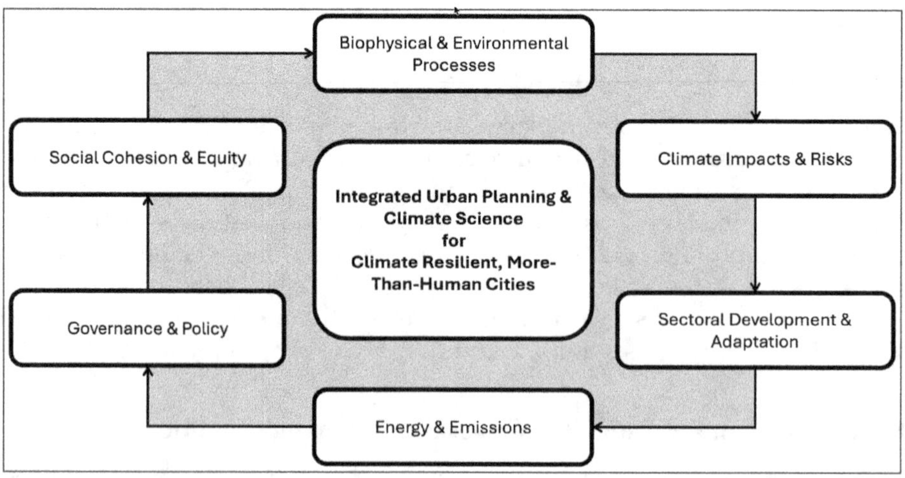

FIGURE 4.1
Multi-scalar governance framework.

Urban governance capacity determines whether biophysical insights and risk assessments translate into adaptive planning decisions, such as zoning reform, building code updates, and strategic infrastructure investment. The IPCC's focus areas on Governance and Policy, Energy and Emissions, and Sectoral Development and Adaptation intersect most strongly at this scale. Where governance structures remain siloed, resilience strategies tend to be fragmented, prioritising technical solutions over systemic transformation.

Importantly, the analysis confirms that metropolitan-scale planning is essential for addressing climate risks that transcend administrative boundaries, including heat islands, flooding, energy systems, and ecological corridors. This reinforces the argument that climate-resilient cities require governance arrangements that operate beyond municipal borders while remaining locally responsive.

4.4.3 Local Systems and Sectors as Sites of Implementation

At the local scale, resilience materialises through sectoral systems and spatial interventions, including land use, transport, energy, water, housing, and ecosystems. The findings indicate that climate impacts and vulnerabilities are most acutely experienced at this level, where infrastructure performance, urban form, and environmental conditions directly affect everyday life.

The IPCC's focus on Biophysical and Environmental Factors and Impacts and Risks is most tangible at this scale, where exposure to heat, flooding, and environmental degradation varies significantly across neighbourhoods. The analysis shows that sectoral adaptation measures (such as nature-based solutions, water-sensitive urban design, and energy-efficient buildings) are most effective when guided by integrated planning frameworks rather than implemented as isolated projects.

Nevertheless, local implementation capacity is uneven, particularly in disadvantaged or rapidly urbanising areas. Without supportive metropolitan governance and enabling national policies, local systems risk reinforcing existing vulnerabilities rather than reducing them. This underscores the importance of aligning sectoral adaptation efforts with broader governance and equity objectives.

4.4.4 Communities, Equity, and More-Than-Human Systems

The findings reaffirm that Social Cohesion and Equity is not supplementary to resilience but constitutive of it. Vulnerable populations often face disproportionate climate risks due to cumulative disadvantages in housing quality, access to green space, and political representation. The multi-scalar framework reveals that these inequities are shaped by decisions made across all governance levels, from national investment priorities to local zoning practices.

At the community scale, inclusive governance and participatory planning enhance adaptive capacity by incorporating local knowledge and lived experience into resilience strategies. The analysis further demonstrates that integrating more-than-human perspectives strengthens urban resilience by recognising the role of ecosystems in regulating climate impacts. Biodiversity, urban forests, wetlands, and hydrological systems function as critical resilience infrastructures, yet their governance often remains fragmented across sectors and scales.

Embedding ecological considerations within governance frameworks (through biodiversity action plans, habitat protection, and ecosystem-based adaptation) emerges as a key mechanism for aligning social and environmental resilience outcomes.

4.4.5 Convergence across Scales: From Fragmentation to Coherence

Taken together, the findings illustrate that climate-resilient cities are produced through convergence across scales rather than optimisation within any single scale. Fragmentation (whether between global commitments and local action, between urban planning and climate science, or between social and environmental objectives) undermines resilience outcomes. Conversely, cities that function as connective nodes within multi-scalar governance systems are better positioned to translate climate knowledge into equitable and durable urban transformation.

This convergence-based understanding reframes urban resilience as a governance process rather than a technical endpoint. It highlights the need for institutional arrangements that support vertical coordination, cross-sector integration, and continuous learning. In this sense, cities are not merely sites of climate impact but active agents capable of aligning global climate ambitions with local realities.

4.5 Conclusion

Cities are indispensable to building a resilient future. The IPCC's *Special Report on Climate Change and Cities* provides a comprehensive guide to urban resilience, emphasising the role of cities as integrators, policy enablers, and agents of social equity in climate adaptation. By adopting updated building codes, modernised governance, and equitable resilience policies, cities can protect their populations, infrastructure, and environments from the escalating impacts of climate change.

The convergence of urban planning, climate science, and social equity (enhanced by the perspective of more-than-human cities) establishes a powerful blueprint for advancing climate resilience at both local and global scales. By embracing these principles, cities can turn climate challenges into opportunities for resilience, building adaptive urban systems that stand as models of sustainability. Through these integrative efforts, cities not only safeguard their communities and urban ecosystems but also ignite a global movement towards a resilient and sustainable future.

As cities worldwide navigate the complexities of climate adaptation, the need for adaptable, evidence-based frameworks becomes increasingly apparent. By leveraging advancements in technology and data analytics, cities can continuously update and refine their resilience strategies to reflect emerging climate risks. For instance, the integration of urban artificial intelligences, urban digital twins and real-time climate monitoring allows cities to model and simulate the impacts of various climate scenarios, empowering local governments to make proactive and data-informed and evidence-based decisions (Shaamala et al., 2024). This dynamic approach not only enhances the precision of resilience planning but also fosters a culture of innovation that can drive sustainable development efforts beyond city boundaries.

Moreover, fostering global collaboration among cities is crucial for amplifying resilience practices. Through networks such as the Global Covenant of Mayors and the C40 Cities Climate Leadership Group, cities can share knowledge, resources, and strategies, creating a global coalition focused on urban climate resilience. This collaborative approach ensures that best practices and lessons learned are accessible to all cities, regardless of their economic capacity. By supporting each other through shared experiences and collective action, cities can accelerate the shift towards sustainable urban development, fortifying their role as pivotal agents in the global fight against climate change. This united front underscores the potential of cities not only to endure climate challenges but also to lead the way in building a resilient and equitable future for all (for humans and non-human urban dwellers).

Acknowledgements

This chapter, with permission from the copyright holder, is a reproduced version of the following journal article: Ye, X., Yigitcanlar, T., Xu, Y., Head, M., Hang, Y., Sanchez, T., Gong, W., & Niyogi, D. (2025). Urban planning and IPCC-like city assessments integration for climate-resilient cities. *Environment and Planning B: Urban Analytics and City Science, 52*(7), 1549–1556.

References

IPCC (Intergovernmental Panel on Climate Change). (2024). *Special report on climate change and cities*. IPCC.

Maller, C. (2018). *Healthy urban environments: More-than-human theories.* Routledge.
Maller, C. (2021). Re-orienting nature-based solutions with more-than-human thinking. *Cities, 113*, 103155.
Shaamala, A., Yigitcanlar, T., Nili, A., & Nyandega, D. (2024). Algorithmic green infrastructure optimisation: Review of artificial intelligence driven approaches for tackling climate change. *Sustainable Cities and Society, 101*, 105182.
Solecki, W., Roberts, D., & Seto, K. C. (2024). Strategies to improve the impact of the IPCC special report on climate change and cities. *Nature Climate Change, 14*(7), 685–691.
Steele, W., Wiesel, I., & Maller, C. (2019). More-than-human cities: Where the wild things are. *Geoforum, 106*, 411–415.
Ye, X., & Niyogi, D. (2022). Resilience of human settlements to climate change needs the convergence of urban planning and urban climate science. *Computational Urban Science, 2*(1), 6.
Yigitcanlar, T. (2018). Post-anthropocentric urbanism – Is more-than-human city just a utopia? *International Journal of Knowledge-Based Development, 9*(4), 337–342.
Yigitcanlar, T., Foth, M., & Kamruzzaman, M. (2019). Towards post-anthropocentric cities: Reconceptualizing smart cities to evade urban ecocide. *Journal of Urban Technology, 26*(2), 147–152.

Part II

Green Infrastructure and Design Innovation

This part of the book highlights the critical role of green infrastructure and design interventions in reducing urban heat and enhancing climate resilience. Focusing on vegetation strategies, tree placement, and cooling microclimates, it shows how urban nature can be strategically deployed to buffer extreme heat. The chapters also explore AI-driven optimisation techniques for green infrastructure, demonstrating how computational models can guide resource-efficient and high-impact interventions. By integrating ecological design with digital intelligence, this part illustrates how urban planning can combine environmental benefits with technological innovation, creating more liveable, healthy, and heat-ready cities.

DOI: 10.1201/9781003737056-6

5

Vegetation Strategies for Urban Heat Mitigation

5.1 Introduction

The rapid pace of global urbanisation has heightened concerns among city planners and policymakers regarding the growing prevalence of urban heat islands (UHIs)—urban areas that exhibit higher temperatures than their surrounding rural regions (Degirmenci et al., 2021; Gunawardena et al., 2017; Kamruzzaman et al., 2018; Levermore et al., 2018; Li et al., 2019; Stone et al., 2012). This temperature disparity arises primarily from the replacement of natural landscapes, such as vegetation and water bodies, with impervious materials that possess greater heat capacity and thermal conductivity, including pavements and buildings (Estrada et al., 2017; Senevirathne et al., 2021).

The UHI effect is most pronounced in densely developed urban cores (García & Díaz, 2021; Ho et al., 2016; Noro et al., 2015), where rapid growth exacerbates vulnerabilities to challenges such as population expansion and the increasing frequency and severity of environmental hazards (Mortoja & Yigitcanlar, 2020; Selby & Desouza, 2019). Vegetation plays a crucial role in enhancing urban resilience through thermal regulation (Gunawardena et al., 2017; Kabano et al., 2021; Noro & Lazzarin, 2015); thus, maintaining an optimal balance between green space and built form is fundamental to sustainable urban planning.

This study focuses on vegetation management as a key strategy for mitigating the UHI effect, while recognising that UHI formation is influenced by a range of factors beyond vegetation. A comprehensive review of UHI mitigation policies and technologies (Degirmenci et al., 2021) highlights several approaches, including urban design interventions (Connors et al., 2013; Koomen & Diogo, 2017), surface albedo enhancement (Pomerantz, 2018; Qin, 2015), sustainable transportation systems (Kamruzzaman et al., 2018; Kolbe, 2019), and improvements in heating, ventilation, and air conditioning (HVAC) technologies (Kohler et al., 2017; Roxon et al., 2020). Another review identifies critical determinants of UHI intensity—vegetation cover, seasonal variation, built-up density, diurnal patterns, population density, and the proportion of water bodies (Deilami et al., 2018).

Urban decision-makers must balance economic growth and societal development with ecological constraints. Optimisation approaches that integrate these objectives have been explored in earlier studies (Antheaume et al., 2018;

DOI: 10.1201/9781003737056-7

Buratto & D'Alpaos, 2015; Kang et al., 2020; Roxon et al., 2020; Yang, 2018), yet from a UHI perspective, there remains an urgent need for models that support data-informed decision-making for sustainable city development (Bakaeva & Le, 2022; Cârlan et al., 2020).

To address this gap, we develop and simulate an optimisation model using data from two regions in Brisbane, Australia. The model assigns optimal combinations of vegetation and building types to urban sites to maximise revenue while constraining UHI intensity—the temperature difference between urban and nearby rural areas (Levermore et al., 2018; Yang & Yao, 2022). This approach provides planners and policymakers with a practical decision support tool for balancing economic and environmental outcomes in urban development.

The chapter proceeds as follows: Section 5.2 introduces the optimisation model, its assumptions, variables, parameters, and formulation. Section 5.3 presents simulation results based on two case studies in Brisbane. Finally, the discussion highlights key findings and implications for urban heat mitigation and sustainable planning.

5.2 Research Design

5.2.1 Assumptions, Optimisation Model Variables and Parameters

In the optimisation model, the notation is as defined in Table 5.1. The model aims to maximise the revenue of the cells within the region of interest, subject to a set of constraints. Revenue is defined as the monetary units the local government can obtain by selling or leasing land to investors who develop the sites. Revenue maximisation is naturally capped once all available land (cells on the map) has been allocated.

The model consists of two primary sets. Firstly, a set of all cells (or pixels) $p \in P$ on the map is defined; "cell" and "pixel" are used interchangeably. Each cell p is identified by its coordinates KX_p and KY_p. Secondly, a set of alternatives $a \in A$ is introduced, from which the model must select exactly one for each pixel. A subset $b \in B \subseteq A$ represents building types, while A contains both building and vegetation alternatives.

A key constraint ensures that residential demand is satisfied within lower ($D_{\min}$) and upper ($D_{\max}$) bounds for the total population. Each cell $p \in P$ must be assigned exactly one alternative $a \in A$ in a feasible solution. The binary decision variable x_{pa} equals 1 if alternative a is placed on cell p, and 0 otherwise. Each alternative a has a sale price v_a and a number of available residential spaces s_a; for vegetation, $s_a = 0$. The total sum of s_a across all cells must satisfy the residential demand, balancing the need for housing with the allocation of open space, which enhances vegetation cover and mitigates UHI intensity (Kamruzzaman et al., 2018).

TABLE 5.1

Notation of the Optimisation Model

Symbol	Description
Sets	
(a \in A)	Set of alternatives (buildings and vegetation)
(b \in B)	Set of building types
(p \in P)	Set of pixels (cells)
Decision Variables	
(w_{p_1p_2})	Distance between cells (p_1) and (p_2)
(x_{pa})	Binary variable: 1 if alternative (a) is placed on cell (p), 0 otherwise
(x_{pb})	Binary variable: 1 if building (b) is placed on cell (p), 0 otherwise
(x_{p_1b_1})	Binary variable: 1 if building (b_1) is placed on cell (p_1), 0 otherwise
(x_{p_2b_2})	Binary variable: 1 if building (b_2) is placed on cell (p_2), 0 otherwise
(y_{p_1b_1p_2b_2})	Binary variable: 1 if building (b_1) is on cell (p_1), and building (b_2) is on cell (p_2), 0 otherwise
Parameters	
(v_a)	Land value (millions per hectare)
(s_a)	Residential capacity (inhabitants per hectare)
(D_{\min})	Minimum total residential demand
(D_{\max})	Maximum total residential demand
(\text{Dist}_{\min})	Minimum mean distance between building options (centre) in metres
(\text{UHI}_{\max})	Maximum allowable UHI intensity
(\beta_0)	Intercept coefficient for UHI model
(\beta_1)	Coefficient for population density in UHI model
(\beta_2)	Coefficient for porosity in UHI model
(\text{Area}_ {\text{sqkm}})	Total map area in km^2
(\text{PopDens})	Population density (residents per km^2 of living area)
(\text{Porosity})	Share of non-built-up area relative to total area
(KX_p)	X-coordinate of pixel (p)
(KY_p)	Y-coordinate of pixel (p)
(P)	Total amount of cells ≙ Total area of the map in hectare

For UHI compliance, the model builds on the work of Deilami et al. (2016), who employed geographically weighted regression (GWR) to quantify the spatial heterogeneity of UHI. GWR has been widely used to analyse urban morphology, land use/land cover change, and their effects on surface temperatures and urban thermal comfort (Gao et al., 2022; Shahfahad et al., 2022; Shaker et al., 2019; Taghipour Javi et al., 2014; Tu & Xia, 2008; Zhao et al., 2015; Zhou & Wang, 2011). In this study, UHI effects are incorporated as a constraint to guide strategic city development from an economic perspective. Following Deilami et al., UHI is modelled as

$$\text{UHI} = \beta_0 + \beta_1 \cdot \text{PopDens} + \beta_2 \cdot \text{Porosity} \tag{5.1}$$

where population density (*PopDens*) is the number of residents per km² of living area, and porosity (*Porosity*) is the share of non-built-up area. Increasing porosity reduces UHI, while higher population density intensifies it. The coefficients β_0, β_1, and β_2 are set according to the specific district. An upper limit $UHI_{\max}$ ensures the solution remains feasible.

To incorporate spatial distribution, a distance matrix $w_{p_1p_2}$ is computed from cell coordinates KX_p and KY_p. Distances are multiplied by decision variables $x_{p_1b_1}$ and $x_{p_2b_2}$ for built-up cells and summed over all cells, capturing only the distances between assigned buildings. The average distance is obtained by dividing the total by the number of links between built-up cells. To maintain linearity, an additional decision variable $y_{p_1b_1p_2b_2}$ and three auxiliary constraints are introduced, preserving the logic of the distance calculation.

5.2.2 Formal Optimisation Model Description

Based on the assumptions and notations introduced, the optimisation model is formally defined as follows:

Objective Function:

$$\text{maximise} \sum_{p \in P} \sum_{a \in A} x_{pa} v_a \tag{5.2}$$

This maximises the total revenue obtained by assigning alternatives $a \in A$ to cells $p \in P$.

Subject to the following constraints:

Residential demand constraints:

$$\sum_{p \in P} \sum_{a \in A} x_{pa} s_a \geq D_{\min} \tag{5.3}$$

$$\sum_{p \in P} \sum_{a \in A} x_{pa} s_a \leq D_{\max} \tag{5.4}$$

These ensure that the total residential capacity across all cells satisfies the lower ($D_{\min}$) and upper ($D_{\max}$) bounds of population demand.

Assignment constraint:

$$\sum_{a \in A} x_{pa} = 1 \forall p \in P \tag{5.5}$$

Each cell must be assigned exactly one alternative, whether a building or vegetation.

UHI constraint:

$$\beta_0 + \beta_1 \frac{\sum_{p \in P} \sum_{a \in A} x_{pa} s_a}{\text{Area}_{\text{km}^2}} + \beta_2 \left(1 - \frac{\sum_{p \in P} \sum_{b \in B} x_{pb}}{|P|} \right) \cdot 100 \leq UHI_{\max} \tag{5.6}$$

This constraint limits UHI intensity based on population density and urban porosity, following the formulation by Deilami et al. (2016).

Distance constraints between buildings:

$$\frac{\sum_{p_1 \in P} \sum_{b_1 \in B} \sum_{p_2 \in P} \sum_{b_2 \in B} w_{p_1 p_2} y_{p_1 b_1 p_2 b_2} \cdot 100}{\sum_{p \in P} \sum_{b \in B} x_{pb} \cdot \left(\sum_{p \in P} \sum_{b \in B} x_{pb} - 1 \right)} \geq \text{Dist}_{\text{min}} \tag{5.7}$$

This ensures that the average distance between built-up cells satisfies a minimum spacing requirement.

Linearisation constraints for distance variable $y_{p_1 b_1 p_2 b_2}$:

$$y_{p_1 b_1 p_2 b_2} \leq x_{p_1 b_1} \forall p_1, p_2 \in P, b_1, b_2 \in B \tag{5.8}$$

$$y_{p_1 b_1 p_2 b_2} \leq x_{p_2 b_2} \forall p_1, p_2 \in P, b_1, b_2 \in B \tag{5.9}$$

$$y_{p_1 b_1 p_2 b_2} \geq x_{p_1 b_1} + x_{p_2 b_2} - 1 \forall p_1, p_2 \in P, b_1, b_2 \in B \tag{5.10}$$

These constraints linearise the product of decision variables in the distance calculation while preserving the logical relationships between cell assignments.

5.2.3 Verbal Elaboration

For calculating the total revenue within the objective function (5.2), the total revenue in the objective function (5.2) is calculated as the sum over all cells $p \in P$ and alternatives $a \in A$ of the decision variable x_{pa} multiplied by the land value v_a.

Constraints (5.3) and (5.4) address the demand for residential space in the region. A feasible solution must provide enough living space to meet the minimum requirement D_{min} without exceeding the maximum D_{max}. The upper bound is important because, without it, all available land could be assigned to buildings to maximise revenue, ignoring UHI restrictions. Total residential space is calculated as the sum over all cells and alternatives of x_{pa} multiplied by the residential capacity s_a of each alternative. For non-building options, $s_a = 0$, so only building types ($b \in B$) contribute to residential capacity. Constraint (5.5) ensures that each cell is assigned exactly one alternative, meaning that for a given cell p, one $x_{pa} = 1$ and all other variables for that cell are 0.

The UHI effect is incorporated through constraint (5.6), which limits the maximum allowable UHI intensity (UHI_{max}). Calculating UHI intensity requires population density and land porosity. Population density is obtained from the total number of residents, as determined by the objective function, divided by the area of the region in km^2. Porosity is calculated as the proportion of non-built-up cells: the number of built-up cells is divided by the total number of cells to obtain the built-up ratio, which is then subtracted from 1 to give the non-built-up ratio and multiplied by 100 according to Deilami et al. (2016).

These values are then inserted into the GWR-based regression equation (5.1) to determine the UHI intensity, which must not exceed UHI_{max}.

Function (5.7) computes the mean distance between built-up cells. For this, all pairs of cells $p_1, p_2 \in P$ and buildings $b_1, b_2 \in B$ are considered. The distance $w_{p_1p_2}$ between the cells is multiplied by the binary variable $y_{p_1b_1p_2b_2}$, which equals 1 only if both buildings b_1 and b_2 are placed on the corresponding cells in the optimal solution. The total distance is converted to metres and divided by the number of links between built-up cells, calculated as the total number of buildings multiplied by one less than this number. The resulting average distance must not fall below the minimum distance parameter Dist_{min}, ensuring adequate spacing between buildings.

Finally, constraints (5.8)–(5.10) are included to linearise the model. They are part of the total distance calculation in constraint (5.7) and preserve the logical relationships between cell assignments while maintaining linearity.

5.3 Analysis and Results

5.3.1 Case Description for the Simulation

To test the theoretical predictions and underlying assumptions of the model, we selected Brisbane as the focus city for our simulations. This choice is well-justified. According to the Köppen climate classification, Brisbane has a humid subtropical climate and is increasingly exposed to extreme temperatures (Deilami & Kamruzzaman, 2017), a situation that is further intensified by the UHI effect (Queensland Fire and Emergency Services, 2019). Incorporating these temperature trends into future urban planning and development is therefore essential. Our research aims to enhance the accuracy and precision of predicting and assessing UHI impacts, which are an escalating concern across Australia (Commonwealth of Australia, 2018; Earl et al., 2016). For the simulations, we focused on two regions within the Brisbane Local Government Area: Brisbane City, representing a densely populated and highly built-up core, and Hamilton, a suburb located approximately 6 km from the city centre that includes the Northshore Development—Queensland's largest waterfront urban renewal project (Queensland Government, 2021b).

5.3.2 Simulation Input Parameters

The land value in millions per hectare (v_a) was calculated through the following procedure. Firstly, a spatial layer representing the valuation roll from the Queensland Valuation and Sales System (QVAS) was obtained via the Queensland Spatial Catalogue (Queensland Government, 2021a). Secondly, this layer was used to create property polygons in ArcGIS Pro, mapping valuation property boundaries for Brisbane City and Hamilton (see Figure 5.1).

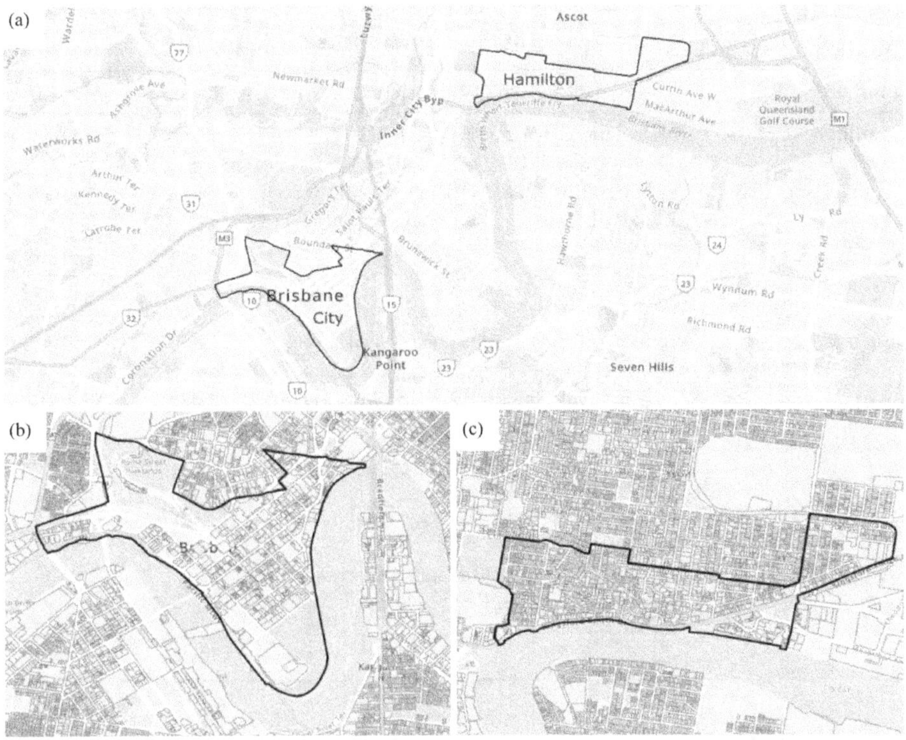

FIGURE 5.1
Map of Brisbane (a) and valuation property boundaries for Brisbane City (b) and Hamilton (c).

Thirdly, the mapping enabled the identification of property IDs within the boundaries of both regions, which were then used to obtain the latest annual land valuations from the Queensland Government website (Queensland Government, 2019). Finally, the average land value per hectare for each region was calculated from all identified property valuations and used as input for the simulations (see Table 5.2).

Resident supply, in inhabitants per hectare, and the total area of the map in km^2 were obtained from the Australian Bureau of Statistics (Australian Bureau of Statistics, Regional population growth, 2018). The maximum UHI intensity (UHI_{max}) was determined using the coefficients β_0, β_1, and β_2 from the GWR analysis conducted by Deilami et al. (2016). Table 5.2 provides a summary of all input parameters used for the simulations in both regions.

5.3.3 Simulation Results

The simulations were performed using a heuristic approach implemented in LocalSolver. Table 5.3 presents the output values. In Brisbane City, increasing UHI_{max} from 4°C to 5°C results in a revenue increase from AUD 9.06 billion to

TABLE 5.2

Simulation Parameters

Parameter	Reference/Calculation	Brisbane City	Hamilton
v_a (million AUD/ha)	Queensland Government (Queensland Government, 2019)	84.68	16.97
s_a (inhabitants/ha)	Australian Bureau of Statistics (Australian Bureau of Statistics, Regional population growth, 2018)	51.84	45.29
D_{min}	Entire surface as vegetation	0	0
D_{max}		P	$\times s_a$
$Dist_{min}$ (m)	Average distance of the centres of the built-up pixels	100	100
β_0	Geographically weighted regression (Deilami et al., 2016)	4.532	4.532
β_1	Geographically weighted regression (Deilami et al., 2016)	0.0003920	0.0001785
β_2	Geographically weighted regression (Deilami et al., 2016)	−0.0260	−0.0145
$Area_{sqkm}$	Australian Bureau of Statistics (Australian Bureau of Statistics, Regional population growth, 2018)	2.4	1.6
P	$Area_{sqkm} \times 100$	240	100

AUD 13.38 billion, achieved by raising the number of buildings from 107 to 158. However, porosity, defined as the share of vegetation, decreases from 55.42% to 34.17%, while population density rises from 2,311.2 to 3,412.8 people per km^2.

For Hamilton, an increase in UHI_{max} from 4°C to 5°C increases revenue from AUD 1.10 billion to AUD 2.29 billion, with the number of buildings rising from 65 to 135. Porosity declines from 59.38% to 15.63%, and population density increases from 1,839.9 to 3,821.3 people per km^2.

Figure 5.2 visualises the distribution of vegetation and buildings, where green cells represent vegetation and yellow cells represent buildings. The figure illustrates that a 1°C increase in UHI_{max} corresponds to a larger area occupied by buildings (yellow cells) in both Brisbane City and Hamilton.

TABLE 5.3

Output Values

Output	Brisbane City (UHI_{max} = 4°C)	Brisbane City (UHI_{max} = 5°C)	Hamilton (UHI_{max} = 4°C)	Hamilton (UHI_{max} = 5°C)
Revenue (billion AUD)	9.06	13.38	1.10	2.29
Buildings	107	158	65	135
UHI_{max} final (°C)	3.9972	4.9815	3.9995	4.9875
Porosity (%)	55.42	34.17	59.38	15.63
PopDens (ppl/km^2)	2,311.2	3,412.8	1,839.91	3,821.34

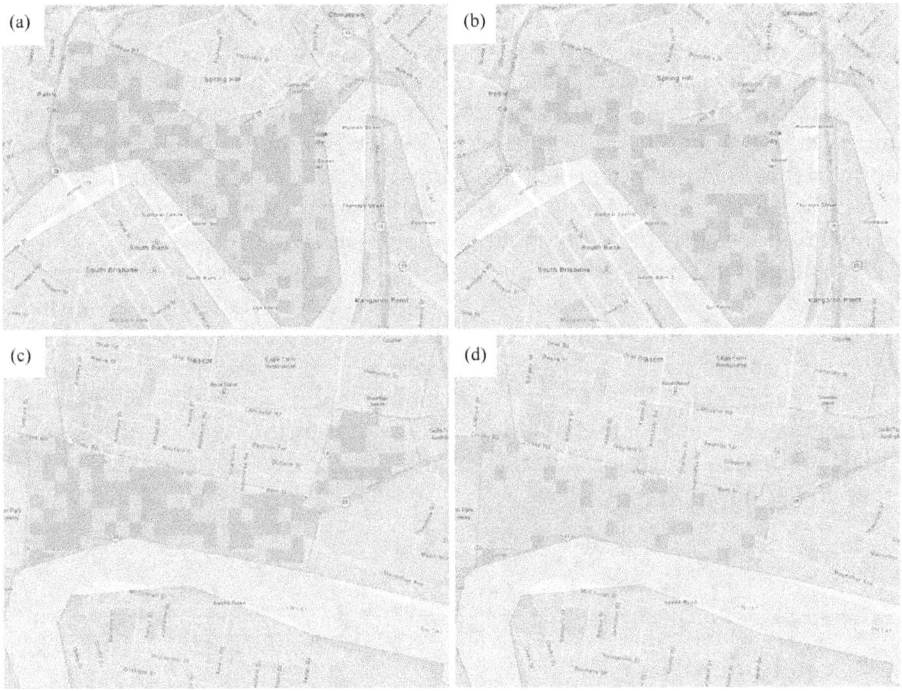

FIGURE 5.2
Result map plot: Brisbane City UHI_{max} = 4°C (a); Brisbane City UHI_{max} = 5°C (b); Hamilton UHI_{max} = 4°C (c); Hamilton UHI_{max} = 5°C (d).

5.4 Findings and Discussion

The simulation results reveal clear trade-offs between economic gains and ecological outcomes in urban development when vegetation strategies are employed to mitigate the UHI effect. In both Brisbane City and Hamilton, increasing the allowable UHI intensity (UHI_{max}) by 1°C led to substantial increases in total revenue—AUD 4.32 billion in Brisbane City and AUD 1.19 billion in Hamilton. These gains were accompanied by an expansion in the number of buildings (Brisbane City: 107 → 158; Hamilton: 65 → 135) and a corresponding increase in population density (Brisbane City: 2,311 → 3,413 ppl/km^2; Hamilton: 1,840 → 3,821 ppl/km^2).

However, this economic growth comes at a considerable ecological cost. Porosity, representing the share of non-built-up, vegetated area, declined sharply in both regions as UHI_{max} increased (Brisbane City: 55.4% → 34.2%; Hamilton: 59.4% → 15.6%), highlighting a reduction in vegetation cover. This indicates that relaxing UHI constraints in favour of maximising revenue may exacerbate thermal stress in urban areas, undermining the benefits of vegetation for heat mitigation and urban liveability.

The spatial distribution of buildings and vegetation further underscores the importance of strategic planning. Figure 5.2 shows that even a 1°C increase in UHI_{max} shifts the landscape significantly towards built-up cells, reducing green space continuity and potentially affecting microclimatic conditions, stormwater management, and overall urban resilience. The optimisation model demonstrates the potential to balance residential demand, revenue generation, and UHI mitigation, providing planners with actionable insights on how to allocate vegetation effectively across urban areas.

Overall, the findings highlight that optimal vegetation management is crucial for controlling UHI intensity while accommodating urban growth. Decision-makers must consider these trade-offs in local planning strategies, using quantitative tools such as the presented optimisation model to ensure that economic objectives do not compromise environmental sustainability. The case studies illustrate that context-specific solutions are required, as the balance between development and vegetation differs between dense urban cores and suburban renewal areas.

5.5 Conclusion

The results demonstrate that economic gains come at a substantial ecological cost. In the simulated scenario, increasing UHI_{max} from 4°C to 5°C leads to revenue increases of AUD 4.32 billion in Brisbane City and AUD 1.19 billion in Hamilton. However, this is accompanied by a 1°C rise in the maximum temperature difference between developed and undeveloped sites, driven by an increase in buildings, a corresponding decrease in porosity, and higher population density.

The findings highlight the trade-offs between economic and ecological objectives, which vary across regions, as illustrated by the differences between Brisbane City and Hamilton. Optimisation models such as the one presented here can support city planners and policymakers in balancing economic outcomes with environmental sustainability goals, providing a quantitative basis for informed decision-making in urban development.

Acknowledgements

This chapter, with permission from the copyright holder, is a reproduced version of the following journal article: Tuczek, M., Degirmenci, K., Desouza, K., Watson, R., Yigitcanlar, T., & Breitner, M. (2022). Mitigating urban heat with optimal distribution of vegetation and buildings. *Urban Climate*, 44(1), 101208.

References

Antheaume, N., Thiel, D., De Corbière, F., Rowe, F., & Takeda, H. (2018). An analytical model to investigate the economic and environmental benefits of a supply chain resource-sharing scheme based on collaborative consolidation centres. *Journal of the Operational Research Society, 69*(12), 1888–1902.

Australian Bureau of Statistics. (2020). *Regional population growth, Australia, 2018-19.* https://www.abs.gov.au/AUSSTATS/abs@.nsf/DetailsPage/3218.02018-19?OpenDocument (accessed 21.12.21).

Bakaeva, N., & Le, M. T. (2022). Determination of urban pollution islands by using remote sensing technology in Moscow, Russia. *Ecological Informatics, 67,* 101493.

Buratto, A., & D'Alpaos, C. (2015). Optimal sustainable use of drinking water sources and interactions between multiple providers. *Operations Research Letters, 43*(4), 389–395.

Cârlan, I., Mihai, B.-A., Nistor, C., & Große-Stoltenberg, A. (2020). Identifying urban vegetation stress factors based on open access remote sensing imagery and field observations. *Ecological Informatics, 55,* 101032.

Commonwealth of Australia. (2018). *Current and future impacts of climate change on housing, buildings and infrastructure.* https://www.aph.gov.au/~/media/Committees/ec_ctte/CCInfrastructure/report.pdf (accessed 21.12.21).

Connors, J. P., Galletti, C. S., & Chow, W. T. (2013). Landscape configuration and urban heat island effects: Assessing the relationship between landscape characteristics and land surface temperature in Phoenix, Arizona. *Landscape Ecology, 28*(2), 271–283.

Degirmenci, K., Desouza, K. C., Fieuw, W., Watson, R. T., & Yigitcanlar, T. (2021). Understanding policy and technology responses in mitigating urban heat islands: A literature review and directions for future research. *Sustainable Cities and Society, 70,* 1–35.

Deilami, K., & Kamruzzaman, M. (2017). Modelling the urban heat island effect of smart growth policy scenarios in Brisbane. *Land Use Policy, 64,* 38–55.

Deilami, K., Kamruzzaman, M., & Hayes, J. F. (2016). Correlation or causality between land cover patterns and the urban heat island effect? Evidence from Brisbane, Australia. *Remote Sensing, 8*(9), 1–28.

Deilami, K., Kamruzzaman, M., & Liu, Y. (2018). Urban heat island effect: A systematic review of spatio-temporal factors, data, methods, and mitigation measures. *International Journal of Applied Earth Observation and Geoinformation, 67,* 30–42.

Earl, N., Simmonds, I., & Tapper, N. (2016). Weekly cycles in peak time temperatures and urban heat island intensity. *Environmental Research Letters, 11*(7), 074003.

Estrada, F., Botzen, W. W., & Tol, R. S. (2017). A global economic assessment of city policies to reduce climate change impacts. *Nature Climate Change, 7*(6), 403–406.

Gao, Y., Zhao, J., & Han, L. (2022). Exploring the spatial heterogeneity of urban heat island effect and its relationship to block morphology with the geographically weighted regression model. *Sustainable Cities and Society, 76,* 103431.

García, D. H., & Díaz, J. A. (2021). Modeling of the urban heat island on local climatic zones of a city using Sentinel 3 images: Urban determining factors. *Urban Climate, 37,* 100840.

Gunawardena, K. R., Wells, M. J., & Kershaw, T. (2017). Utilising green and bluespace to mitigate urban heat island intensity. *Science of the Total Environment, 584–585,* 1040–1055.

Ho, H. C., Knudby, A., Xu, Y., Hodul, M., & Aminipouri, M. (2016). A comparison of urban heat islands mapped using skin temperature, air temperature, and apparent temperature (Humidex), for the greater Vancouver area. *Science of the Total Environment, 544*, 929–938.

Kabano, P., Lindley, S., & Harris, A. (2021). Evidence of urban heat island impacts on the vegetation growing season length in a tropical city. *Landscape and Urban Planning, 206*, 103989.

Kamruzzaman, M., Deilami, K., & Yigitcanlar, T. (2018). Investigating the urban heat island effect of transit-oriented development in Brisbane. *Journal of Transport Geography, 66*, 116–124.

Kang, J., Ng, T. S., & Su, B. (2020). Optimising electricity mix for CO_2 emissions reduction: A robust input-output linear programming model. *European Journal of Operational Research, 287*(1), 280–292.

Kohler, M., Tannier, C., Blond, N., Aguejdad, R., & Clappier, A. (2017). Impacts of several urban-sprawl countermeasures on building (space heating) energy demands and urban heat island intensities: A case study. *Urban Climate, 19*, 92–121.

Kolbe, K. (2019). Mitigating urban heat island effect and carbon dioxide emissions through different mobility concepts: Comparison of conventional vehicles with electric vehicles, hydrogen vehicles and public transportation. *Transport Policy, 80*, 1–11.

Koomen, E., & Diogo, V. (2017). Assessing potential future urban heat island patterns following climate scenarios, socio-economic developments and spatial planning strategies. *Mitigation and Adaptation Strategies for Global Change, 22*(2), 287–306.

Levermore, G., Parkinson, J., Lee, K., Laycock, P., & Lindley, S. (2018). The increasing trend of the urban heat island intensity. *Urban Climate, 24*, 360–368.

Li, D., Liao, W., Rigden, A. J., Liu, X., Wang, D., Malyshev, S., & Shevliakova, E. (2019). Urban heat island: Aerodynamics or imperviousness?. *Science Advances, 5*(4), 1–4.

Mortoja, M. G., & Yigitcanlar, T. (2020). Local drivers of anthropogenic climate change: Quantifying the impact through a remote sensing approach in Brisbane. *Remote Sensing, 12*(14), 1–24.

Noro, M., Busato, F., & Lazzarin, R. (2015). Urban heat island in Padua, Italy: Experimental and theoretical analysis. *Indoor and Built Environment, 24*(4), 514–533.

Noro, M., & Lazzarin, R. (2015). Urban heat island in Padua, Italy: Simulation analysis and mitigation strategies. *Urban Climate, 14*, 187–196.

Pomerantz, M. (2018). Are cooler surfaces a cost-effective mitigation of urban heat islands? *Urban Climate, 24*, 393–397.

Qin, Y. (2015). A review on the development of cool pavements to mitigate urban heat island effect. *Renewable and Sustainable Energy Reviews, 52*, 445–459.

Queensland Fire and Emergency Services. (2019). *Queensland state heatwave risk assessment 2019*. https://www.disaster.qld.gov.au/__data/assets/pdf_file/0020/339311/QFES-State-Heatwave-Risk-Assessment-Executive-Summary.pdf (accessed 21.12.21).

Queensland Government. (2019). *Find your land valuation*. https://www.qld.gov.au/environment/land/title/valuation/find-your-land-valuation (accessed 21.12.21).

Queensland Government. (2021a). *Queensland spatial catalogue*. https://qldspatial.information.qld.gov.au/catalogue (accessed 21.12.21).

Queensland Government. (2021b). *Welcome to Northshore: Brisbane's immersive riverfront.* https://northshorebrisbane.com.au (accessed 21.12.21).

Roxon, J., Ulm, F.-J., & Pellenq, R.-M. (2020). Urban heat island impact on state residential energy cost and CO_2 emissions in the United States. *Urban Climate, 31,* 100546.

Selby, J. D., & Desouza, K. C. (2019). Fragile cities in the developed world: A conceptual framework. *Cities, 91,* 180–192.

Senevirathne, D., Jayasooriya, V., Dassanayake, S., & Muthukumaran, S. (2021). Effects of pavement texture and colour on urban heat islands: An experimental study in tropical climate. *Urban Climate, 40,* 101024.

Shahfahad, Naikoo, M. W., Islam, A. R. M. T., Mallick, J., & Rahman, A. (2022). Land use/land cover change and its impact on surface urban heat island and urban thermal comfort in a metropolitan city. *Urban Climate, 41,* 101052.

Shaker, R. R., Altman, Y., Deng, C., Vaz, E., & Forsythe, K. W. (2019). Investigating urban heat island through spatial analysis of New York City streetscapes. *Journal of Cleaner Production, 233,* 972–992.

Stone, B., Vargo, J., & Habeeb, D. (2012). Managing climate change in cities: Will climate action plans work? *Landscape and Urban Planning, 107*(3), 263–271.

Taghipour Javi, S., Malekmohammadi, B., & Mokhtari, H. (2014). Application of geographically weighted regression model to analysis of spatiotemporal varying relationships between groundwater quantity and land use changes (case study: Khanmirza Plain, Iran). *Environmental Monitoring and Assessment, 186*(5), 3123–3138.

Tu, J., & Xia, Z.-G. (2008). Examining spatially varying relationships between land use and water quality using geographically weighted regression I: Model design and evaluation. *Science of the Total Environment, 407*(1), 358–378.

Yang, C.-H. (2018). An optimization portfolio decision model of life cycle activity-based costing with carbon footprint constraints for hybrid green power strategies. *Computers & Operations Research, 96,* 256–271.

Yang, X., & Yao, L. (2022). Reexamining the relationship between surface urban heat island intensity and annual precipitation: Effects of reference rural land cover. *Urban Climate, 41,* 101074.

Zhao, Z., Gao, J., Wang, Y., Liu, J., & Li, S. (2015). Exploring spatially variable relationships between NDVI and climatic factors in a transition zone using geographically weighted regression. *Theoretical and Applied Climatology, 120*(3), 507–519.

Zhou, X., & Wang, Y.-C. (2011). Dynamics of land surface temperature in response to land-use/cover change. *Geographical Research, 49*(1), 23–36.

6

Tree Placement for Cooling Microclimates

6.1 Introduction

The impact of climate change on global weather patterns is becoming more pronounced, resulting in noticeable changes in urban microclimates that pose significant challenges to human well-being and the sustainability of biodiversity (Mortoja & Yigitcanlar, 2020a, 2020b; Shaamala et al., 2024; Shivanna, 2022). As urban areas continue to expand and human activities including transport compromise environmental quality, creating and sustaining comfortable microclimates became an increasingly important act (Dizdaroglu & Yigitcanlar, 2014; Yigitcanlar et al., 2008). Microclimates, localised atmospheric zones which differ from surrounding areas, are significantly influenced by urban infrastructure and green spaces (Aram et al., 2019; Degirmenci et al., 2021).

Considering the above, urban areas are increasingly challenged by the impacts of climate change, which affect outdoor thermal comfort and overall quality of life (Degirmenci et al., 2021; Yin et al., 2023). These challenges negatively affect the quality of life and pose health risks, particularly during periods of extreme heat (Tuczek et al., 2022; Yigitcanlar, 2010). To address these issues, researchers introduced an urban thermal comfort index (UTCI), a bioclimatic index for describing the physiological comfort of the human body under specific meteorological conditions (Bröde et al., 2012). This index was chosen for its comprehensive approach, which considers air temperature, wind speed, humidity, and solar radiation, providing a detailed evaluation of thermal comfort (Guo & Sun, 2024). Moreover, while various metrics are available for evaluating outdoor thermal comfort, UTCI is preferred in this study owing to its comprehensive approach and global acceptance (Mahdavinejad et al., 2024; Nazarian et al., 2019).

In this context, modulation of microclimate challenges is vital for reducing the adverse consequences of climate change in densely populated areas (Brigham & Suding, 2023; Irfeey et al., 2023). Central to addressing these challenges is the strategic use of urban green infrastructure, particularly trees, which play a significant role in improving urban microclimate (Deng et al., 2022; Meili et al., 2021; Sanusi et al., 2017). The incorporation of shading and evapotranspiration functions in urban planning and design helps mitigate the severity of urban heat island issues and improves the thermal comfort of city residents (Armson et al., 2012; Cheela et al., 2021), including the non-human

 DOI: 10.1201/9781003737056-8

animals in the city. Thus, incorporating trees into urban planning and design is essential to improve the visual appeal and, more importantly, to facilitate the implementation of adaptation and mitigation measures against climate change's effects on urban microclimates (Chen et al., 2019; Zhang & Brack, 2021).

Determining the locations of trees to promote cooling, however, reveals a significant gap in current urban planning policies and practices (Bajsanski et al., 2016; Hao et al., 2023; Milošević et al., 2017; Shaamala et al., 2024). Current guidelines often prioritise the quantity of trees by considering factors such as space and maintenance budgets without instructions on where to plant trees for optimal cooling benefits (Hao et al., 2023). Moreover, the academic literature presents varying perspectives, with some studies suggesting clustering trees in locations for cooling effects, whereas others propose spreading them over a wider area (Morakinyo & Lam, 2016; Zhao et al., 2018). For example, recent studies have emphasised the advantages of placing trees on pathways to alleviate heat stress on pedestrians in urban areas (Azcarate et al., 2021; Lachapelle et al., 2023; Ren et al., 2022). Concurrently, other studies advocate for sparse planting of trees and ensure minimal canopy overlap (Zhang et al., 2018; Zhao et al., 2018).

Recently, the optimal location of trees within urban areas has emerged as a viable intervention for mitigating such issues. The term "tree location" in the context of urban planning and landscape architecture, as used in this study, involves several key components that help in designing green areas effectively (Gkatsopoulos, 2017; Li et al. 2023a, 2023b; Liu et al., 2024). These components include tree spacing, tree massing, and positioning relative to other land uses (Back et al., 2023; Wallenberg et al., 2022, 2022; Zhao et al., 2018). By optimising tree locations, it is possible to maximise cooling effects, thereby enhancing UTCI (Bajsanski et al., 2016; Milošević et al., 2017; Zhao et al., 2018). Nevertheless, it is crucial to acknowledge that besides tree location other factors such as tree species, foliage type and density, seasonality of tree foliage, and tree height influence the tree's capability of cooling enhancement (Irmak et al., 2018; Meili et al., 2021; Morakinyo et al., 2020).

Against this backdrop, this study introduces a novel methodology aimed at the optimal positioning of trees to enhance urban microclimates, particularly targeting the mitigation of urban thermal comfort. Specifically, this study seeks to answer three key research questions: (a) How can we determine the optimal location of trees to improve urban thermal comfort? (b) How does the location of trees influence the UTCI in urban areas? (c) Which optimisation algorithms are most effective for optimising urban tree locations to maximise heat mitigation?

Building on this, our approach systematically exploits three distinct optimisation algorithms, including "Non-dominated Sorting Genetic Algorithm II" (NSGA-II), "Particle Swarm Optimisation" (PSO), and "Ant Colony Optimisation" (ACO), to establish the most effective arboreal configurations that contribute to the microclimate. This comprehensive analysis lays the foundation for enhancing the precision of urban planning and design and contributes to the development of robust data-driven solutions for sustainable urban living environments.

This research is underscored by the need for effective urban heat mitigation strategies in the face of climate change. Although previous studies have explored the benefits of urban vegetation, there is a lack of comprehensive frameworks for optimising tree placement. This approach offers new insights by combining well-established questions with advanced optimisation techniques to develop practical urban planning strategies, particularly for tree placement and its impact on urban microclimates. By considering environmental factors such as air temperature, humidity, wind speed, and mean radiant temperature (MRT), we adopted a holistic approach to address urban heat mitigation. This study on the innovative use of AI tools enhances the precision and efficacy of these strategies, providing a nuanced understanding of environmental interactions and offering timely and actionable recommendations for optimising urban environments, along with helping planning decisions (Son et al., 2023).

Following this introduction section, Section 6.2 delves into the literature. Section 6.3 outlines the methodological approach used to investigate the effects of tree location on microclimate. Section 6.4 presents the results of the study and analyses the performance of different optimisation algorithms. Section 6.5 discusses the implications of the findings and suggests directions for future research. Finally, Section 6.6 concludes the study and summarises the key insights.

6.2 Literature Background

Urban microclimates refer to localised atmospheric conditions found within cities that display variations in temperature, humidity, wind patterns, and other climatic factors compared to the regional climate (Yang et al., 2023). These differences are primarily influenced by the layout and features of the environment, such as buildings, roads, water bodies, and green areas (Upreti et al., 2024). The urban microclimate concept is crucial for understanding the environmental impacts of urbanisation, particularly UTCI (Soltani & Sharifi, 2017). The UTCI was developed through comprehensive climatological research, which accounts for the multilayered nature of the thermal environment and human physiology, offering a unified standard to assess perceived outdoor temperature (Bröde et al., 2012; Skutecki et al., 2019). However, it is important to acknowledge that the UTCI is specifically designed from a human-centric perspective and does not account for the thermal comfort or stress responses of other species, which may experience the thermal environment differently (Park et al., 2014).

Urban heat island phenomenon is attributed to the capacity of buildings, paved surfaces, and imitated vegetation to absorb heat from anthropogenic emissions, vehicles, industrial processes, and air conditioning while retaining heat (Faragallah & Ragheb, 2022; Kamruzzaman et al., 2018; Tuczek et al., 2022). The impact of trees on the urban microclimate has been thoroughly examined, with emphasis on their cooling contributions via shading, evapotranspiration, and the alteration of wind patterns (Hao et al., 2023; Li et al., 2022; Li et al., 2024; Milošević et al., 2017; Park et al., 2020). In addition, beyond the acknowledged benefits of trees, their location is crucial, as it directly influences the amount of solar radiation an area receives (Bajsanski et al., 2016). The following sections focus on the role of trees in urban microclimate regulation and examine their cooling effects and significance for urban sustainability and human comfort.

6.2.1 Trees Mechanism in Urban Microclimate Regulation

Trees contribute to urban cooling and microclimate improvement through several mechanisms (Figure 6.1). These natural processes not only enhance local environmental conditions but also provide a sustainable means of mitigating urban heat effects (Balany et al., 2020; Irfeey et al., 2023). Understanding these mechanisms is crucial for urban planning, landscape, and environmental management (Gago et al., 2013; Woodward et al., 2023; Zheng et al., 2023). Trees significantly enhance urban microclimates and cooling through shading, evapotranspiration, and windbreaking (Coutts et al., 2016). They

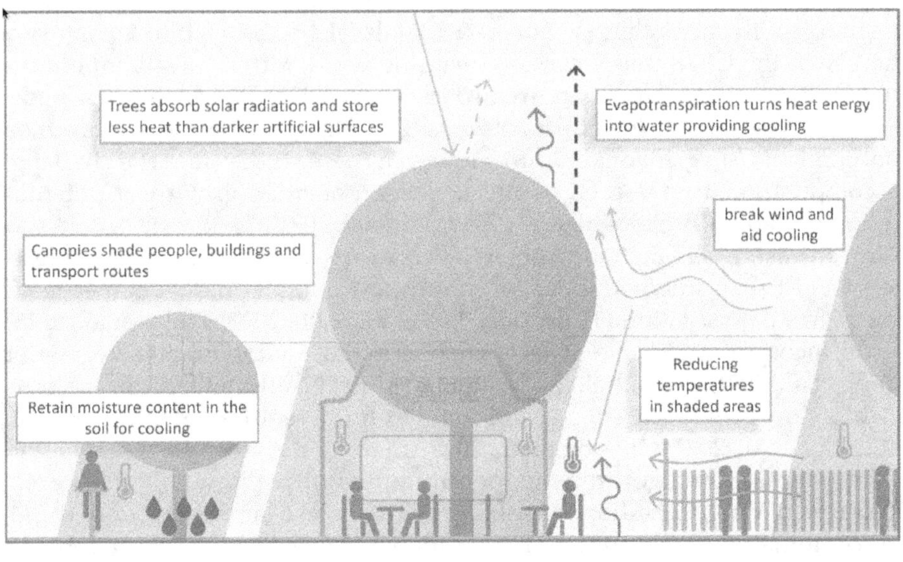

FIGURE 6.1
Tree cooling mechanism.

lower surface and air temperatures by reducing solar radiation, especially in dense urban areas where hard surfaces exacerbate heat (Jareemit & Srivanit, 2022; Speak et al., 2020; Yu et al., 2020). By shading surfaces, particularly low-albedo materials such as asphalt, trees significantly reduce heat absorption and effectively cool the area (Kaluarachichi et al., 2020).

Trees also release water vapour, which cools air through evaporation and transpiration (Pace et al., 2021; Speak et al., 2020; Tams et al., 2023). Furthermore, trees help to maintain cooler temperatures and higher humidity by shading the soil, which reduces ground moisture evaporation (Hwang et al., 2023; Winbourne et al., 2020). Additionally, by altering wind patterns and reducing speed, trees can enhance air cooling (Lu et al., 2023; Ren et al., 2023). In other contexts, optimal tree locations can channel breezes into urban areas, thereby augmenting cooling (Park et al., 2012). The turbulence caused by wind interacting with tree canopies results in enhanced evaporation and transpiration rates, contributing to enhanced cooling (Giometto et al., 2017; Gunawardena et al., 2017).

6.2.2 Tree Computational Modelling

Tree computational modelling is an essential process in tree location optimisation. This involves simulating the impact of each potential tree location or trees configuration and providing feedback to the optimisation algorithm, creating a loop that fine-tunes the placement for maximum environmental benefit. Tree mechanisms have been extensively modelled using a variety of methods (Hao et al., 2023). For instance, computational fluid dynamics (CFD) is utilised to model the behaviour of fluids and gases, as well as their relationships with each other (Versteeg & Malalasekera, 2007). In the context of the UTCI, the CFD model can simulate the wind pattern and temperature distribution within an urban area (Back et al., 2023). This model provides detailed insights into the effects of trees (Zeng et al., 2020) and surface thermal comfort (Allegrini et al., 2015). However, these studies face the challenge of computational feasibility, owing to the exponential growth of potential design variables (Broadbent et al., 2019; Mirzaei, 2021).

In contrast, ENVI-met, a commercial software, has been used to simulate vegetated urban microclimates by inputting tree data in 3D geometries to calculate solar radiation absorption (Liu et al., 2018, 2023; Tsoka et al., 2018). It can incorporate diverse aspects such as wind, radiation, and vegetation (Yang et al., 2021). Nonetheless, despite its utility, ENVI-met faces limitations, such as slow processing speeds and complex input requirements (Alves et al., 2022; Liu et al., 2021; Ouyang et al., 2022), making it challenging to integrate with optimisation algorithms to identify optimal tree locations (Hao et al., 2023). In contrast, Grasshopper software can analyse large number of different combinations to optimise tree locations. However, it does not consider the physical characteristics of trees.

RayMan is another modelling tool that is particularly renowned for its application in the assessment of UTCI (Matzarakis et al., 2010). Despite its extraordinarily high sensitivity, which allows the generation of results using simple parameters, RayMan's focus on spot-related simulations of radiant flux densities and thermos-physiological evaluation metrics limits its ability to provide a broader analysis, RayMan's focus on spot-related simulations of radiant flux densities and thermos-physiological evaluation metrics limits its ability to provide a broader analysis (Jänicke et al., 2021).

In contrast, ray-tracing algorithms and commercial software such as SOLWEIG and CityComfort+ have been used to model shading effects that account for solar radiation dynamics (Briegel et al., 2023; Huang et al., 2014; Jänicke et al., 2021; Lindberg et al., 2008). Although SOLWEIG and CityComfort+ are effective for specific applications, ENVI-met was selected for the current study because of its effectiveness in preparing numerical data in open spaces in urban areas (Mahdavinejad et al., 2024). The ENVI-met is particularly suitable for this study because it provides detailed microclimate modelling, which is essential for evaluating outdoor thermal comfort and optimising tree configurations in urban environments.

6.2.3 Tree Optimisation

Despite advancements in simulation-based optimisation, the application of such methods for tree allocation, particularly to enhance the cooling performance, remains a developing area. Initial studies employed CFD models and heuristic methods to explore designs aimed at mitigating urban heat stress. These early studies provided a foundation but also encountered computational challenges (Chen et al., 2008; Mochida et al., 2008). Subsequent research efforts have applied brute-force methods to determine optimal tree layouts within urban spaces, focusing on maximising shade coverage and minimising solar intensity, which are key factors in mitigating heat. Still, these studies were limited by the exponential increase in computational load as the number of design variables increased (Bajsanski et al., 2016; Milošević et al., 2017; Zhao et al., 2017).

Other studies have sought to achieve optimisation by partially simulating the tree-cooling effects or by using modified optimisation approaches. For example, Stojakovic et al. (2020) prioritised MRT when optimising tree locations, whereas Park et al. (2020) and Wallenberg et al. (2022) noted limitations in their methodologies, which are often adjusted based on the tools available and concentrate on particular facets of urban cooling, such as shade provision. Hao et al. (2023) integrated a rapid simulation model with a genetic algorithm, concentrating on location of trees within urban park to optimise cooling effects.

Existing research in the field of urban greenery optimisation has primarily concentrated on either the cooling effects of tree coverage ratios and species or on the computational modelling of tree-induced microclimate changes without adequately addressing the optimisation of tree locations (Li & Song, 2019; Morakinyo et al., 2020; Yang et al., 2017; Zhang & Gou, 2021). Several studies have attempted to integrate tree simulation models with optimisation algorithms (Hao et al., 2023; Stojakovic et al., 2020; Tuczek et al., 2022; Wallenberg et al., 2022; Zhao et al., 2017). Yet, these efforts often encounter challenges in balancing computational efficiency with the complexity of tree-cooling mechanisms. Additionally, research has largely overlooked the spatial dynamics of tree placement in relation to each other, which is crucial for maximising urban cooling benefits.

The study reported in this chapter aims to overcome these challenges by testing three distinct optimisation algorithms: NSGA-II, PSO, and ACO. These algorithms have demonstrated their potential in solving intricate optimisation problems. NSGA-II is a popular multi-objective optimisation algorithm that excels in handling multiple conflicting objectives (Ishibuchi et al., 2009; Verma et al., 2021). It works based on the principles of natural selection and genetic algorithms. The core mechanism of the NSGA-II involves generating an initial population of potential solutions and then classifying this population into different fronts based on Pareto dominance (Imene et al., 2022; Song & Chen, 2018). NSGA-II is particularly effective for problems that require a diverse set of solutions along the Pareto front (Vachhani et al., 2016). It has been employed in a planning model to identify the optimal locations and types of green spaces based on their varied effects (Yoon et al., 2019). This algorithm was selected for its robustness in providing a range of optimal solutions that balance multiple objectives.

Concurrently, PSO is inspired by the social behaviour of flocking birds or schooling fish (Ma et al., 2011). It optimises a problem by iteratively improving candidate solutions based on a given measure of quality (Vafaeinejad & Mahmoudi Jam, 2021; Zhang et al., 2019). PSO is known for its simplicity, ease of implementation, and fast convergence (Noel, 2012; Zheng et al., 2021). It was used to strategically select park locations, with the aim of minimising walking distances for urban residents (Zhao & Gong, 2024). This algorithm was chosen for its ability to quickly converge to high-quality solutions, making it suitable for optimising tree placement to enhance thermal comfort efficiently.

In addition, ACO is inspired by the foraging behaviour of ants and their ability to find the shortest path between their colony and the food source (Sudholt & Thyssen, 2012). It is a probabilistic technique for solving computational problems that can be reduced to finding good paths through graphs (Liao et al., 2011). In ACO, pheromone levels and parameters are initially set up, and ants construct solutions by moving through the problem's graph based on pheromone levels and heuristic information. After constructing a solution, pheromones are updated to reinforce good paths and evaporate less attractive ones (Kozłowska, 2014; Li et al., 2019). This process is repeated for several iterations or until convergence. ACO was used to efficiently enhance decision support for green area maintenance (Lee & Lee, 2010). This algorithm was selected for its strength in solving combinatorial optimisation problems and its flexibility in adapting to various problem constraints.

These three algorithms were chosen for their complementary strengths in solving multi-objective optimisation problems. Undertaking individual evaluations of these algorithms enables the assessment of their computational requirements and optimisation effectiveness. This approach not only identifies the most computationally efficient algorithm but also evaluates the quality of optimisation in terms of improving the urban microclimate. It is anticipated that this detailed analysis will reveal the unique strengths and limitations of each algorithm and guide their application in urban planning and environmental management.

The importance of this research lies in addressing the gap in current urban planning practices regarding the optimisation of tree placement for urban cooling. By systematically comparing the effectiveness of the three advanced optimisation algorithms, this study provides insights into how urban planners can maximise the cooling benefits of trees. The outcomes of this research have the potential to inform policies, improve urban design, and contribute to the development of sustainable cities capable of mitigating the adverse effects of climate change. This approach ensures that the results are not only academically significant, but also practically applicable, offering data-driven solutions for enhancing urban microclimates.

6.3 Research Design

In this study, we introduced a novel methodology aimed at optimising tree locations to enhance urban microclimates, specifically targeting the mitigation of urban heat. Our approach systematically exploits three distinct optimisation algorithms, NSGA-II, PSO, and ACO, to establish the most effective tree-massing configurations that contribute to the microclimate. These algorithms were integrated with a simulation model for microclimate evaluation. This integration is pivotal for examining the influence of strategic tree placement on the Universal Thermal Comfort Index (UTCI) (Figure 6.2).

6.3.1 Initialisation Phase

6.3.1.1 Study Area

The methodology was initiated with the selection of a hypothetical study area and initial conditions. For this study, the plot was assumed to have a 50 × 50 m, which was deemed an appropriate size for an urban park in which to perform experiments on tree optimisation. The terrain was generally flat, and the vegetation cover was limited (Figure 6.3). The soil type was

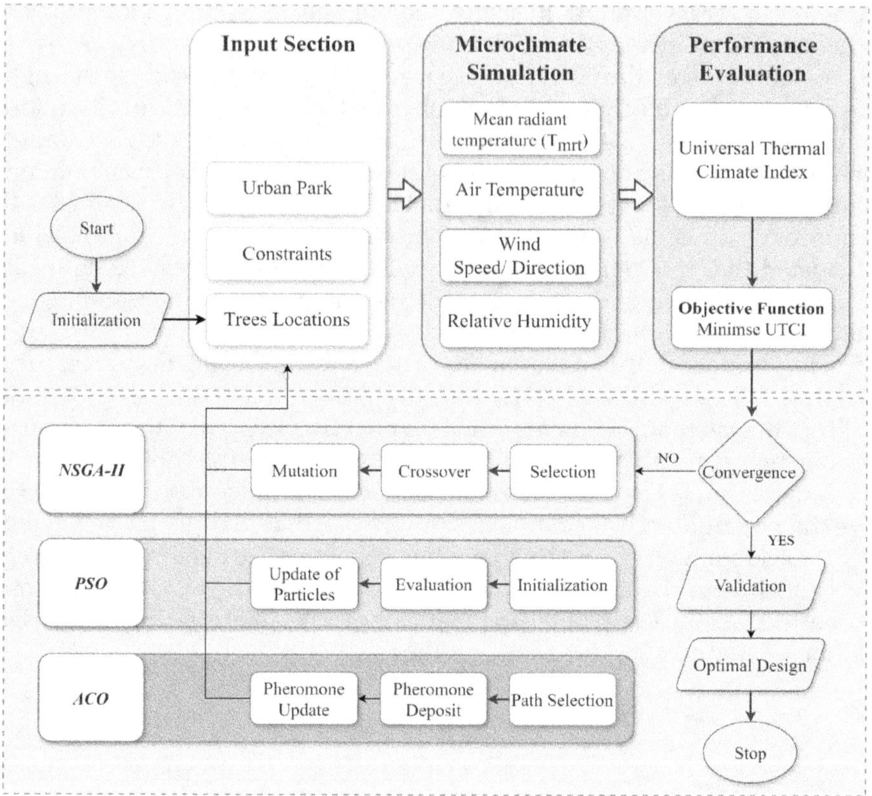

FIGURE 6.2
Flowchart of the proposed methodology.

considered loamy, offering suitable conditions and deep soil for tree growth without requiring significant modification.

6.3.1.2 Simulation Period Selection

To assess the cooling effects under extreme heat conditions, simulations were run over the hottest week in the capital city of Queensland-Brisbane (17th to 23rd of February 2023), which ranges in temperature from 23°C to 37°C. These periods typically fall within the summer months, and the effects of the urban heat island tend to increase temperature. In Brisbane, a subtropical climate, the high relative humidity (RH) exacerbates this heat, making it feel hotter than the actual temperature. This combination significantly affects human comfort, energy usage, and overall liveability of the area. The selected week will allow us to look at tree placement optimisation under difficult yet increasingly common conditions.

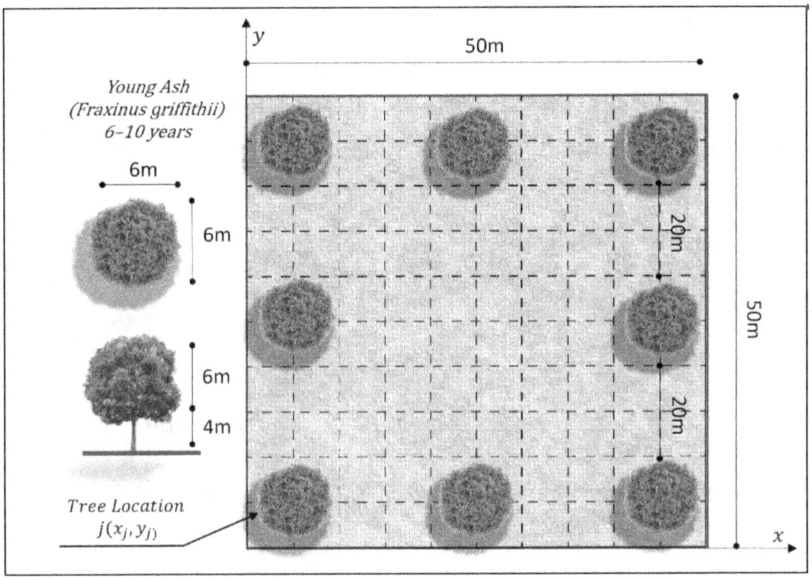

FIGURE 6.3
Study area.

6.3.1.3 Constraint Considerations

In line with the objective of optimising the tree location, we accounted for several constraints.

- Area Constraints: The placement of trees was restricted within the confines of the predetermined urban plot area.
- Tree Count: A total of eight trees were to be placed within the plot.
- Canopy Overlap: The simulation model ensured that the canopies of the selected trees did not intersect, thus mimicking realistic urban planning and design conditions with space for each tree.

6.3.2 Tree Location Optimisation

Optimising the location of trees within an urban setting involves a complex decision-making process that balances ecological benefits with urban space constraints. In this study, we formulated this as a mathematical optimisation problem within a Cartesian Coordinate System. The solution to this problem will determine the optimal locations for tree planting that enhance the microclimate and thermal comfort for park users, as assessed by UTCI.

6.3.2.1 Notions and Definitions

i: index representing potential tree locations.

I: the set comprising all viable tree locations within the defined area.

p: predetermined number of trees to be distributed within a defined area.

6.3.2.1.1 Objective Function

An objective function was constructed to minimise the UTCI. It is articulated as follows:

$$Minimize\ f(x_1, x_2, \ldots, x_i)$$

where x_i indicates the binary decision of placing a tree at the ith potential site.

Constraints

The optimisation is subject to constraints ensuring the defined number of trees and feasibility of their location:

$$\sum_{i \in I} x_i = p$$

$$x_i = \in \{0, 1\} \ \forall_i \in I$$

where

$$x_i = \begin{cases} 1 \ if \ a \ tree \ is \ allocated \ to \ site \ i, \ and \ x_i \\ 0 \ otherwise \end{cases}$$

6.3.2.2 Iteration to Convergence

6.3.2.2.1 Convergence Criteria

For the convergence criterion, a predefined iteration limit of 200 cycles is adopted for each algorithm. Convergence is determined when the algorithms no longer yield significant improvements in the objective function, which, in this case, is the optimisation of the microclimate, as quantified by UTCI.

6.3.2.2.2 Iterative Process

The iterative process begins with each algorithm, NSGA-II, PSO, and ACO, starting from a random initial configuration of tree locations within a 50*50 m area. At each iteration, the algorithms perform their respective operations as follows:

- NSGA-II performs mutation, crossover, and selection to explore and exploit the search space, thereby generating new populations of tree configurations.

- PSO updates the particle positions, with each particle's movement informed by its own experience and the swarm's collective intelligence.
- ACO simulates ants laying down pheromones, which guide the search for more promising areas of the solution space.

After each iteration, the newly generated tree configurations were input into the microclimate simulation model to evaluate their impact on the microclimate. The shading of the tree was modelled physically using a 3D model to calculate shadows. Simultaneously, evapotranspiration was numerically represented by estimating the cooling effect of water vapour released by the tree. The simulation results were fed back into the optimisation algorithms, informing the next iteration of adjustments. This feedback loop continued until the algorithm reached the 200th iteration. At this point, the solutions are analysed to extract the configuration that best mitigates microclimate effects according to the UTCI. The iterative process is designed to harmonise the exploration of the search space by exploiting the best-found solutions to effectively navigate toward an optimal design.

6.3.3 Universal Thermal Climate Index Modelling

The Universal Thermal Climate Index (UTCI) serves as a comprehensive measure to evaluate the combined effect of these cooling mechanisms on human thermal comfort. By simulating changes in shading, evapotranspiration rates, and wind patterns, the model provides insights into how strategic tree placement can optimise thermal comfort in urban outdoor spaces. UTCI is used to quantify the perceived temperature experienced by humans, integrating the multifaceted influences of the thermal environment into a single, understandable metric.

This section underlines the complexity and importance of every cooling mechanism given by trees, showing the approach the simulation model has taken by the simulation model to evaluate their contributions to urban microclimate improvement. The influence of these cooling mechanisms on the UTCI can be conceptualised through an equation developed by Brode and Wojtach (2010), which incorporates changes in temperature, wind speed, and humidity due to shading, evapotranspiration, and wind modification as follows:

$$UTCI = f\left(T_{mrt}, T_a, v, RH\right)$$

where

T_{mrt}: is the MRT affected by shading (°C),

T_a: is the air temperature adjusted for evapotranspiration cooling (°C),

v: is the wind speed modified by tree placement (m/s),

RH: is the RH adjusted for evapotranspiration effects (%).

In the comprehensive examination of the effects on UTCI through tree allocation, tree placements were analysed using Ladybug plugin in Grasshopper. Ladybug was used (Roudsari and Pak (2013) to calculate UTCI considering T_a, RH, and v imported from .epw file as well as mean radiant temperature (T_{mrt}). The analysis period was on the 17th of February during the hottest hour at 3 PM taking in account the consideration of the solar path. Moreover, the model was based on eight sphere-shaped trees with dense foliage (6 m tree crown diameter, 4 m trunk height).

6.3.4 Model Validation

To ensure the accuracy and applicability of our novel methodology, we used a baseline scenario comparison as the primary method for the model verification. The simulated baseline scenario represented the hypothetical study area. This provided a control set of data against which we compared the effects of different tree configurations on urban microclimate. By comparing these baseline conditions with scenarios where tree placements were optimised using NSGA-II, PSO, and ACO algorithms, we were able to assess the relative impact of these optimisations on factors such as air temperature, RH, wind speed, and MRT. We conducted validation using ENVI-met 5.0, a sophisticated microclimate simulation software. Consistency in the tree characteristics was maintained throughout the study to ensure validity. The tree used in the simulation and validation phases was a young ash tree (Fraxinus Griffithii), aged six to ten years, selected for its prevalence in urban areas in Brisbane and its suitability for achieving short-term goals.

6.4 Analysis and Results

This section presents the empirical outcomes derived from applying our novel methodology, which utilises the optimisation algorithms NSGA-II, ACO, and PSO to optimise tree locations to enhance urban microclimates. This methodology was systematically applied to identify the most effective tree configuration that enhanced the UTCI. The validation phase of our study tested the tree configurations recommended by the NSGA-II, ACO, and PSO algorithms against their impact on UTCI and key microclimatic variables. This test demonstrated significant impacts on the RH, mean radiant temperature, potential air temperature, and wind speed, confirming the efficacy of our novel optimisation methodology.

The graphical data extracted from the simulation models offers a detailed representation of the thermal dynamics within the urban fabric. This section discusses the nuanced performance of each algorithmic configuration, showing their individual and collective influence on the urban microclimate.

By examining the trends across different times of the day, we gained insight into the daily efficacy of tree location.

A detailed analysis of these results highlights not only the potential for optimised tree locations to mitigate urban heat, but also the relative strengths and weaknesses of each optimisation approach. This detailed evaluation serves as the foundation for understanding the practical implications of leveraging computational algorithms in urban forestry and planning, particularly focusing on enhancing environmental conditions and human comfort, which are increasingly being challenged by climate change. The findings are pivotal, as they go beyond theoretical modelling to suggest practical interventions for urban planners and policymakers seeking data-driven methods to counteract the urban heat effect and move toward more sustainable and liveable urban environments.

6.4.1 Spatial Configurations from Optimisation Algorithms

The analysis commenced with an assessment of the spatial arrangements of the trees as proposed by each algorithm. The three configurations are illustrated in Figure 6.4. Across all configurations, the algorithms considered the need to enhance urban thermal comfort through strategic tree placement. While NSGA-II aims for broad coverage, ACO focuses on creating microclimate havens and PSO targets critical urban areas for heat mitigation. Each pattern reflects a thoughtful approach to leveraging green infrastructure for climate adaptation and urban sustainability.

6.4.2 Optimisation Trajectory and Universal Thermal Climate Index Reduction

The results showed a pattern of convergence among NSGA-II, ACO, and PSO to optimise tree locations for urban microclimate enhancement (Figure 6.5). Specifically, NSGA-II suggests a strong and consistent pursuit toward an optimal solution with a robust and continuous decrease in UTCI, implying high reliability and stability in its performance owing to minimal fluctuations.

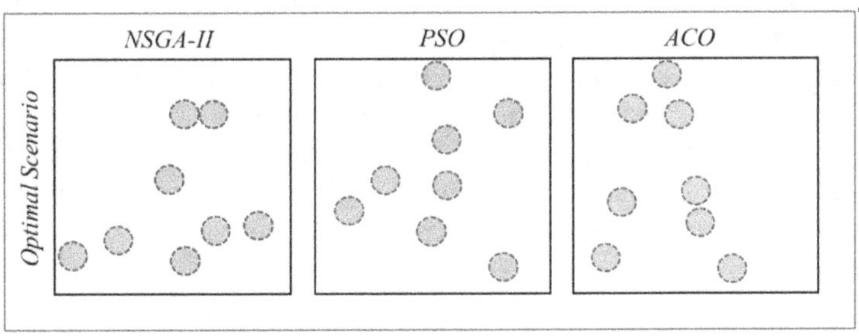

FIGURE 6.4
Tree configurations as determined by the optimisation algorithms NSGA-II, PSO, and ACO.

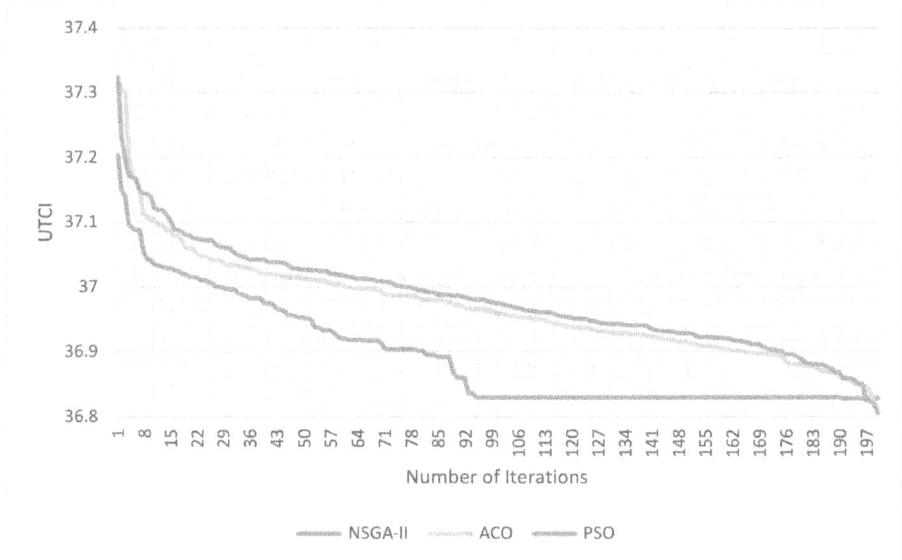

FIGURE 6.5
Comparative performance of optimisation algorithms NSGA-II, ACO, and PSO over iterations in reducing UTCI.

The performance of ACO was marked by a more cautious and methodical reduction in UTCI values, demonstrating a systematic optimisation process, albeit less quickly converging than NSGA-II. PSO, on the other hand, exhibited a strikingly different performance, with an initially fast reduction in UTCI, indicating an aggressive search strategy that quickly discovered significant microclimate improvement areas. Nonetheless, this was followed by marked variations in performance, which might signal either a strategy of thorough investigation of the search space or the inherent variability of solution quality.

The results obtained using all three algorithms revealed a marked decrease in UTCI, indicating an improved urban microclimate from placing strategic trees in place. Notably, while all algorithms showed a reduction in UTCI, PSO consistently identified effective tree configurations more rapidly, showcasing its advantage in quick decision-making. Although NSGA-II plateaued earlier, suggesting its efficiency in achieving timely solutions, ACO provided the most substantial reduction in air temperature. This demonstrates that each algorithm has distinct strengths depending on the optimisation goals: rapid identification of solutions with PSO, efficiency in timely solutions with NSGA-II, and maximum temperature reduction with ACO.

6.4.3 Microclimate Simulation for Optimal Scenarios

To validate the effectiveness of the proposed method for optimising tree locations, we conducted a series of microclimate simulations. These simulations

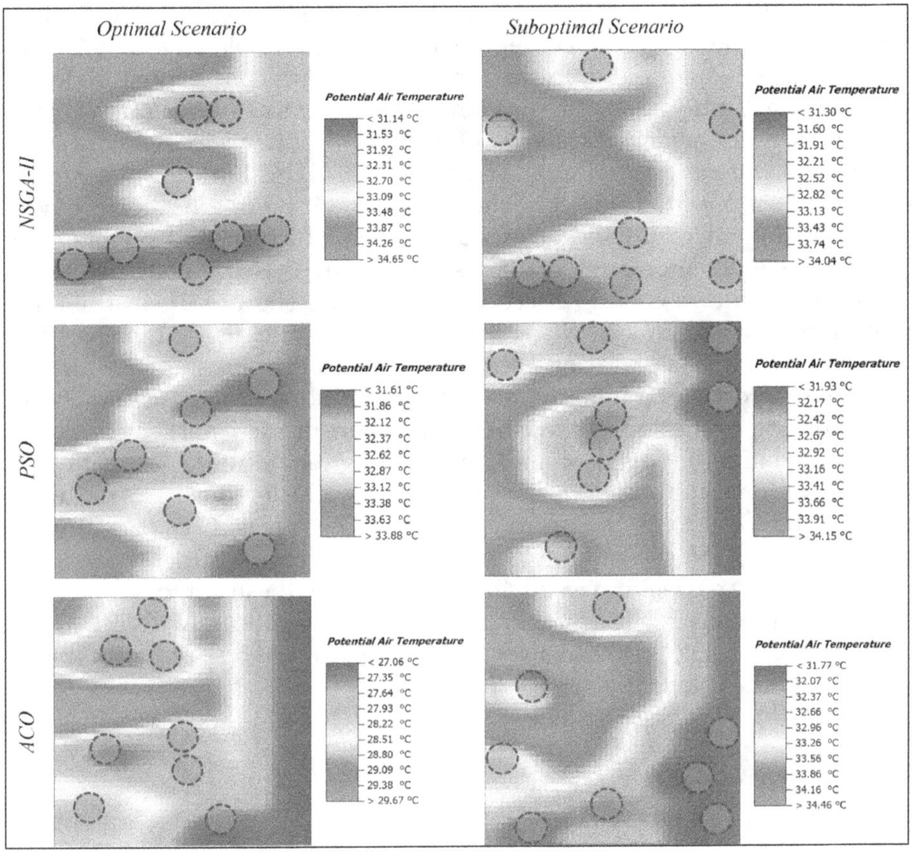

FIGURE 6.6
Variation in potential air temperature with different optimisations of tree location configurations.

projected the impact of the optimal tree configurations generated by the NSGA-II, ACO, and PSO algorithms on critical urban microclimate variables that affect the UTCI. The optimal scenarios were defined as configurations that yielded the most favourable results in terms of microclimatic variables, namely, RH, mean radiant temperature, potential air temperature, and wind speed.

The simulation results (Figure 6.6) indicate tree distribution effects on the urban microclimate, as visualised in the thermal maps, revealing notable variations in air temperature associated with different tree locations. Each thermal map represents a unique scenario generated by the optimisation algorithms. The configuration of tree locations significantly influenced the potential air temperature within the urban area. The maps show a range of temperatures with a variation of 5°C from below 27°C to above 34°C. Notably, areas under tree canopies generally display cooler temperatures. This cooling effect is most pronounced in scenarios where the trees are located in locations that appear to be hotspots in other scenarios. For example, in the ACO optimal scenario, temperatures in the central area were considerably reduced, suggesting an optimised configuration for tree location to maximise shade.

Conversely, in the non-optimal ACO scenario, despite the presence of trees, some areas still experienced higher temperatures. This indicates that simply having trees is not sufficient for cooling; rather, their precise location is crucial for optimising microclimate effects. Furthermore, the scenarios exhibited different degrees of temperature gradients between trees, with some demonstrating a more uniform cooling effect, whereas others showed distinct hot and cool zones, indicating the importance of tree density and distribution patterns.

After the analysis of various scenarios, it was evident that certain configurations produced a more consistent distribution of lower temperatures. These findings indicate a synergistic effect, where one tree not only promotes localised cooling but also enhances the cooling capabilities of surrounding trees. Additionally, optimal scenarios show trees clustered in pairs or threes, which enhance cooling by forming larger shaded areas. Consequently, the overall range of effects was amplified.

Subsequent analyses revealed the outcomes of these interventions. We examined the microclimatic parameters that significantly affect UTCI, including RH, MRT, potential air temperature, and wind speed, in response to nuanced alterations in arboreal layouts.

6.4.4 Model Validation

6.4.4.1 Relative Humidity (RH, %)

The chart presents the RH over a 24-hour period for four scenarios: NSGA-II-optimal, ACO-optimal, PSO-optimal, and baseline. The baseline scenario followed a typical daily RH cycle, peaking in the early morning, dropping in the afternoon, and rising again in the evening. The NSGA-II-optimal scenario closely followed the baseline but with a slightly lower daytime RH, indicating a modest reduction in humidity due to tree placement. The ACO-optimal scenario showed a significant reduction in daytime RH, particularly from 8 AM to 5 PM, suggesting highly effective tree placement for reducing humidity and enhancing thermal comfort. The PSO-optimal scenario also reduced daytime RH, although it was less significant than ACO, indicating moderate effectiveness. In summary, ACO-optimal achieved the greatest reduction in daytime RH, followed by PSO-optimal and NSGA-II-optimal, demonstrating the varying effectiveness of these optimisation algorithms in enhancing urban thermal comfort (Figure 6.7).

6.4.4.2 Potential Air Temperature (°C)

Figure 6.8 indicates that the baseline scenario peaks at approximately 3 PM and follows a typical daily temperature cycle. The NSGA-II-optimal scenario

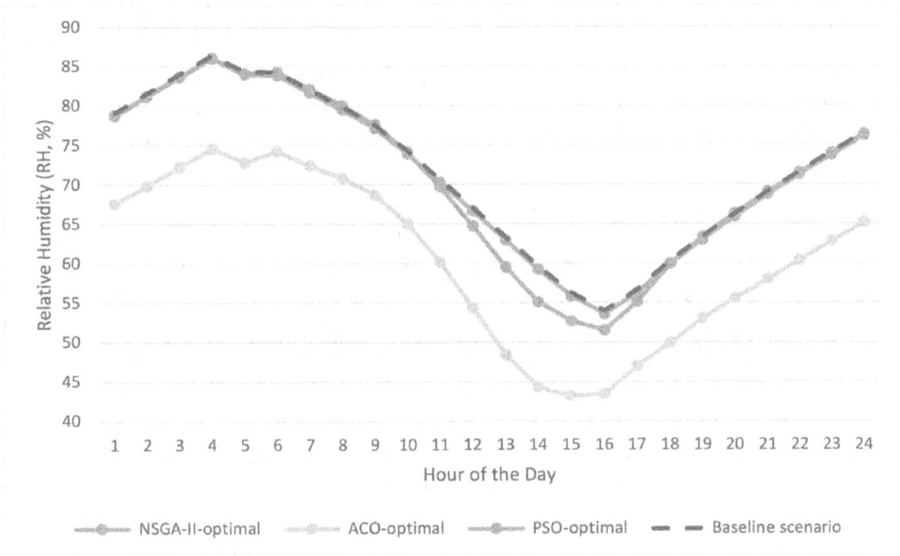

FIGURE 6.7
Tree location impact on relative humidity: Optimal versus suboptimal configurations.

shows slightly lower temperatures throughout the day, especially during peak hours, indicating an effective temperature reduction. The ACO-optimal scenario achieves a greater reduction in daytime temperatures than the NSGA-II-optimal scenario, demonstrating strong effectiveness in cooling.

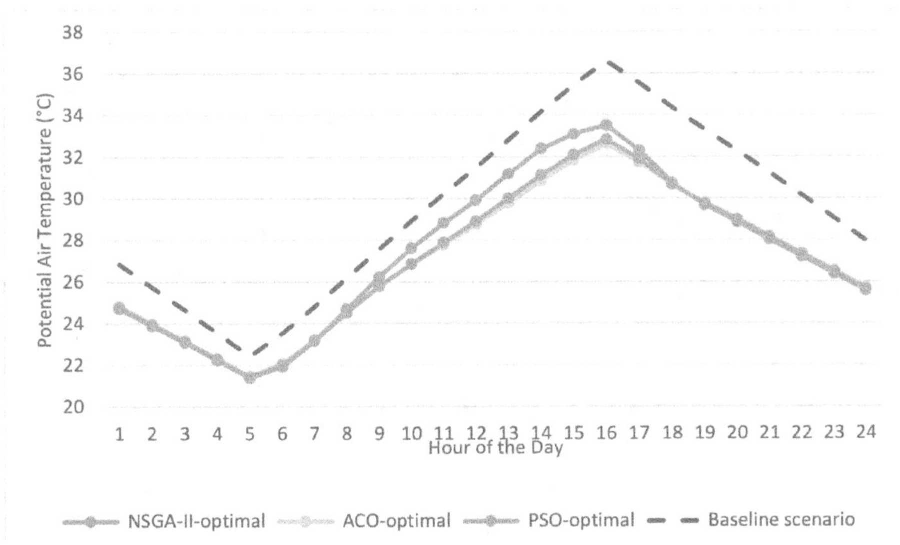

FIGURE 6.8
Tree location impact on potential air temperature: Optimal versus suboptimal configurations.

The PSO-optimal scenario followed a similar pattern to the ACO-optimal scenario, with comparable temperature reductions. In summary, ACO-optimal and PSO-optimal achieved the greatest daytime temperature reduction, followed by NSGA-II-optimal. All three algorithms had significantly lower air temperatures than the baseline, enhancing urban thermal comfort.

6.4.4.3 Mean Radiant Temperature (°C)

Figure 6.9 shows that the baseline scenario exhibited a typical MRT cycle (range error of +6) with a sharp increase in the morning, peaking around noon, and then gradually decreasing. The NSGA-II-optimal scenario showed a reduction in MRT compared to the baseline, especially during peak hours, indicating effective temperature management through tree placement. The ACO-optimal scenario achieved a greater reduction in MRT, particularly in the late morning and early afternoon, demonstrating a strong cooling effect. The PSO-optimal scenario also effectively reduced the MRT, similar to ACO-optimal, but was slightly less pronounced. These reductions in MRT are attributed to tree configurations creating large canopy areas that provide extensive shading. In summary, the ACO-optimal and PSO-optimal scenarios showed the greatest reduction in MRT, followed by NSGA-II-optimal, with all three optimisation algorithms significantly lowering MRT compared to the baseline and enhancing urban thermal comfort.

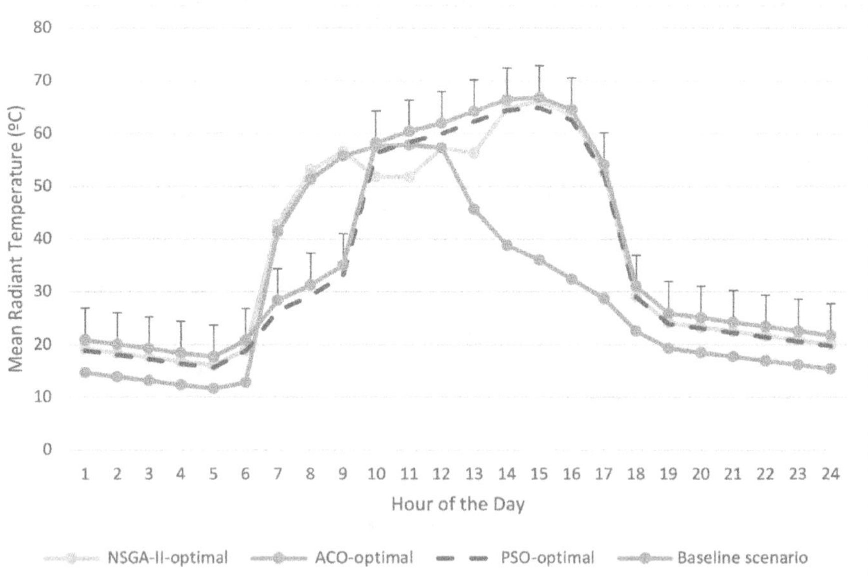

FIGURE 6.9
Tree location impact on mean radiant temperature: Optimal versus suboptimal configurations.

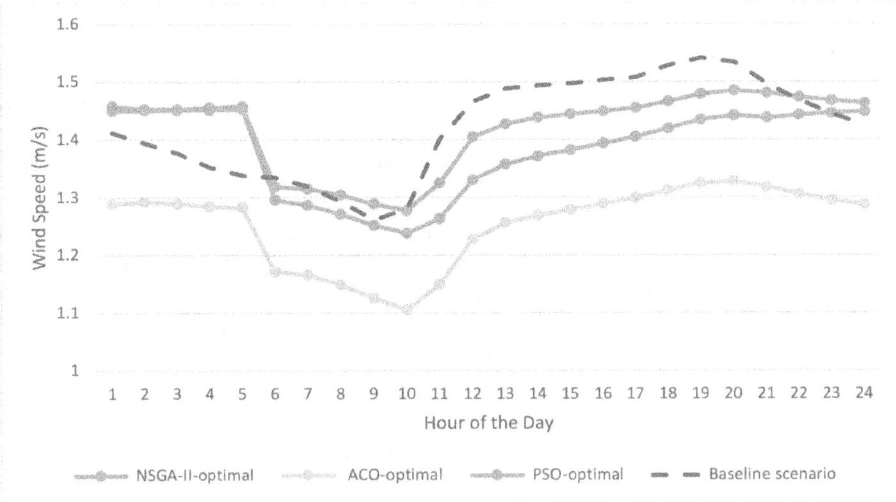

FIGURE 6.10
Effect of tree location on wind speed: Optimal versus suboptimal configurations.

6.4.4.4 *Wind Speed (m/s)*

The baseline scenario had higher wind speeds throughout the day than the other scenarios (Figure 6.10). The NSGA-II-optimal and PSO-optimal scenarios showed moderate reductions in wind speed, indicating that their tree configurations provide some wind buffering. The ACO-optimal scenario had the lowest wind speeds, especially from early morning to late afternoon, suggesting that its tree placement created the most significant wind reduction. These differences in wind speed reductions are due to the varying tree configurations, which shape large canopy areas and act as windbreaks. In summary, all three optimisation algorithms reduce wind speed compared to the baseline, with ACO-optimal showing the most significant reduction, followed by PSO-optimal, and NSGA-II-optimal.

In sum, the analysis of these climatic factors suggests that tree distribution and clustering have a significant impact on the urban microclimate. The best tree location configurations derived from each algorithm showed marked improvements in all measured variables, which contributed to lowering the temperature, increasing the humidity, and promoting air movement. These changes are critical for urban areas to mitigate the heat island effect and improve overall human thermal comfort. Moreover, the validation results emphasise that our improved methodology, based on data-driven insights and algorithmic optimisation, is a powerful tool for urban forestry planning and design. This approach lays the groundwork for optimised urban greening and offers measurable benefits for urban microclimate regulation.

6.5 Findings and Discussion

This study introduced a novel methodology that significantly advances urban planning and design by optimising tree locations to enhance urban microclimates. We utilised three optimisation algorithms, NSGA-II, ACO, and PSO, to optimise the placement of trees. Each algorithm has produced unique spatial arrangements of trees within an urban area, tailored to improve urban climate and comfort levels, as measured by UTCI. This index underscores the profound impact of optimising the tree location on outdoor thermal comfort.

The strength of the proposed method lies in its data-driven approach, which offers precise and actionable insights for urban planners and designers. By implementing tree optimisation adapted to city conditions, the methodology not only optimises thermal comfort but also enhances the liveability of urban areas. The integration of tree placement within urban planning is crucial as it plays a pivotal role in the sustainable development of urban environments.

The findings from the optimisation algorithms revealed that NSGA-II offers consistent and gradual optimisation of tree locations. This leads to a sustained decrease in UTCI. This algorithm can potentially guide long-term urban planning and design strategies. NSGA-II aims to incrementally enhance thermal comfort across urban areas. On the other hand, the ACO cautious optimisation trajectory demonstrates its utility in meticulously enhancing microclimatic conditions. Its effectiveness in slightly elevating RH levels contributes to a nuanced improvement in thermal comfort. This highlights its suitability for urban areas where changes must be methodical and ecologically balanced.

Within this context, the PSO algorithm has demonstrated compelling performance in rapidly reducing the UTCI. The results indicate that, while PSO outperforms NSGA-II and ACO in quickly identifying tree configurations that yield a decrease in UTCI, ACO provides the most substantial reduction in air temperature. NSGA-II, though it plateaus earlier, shows efficiency in achieving timely solutions. This suggests that each algorithm offers unique advantages: PSO for rapid solution identification, ACO for maximum air temperature reduction, and NSGA-II for efficiency in achieving timely solutions for urban cooling.

The practical implications of this study are twofold, encompassing both policy development and landscape planning and design enhancement. At the policymaking level, this study points out that not only is the number of trees particularly important, but also the location and massing configuration of trees. The empirical results presented in this study demonstrate that optimising the location of trees can produce more cooling effects compared with a random distribution of the same number of trees. This points to the potential for refined policy guidelines that go beyond general green coverage ratios, advocating detailed prescriptions for tree placement to maximise urban cooling benefits.

From a landscape design perspective, this study revealed that planting designs that consider tree foliage and the clustering of tree canopies can amplify their cooling effects. Such strategic canopy concentrations can significantly cool urban areas without increasing the number of trees. The variance in UTCI observed between optimal and suboptimal tree locations further quantified this effect, with a significant variation of up to 5°C in potential temperature. This insight propels a shift in landscape planting design practices, encouraging the creation of cluster configurations of canopy formations that can function more effectively than merely focusing on increasing tree count.

Moreover, it will help to improve the tools that landscape architects use to optimise planting designs for urban cooling. Ultimately, this study provides a compelling argument for the intelligent allocation of trees within urban environments, promoting a more efficient and effective approach for urban cooling and sustainable landscape design. However, it acknowledges the limitations of not incorporating the full range of urban complexities within its scope. It also acknowledges the need to reconcile the urban cooling approach to tree optimisation with other considerations landscape architects make in planting design, including defining spaces, creating habitats, aesthetic considerations, noise reduction, and the cultural significance of the tree.

6.5.1 Limitations and Research Directions

The methodology employed in this study, while innovative in its approach to enhancing urban microclimates through optimised tree locations. Nevertheless, it has certain limitations, particularly those related to the software tools available for implementation. Although we utilised advanced simulation software to assess the impact of tree placement on microclimatic variables, there is a notable deficiency in comprehensive tools that can fully capture the complex interactions of trees within urban environments and seamlessly integrate them with optimisation algorithms.

For instance, the Ladybug plugin for Grasshopper software recognises the impact of trees in a concrete manner, as tree crown geometry projects a complete shadow, but it fails to take into account the tangible attributes of trees, such as leaf type and thickness (Bajsanski et al., 2016). Besides, performing simultaneous UTCI calculations at numerous locations is time-consuming and computationally demanding (Hao et al., 2023; Milošević et al., 2017). In contrast, Envi-met facilitates a more detailed input of tree characteristics (Alves et al., 2022; Guerri et al., 2023; Milošević et al., 2017). However, it is a closed source that does not allow it to be coupled with an optimisation algorithm (Hao et al., 2023).

Furthermore, studies have indicated that ENVI-met tends to underestimate air temperature under certain conditions. This consistent underestimation may be due to miscalculations of solar radiation levels reaching the plant surfaces (Dizdaroglu et al., 2012; Liu et al., 2021). Moreover, previous verifications have identified specific errors in predicting MRT, with an error margin of up to 6°C. This creates a need for modelling software that contains the accuracy and flexibility to assess tree locations more accurately. In addition, these findings were based on simulations that may not completely encompass the diversity of real-world urban areas. The generalisability of these results to various urban settings with different climatic conditions, urban fabrics, and social contexts is limited. Furthermore, while UTCI serves as a holistic metric for thermal comfort, the focus of this study on this single metric might omit other relevant environmental or social benefits of urban trees, such as biodiversity, air quality improvement, and psychological well-being.

Prospective research should incorporate the characteristics of different tree species, such as leaf area density, canopy dimension, and trunk height, and whether the species is endemic or exotic, to optimise urban cooling efficacy. Enhancing tree modelling should include precise details such as the minimum distances between trees, buildings, and streets. Additionally, exploring different planting designs and tree-spacing configurations can provide insights into how spatial arrangements affect urban cooling, aesthetics, and the functionality of green spaces.

For validation, the tree simulation model integrated empirical data. We acknowledge the challenges in validating models that combine tree shading, evapotranspiration, and windbreaks. Recent scaled outdoor experiments, such as those conducted by Chen et al. (2021) and Wu et al. (2024), have provided valuable data for assessing the impacts of tree evapotranspiration and shading on urban microclimates. Incorporating these empirical findings into our model validation will enhance the accuracy and reliability of the simulations, ensuring that they reflect real-world conditions more closely.

6.6 Conclusion

The study reported in this chapter advances urban environmental management by demonstrating the utility of optimisation algorithms that include NSGA-II, ACO, and PSO in selecting the most beneficial locations for tree planting to improve the UTCI. The investigation highlighted that the optimised location is crucial for enhancing urban thermal comfort more effectively than merely increasing the number of trees. The challenges persist in accurately simulating the impact of trees on urban microclimates, highlighting the need for tools that can fully integrate complex tree characteristics with optimisation algorithms.

Future research efforts should integrate findings from recent outdoor experiments to validate and improve the tree-effect model. Incorporating these empirical data will refine and enhance the real-world applicability. Additionally, these computational strategies can be developed into comprehensive tools that offer a promising avenue for supporting policymaking, landscape planning, and design practices. Through the application of optimisation algorithms, urban planners and policymakers can aim to develop more habitable, cooler, and environmentally sustainable urban areas, thereby improving the quality of life of urban residents both presently and in the long term.

Acknowledgements

This chapter, with permission from the copyright holder, is a reproduced version of the following journal article: Shaamala, A., Yigitcanlar, T., Nili, A., and Nyandega, D. (2024). Strategic tree placement for urban cooling: a novel optimisation approach for desired microclimate outcomes. *Urban Climate, 56*(1), 102084.

References

Allegrini, J., Dorer, V., & Carmeliet, J. (2015). Coupled CFD, radiation and building energy model for studying heat fluxes in an urban environment with generic building configurations. *Sustainable Cities and Society, 19*, 385–394. https://doi.org/10.1016/j.scs.2015.07.009

Alves, F. M., Gonçalves, A., & del Caz-Enjuto, M. R. (2022). The use of ENVI-met for the assessment of nature-based solutions' potential benefits in industrial parks—A case study of Argales Industrial Park (Valladolid, Spain). *Infrastructures, 7*(6), 85. https://doi.org/10.3390/infrastructures7060085

Aram, F., Higueras García, E., Solgi, E., & Mansournia, S. (2019). Urban green space cooling effect in cities. *Heliyon, 5*(4), e01339. https://doi.org/10.1016/j.heliyon.2019.e01339

Armson, D., Stringer, P., & Ennos, A. R. (2012). The effect of tree shade and grass on surface and globe temperatures in an urban area. *Urban Forestry & Urban Greening, 11*(3), 245–255. https://doi.org/10.1016/j.ufug.2012.05.002

Azcarate, I., Acero, J., Garmendia, L., & Rojí, E. (2021). Tree layout methodology for shading pedestrian zones: Thermal comfort study in Bilbao (Northern Iberian Peninsula). *Sustainable Cities and Society, 72*, 102996. https://doi.org/10.1016/j.scs.2021.102996

Back, Y., Kumar, P., Bach, P. M., Rauch, W., & Kleidorfer, M. (2023). Integrating CFD-GIS modelling to refine urban heat and thermal comfort assessment. *Science of the Total Environment, 858*, 159729. https://doi.org/10.1016/j.scitotenv.2022.159729

Bajsanski, I., Stojakovic, V., & Jovanovic, M. (2016). Effect of tree location on mitigating parking lot insolation. *Computers, Environment and Urban Systems, 56*, 59–67. https://doi.org/10.1016/j.compenvurbsys.2015.11.006

Balany, F., Ng, A. W., Muttil, N., Muthukumaran, S., & Wong, M. S. (2020). Green infrastructure as an urban heat island mitigation strategy—A review. *Water, 12*(12):3577. https://doi.org/10.3390/w12123577

Briegel, F., Makansi, O., Brox, T., Matzarakis, A., & Christen, A. (2023). Modelling long-term thermal comfort conditions in urban environments using a deep convolutional encoder-decoder as a computational shortcut. *Urban Climate, 47,* 101359. https://doi.org/10.1016/j.uclim.2022.101359

Brigham, L. M., & Suding, K. (2023). Cross-scale facilitation: A framework for microclimate moderation of climate change. *Oikos,* e10241. https://doi.org/10.1111/oik.10241

Broadbent, A. M., Coutts, A. M., Nice, K. A., Demuzere, M., Krayenhoff, E. S., Tapper, N. J., & Wouters, H. (2019). The air-temperature response to green/blue-infrastructure evaluation tool (TARGET v1.0): An efficient and user-friendly model of city cooling. *Geoscientific Model Development, 12*(2), 785–803. https://doi.org/10.5194/gmd-12-785-2019

Brode, P., & Wojtach, B. (2010). *UTCI homepage.* https://www.utci.org/index.html

Bröde, P., Fiala, D., Błażejczyk, K., Holmér, I., Jendritzky, G., Kampmann, B., Tinz, B., & Havenith, G. (2012). Deriving the operational procedure for the universal thermal climate index (UTCI). *International Journal of Biometeorology, 56*(3), 481–494. https://doi.org/10.1007/s00484-011-0454-1

Cheela, V. R. S., John, M., Biswas, W., & Sarker, P. (2021). Combating urban heat island effect—A review of reflective pavements and tree shading strategies. *Buildings, 11*(3), 93. https://doi.org/10.3390/buildings11030093

Chen, H., Ooka, R., & Kato, S. (2008). Study on optimum design method for pleasant outdoor thermal environment using genetic algorithms (GA) and coupled simulation of convection, radiation and conduction. *Building and Environment, 43*(1), 18–30. https://doi.org/10.1016/j.buildenv.2006.11.039

Chen, T., Yang, H., Chen, G., Lam, C. K. C., Hang, J., Wang, X., Liu, Y., & Ling, H. (2021). Integrated impacts of tree planting and aspect ratios on thermal environment in street canyons by scaled outdoor experiments. *Science of the Total Environment, 764,* 142920. https://doi.org/10.1016/j.scitotenv.2020.142920

Chen, X., Zhao, P., Hu, Y., Ouyang, L., Zhu, L., & Ni, G. (2019). Canopy transpiration and its cooling effect of three urban tree species in a subtropical city – Guangzhou, China. *Urban Forestry & Urban Greening, 43,* 126368. https://doi.org/10.1016/j.ufug.2019.126368

Coutts, A. M., Harris, R. J., Phan, T., Livesley, S. J., Williams, N. S. G., & Tapper, N. J. (2016). Thermal infrared remote sensing of urban heat: Hotspots, vegetation, and an assessment of techniques for use in urban planning. *Remote Sensing of Environment, 186,* 637–651. https://doi.org/10.1016/j.rse.2016.09.007

Degirmenci, K., Desouza, K. C., Fieuw, W., Watson, R. T., & Yigitcanlar, T. (2021). Understanding policy and technology responses in mitigating urban heat islands: A literature review and directions for future research. *Sustainable Cities and Society, 70,* 102873. https://doi.org/10.1016/j.scs.2021.102873

Deng, L., Jia, X., Wang, W., & Hussain, S. A. (2022). Revealing impacts of trees on modeling microclimate behavior in spaces between buildings through simulation monitoring. *Buildings, 12*(8), 1168. https://doi.org/10.3390/buildings12081168

Dizdaroglu, D., & Yigitcanlar, T. (2014). A parcel-scale assessment tool to measure sustainability through urban ecosystem components: The MUSIX model. *Ecological Indicators, 41,* 115–130. https://doi.org/10.1016/j.ecolind.2014.01.037

Dizdaroglu, D., Yigitcanlar, T., & Dawes, L. (2012). A micro-level indexing model for assessing urban ecosystem sustainability. *Smart and Sustainable Built Environment, 1*(3), 291–315. https://doi.org/10.1108/20466091211287155

Faragallah, R. N., & Ragheb, R. A. (2022). Evaluation of thermal comfort and urban heat island through cool paving materials using ENVI-met. *Ain Shams Engineering Journal, 13*(3), 101609. https://doi.org/10.1016/j.asej.2021.10.004

Gago, E. J., Roldan, J., Pacheco-Torres, R., & Ordóñez, J. (2013). The city and urban heat islands: A review of strategies to mitigate adverse effects. *Renewable and Sustainable Energy Reviews, 25*, 749–758. https://doi.org/10.1016/j.rser.2013.05.057

Giometto, M. G., Christen, A., Egli, P. E., Schmid, M. F., Tooke, R. T., Coops, N. C., & Parlange, M. B. (2017). Effects of trees on mean wind, turbulence and momentum exchange within and above a real urban environment. *Tribute to Professor Garrison Sposito: An Exceptional Hydrologist and Geochemist, 106*, 154–168. https://doi.org/10.1016/j.advwatres.2017.06.018

Gkatsopoulos, P. (2017). A methodology for calculating cooling from vegetation evapotranspiration for use in urban space microclimate simulations. *Sustainable Synergies from Buildings to the Urban Scale, 38*, 477–484. https://doi.org/10.1016/j.proenv.2017.03.139

Guerri, G., Crisci, A., & Morabito, M. (2023). Urban microclimate simulations based on GIS data to mitigate thermal hot-spots: Tree design scenarios in an industrial area of Florence. *Building and Environment, 245*, 110854. https://doi.org/10.1016/j.buildenv.2023.110854

Gunawardena, K. R., Wells, M. J., & Kershaw, T. (2017). Utilising green and bluespace-blue space to mitigate urban heat island intensity. *Science of the Total Environment, 584–585*, 1040–1055. https://doi.org/10.1016/j.scitotenv.2017.01.158

Guo, X., & Sun, Y. (2024). A thermal comfort assessment on semi-outdoor sports stadiums located in 3 different climate zones in China. *Building and Environment, 261*, 111660. https://doi.org/10.1016/j.buildenv.2024.111660

Hao, T., Zhao, Q., & Huang, J. (2023). Optimization of tree locations to reduce human heat stress in an urban park. *Urban Forestry & Urban Greening, 86*, 128017. https://doi.org/10.1016/j.ufug.2023.128017

Huang, J., Cedeño-Laurent, J. G., & Spengler, J. D. (2014). CityComfort+: A simulation-based method for predicting mean radiant temperature in dense urban areas. *Building and Environment, 80*, 84–95. https://doi.org/10.1016/j.buildenv.2014.05.019

Hwang, B., Sou, H.-D., Oh, J.-H., & Park, C.-R. (2023). Cooling effect of urban forests on the urban heat island in Seoul, South Korea. *PLoS ONE, 18*(7), e0288774. https://doi.org/10.1371/journal.pone.0288774

Imene, L., Sihem, S., Okba, K., & Mohamed, B. (2022). A third generation genetic algorithm NSGAIII for task scheduling in cloud computing. *Journal of King Saud University – Computer and Information Sciences, 34*(9), 7515–7529. https://doi.org/10.1016/j.jksuci.2022.03.017

Irfeey, A. M., Chau, H.-W., Sumaiya, M. M., Wai, C. Y., Muttil, N., & Jamei, E. (2023). Sustainable mitigation strategies for urban heat island effects in urban areas. *Sustainability, 15*(14), 10767. https://doi.org/10.3390/su151410767

Irmak, M. A., Yilmaz, S., Mutlu, E., & Yilmaz, H. (2018). Assessment of the effects of different tree species on urban microclimate. *Environmental Science and Pollution Research, 25*(16), 15802–15822. https://doi.org/10.1007/s11356-018-1697-8

Ishibuchi, H., Sakane, Y., Tsukamoto, N., & Nojima, Y. (2009, 11-14 Oct. 2009). Evolutionary many-objective optimization by NSGA-II and MOEA/D with large populations. In *2009 IEEE international conference on systems, man and cybernetics* (pp. 1758–1763). https://doi.org/10.1109/ICSMC.2009.5346628

Jänicke, B., Milošević, D., & Manavvi, S. (2021). Review of user-friendly models to improve the urban micro-climate. *Atmosphere, 12*(10), 1291. https://doi.org/10.3390/atmos12101291

Jareemit, D., & Srivanit, M. (2022). A comparative study of cooling performance and thermal comfort under street market shades and tree canopies in tropical Savanna climate. *Sustainability, 14*(8), 4653. https://doi.org/10.3390/su14084653

Kaluarachichi, T., Tjoelker, M., & Pfautsch, S. (2020). Temperature reduction in urban surface materials through tree shading depends on surface type not tree species. *Forests, 11*(11), 1141. https://doi.org/10.3390/f11111141

Kamruzzaman, M., Deilami, K., & Yigitcanlar, T. (2018). Investigating the urban heat island effect of transit oriented development in Brisbane. *Journal of Transport Geography, 66*, 116–124. https://doi.org/10.1016/j.jtrangeo.2017.11.016

Kozłowska, A. Operation of ant algorithm on different hardware platforms. In *2014 Proceedings of the 21st international conference mixed design of integrated circuits and systems (MIXDES)* (pp. 537–539). https://doi.org/10.1109/MIXDES.2014.6872259

Lachapelle, J. A., Scott Krayenhoff, E., Middel, A., Coseo, P., & Warland, J. (2023). Maximizing the pedestrian radiative cooling benefit per street tree. *Landscape and Urban Planning, 230*, 104608. https://doi.org/10.1016/j.landurbplan.2022.104608

Lee, T. W., & Lee, H. Y. (2010, 19-21 March 2010). Applying ant colony optimization for the decision support of green area maintenance. In *2010 Second international conference on computer engineering and applications* (1, pp. 138–142). https://doi.org/10.1109/ICCEA.2010.35

Li, F., Yigitcanlar, T., Nepal, M., Thanh, K. N., & Dur, F. (2022). Understanding urban heat vulnerability assessment methods: A PRISMA review. *Energies, 15*(19), 6998. https://doi.org/10.1016/j.jtrangeo.2017.11.016

Li, F., Yigitcanlar, T., Nepal, M., Thanh, K. N., & Dur, F. (2024). A novel urban heat vulnerability analysis: Integrating machine learning and remote sensing for enhanced insights. *Remote Sensing, 16*(16), 3032.

Li, R., Zeng, F., Zhao, Y., Wu, Y., Niu, J., Wang, L., Leon, Gao, N., & Shi, X.. (2023a). CFD simulations of the tree effect on the outdoor microclimate by coupling the canopy energy balance model. *Building and Environment, 230*, 109995. https://doi.org/10.1016/j.buildenv.2023.109995

Li, Y., Bao, W., Bongers, F., Chen, B., Chen, G., Guo, K., Jiang, M., Lai, J., Lin, D., Liu, C. & Liu, X., (2019). Drivers of tree carbon storage in subtropical forests. *Science of the Total Environment, 654*, 684–693.

Li, Y., Lin, D., Zhang, Y., Song, Z., Sha, X., Zhou, S., Chen, C., & Yu, Z. (2023b). Quantifying tree canopy coverage threshold of typical residential quarters considering human thermal comfort and heat dynamics under extreme heat. *Building and Environment, 233*, 110100. https://doi.org/10.1016/j.buildenv.2023.110100

Li, Y., & Song, Y. (2019). Optimization of vegetation arrangement to improve microclimate and thermal comfort in an urban park. *International Review for Spatial Planning and Sustainable Development*, 7(1), 18–30. https://doi.org/10.14246/irspsd.7.1_18

Liao, C.-J., Tsai, Y.-L., & Chao, C.-W. (2011). An ant colony optimization algorithm for setup coordination in a two-stage production system. *Applied Soft Computing, 11*(8), 4521–4529. https://doi.org/10.1016/j.asoc.2011.08.014

Lindberg, F., Holmer, B., & Thorsson, S. (2008). SOLWEIG 1.0 – Modelling spatial variations of 3D radiant fluxes and mean radiant temperature in complex urban settings. *International Journal of Biometeorology, 52*(7), 697–713. https://doi.org/10.1007/s00484-008-0162-7

Liu, Y., Yigitcanlar, T., Guaralda, M., Degirmenci, K., & Liu, A. (2024). Spatial modelling of urban wind characteristics: Review of contributions to sustainable urban development. *Buildings, 14*(3), 737. https://doi.org/10.3390/buildings14030737

Liu, Z., Cheng, K. Y., Sinsel, T., Simon, H., Jim, C. Y., Morakinyo, T. E., He, Y., Yin, S., Ouyang, W., Shi, Y., & Ng, E. (2023). Modeling microclimatic effects of trees and green roofs/façades in ENVI-met: Sensitivity tests and proposed model library. *Building and Environment, 244*, 110759. https://doi.org/10.1016/j.buildenv.2023.110759

Liu, Z., Cheng, W., Jim, C. Y., Morakinyo, T. E., Shi, Y., & Ng, E. (2021). Heat mitigation benefits of urban green and blue infrastructures: A systematic review of modeling techniques, validation and scenario simulation in ENVI-met V4. *Building and Environment, 200*, 107939. https://doi.org/10.1016/j.buildenv.2021.107939

Liu, Z., Zheng, S., & Zhao, L. (2018). Evaluation of the ENVI-met vegetation model of four common tree species in a subtropical hot-humid area. *Atmosphere, 9*(5), 198. https://doi.org/10.3390/atmos9050198

Lu, F., Gao, Y., Jiang, L., Chen, Y., & Hao, Z. (2023). The effect of greening layout on microclimate in urban residential areas in hot summer–cold winter zones. *Atmosphere, 14*(12), 1824. https://doi.org/10.3390/atmos14121824

Ma, S., He, J., Liu, F., & Yu, Y. (2011). Land-use spatial optimization based on PSO algorithm. *Geo-Spatial Information Science, 14*(1), 54–61. https://doi.org/10.1007/s11806-011-0437-8

Mahdavinejad, M., Shaeri, J., Nezami, A., & Goharian, A. (2024). Comparing universal thermal climate index (UTCI) with selected thermal indices to evaluate outdoor thermal comfort in traditional courtyards with BWH climate. *Urban Climate, 54*, 101839. https://doi.org/10.1016/j.uclim.2024.101839

Matzarakis, A., Rutz, F., & Mayer, H. (2010). Modelling radiation fluxes in simple and complex environments: Basics of the RayMan model. *International Journal of Biometeorology, 54*(2), 131–139. https://doi.org/10.1007/s00484-009-0261-0

Meili, N., Manoli, G., Burlando, P., Carmeliet, J., Chow, W. T. L., Coutts, A. M., Roth, M., Velasco, E., Vivoni, E. R., & Fatichi, S. (2021). Tree effects on urban microclimate: Diurnal, seasonal, and climatic temperature differences explained by separating radiation, evapotranspiration, and roughness effects. *Urban Forestry & Urban Greening, 58*, 126970. https://doi.org/10.1016/j.ufug.2020.126970

Milošević, D. D., Bajšanski, I. V., & Savić, S. M. (2017). Influence of changing trees locations on thermal comfort on street parking lot and footways. *Urban Forestry & Urban Greening, 23*, 113–124. https://doi.org/10.1016/j.ufug.2017.03.011

Mirzaei, P. A. (2021). CFD modeling of micro and urban climates: Problems to be solved in the new decade. *Sustainable Cities and Society, 69*, 102839. https://doi.org/10.1016/j.scs.2021.102839

Mochida, A., Tabata, Y., Iwata, T., & Yoshino, H. (2008). Examining tree canopy models for CFD prediction of wind environment at pedestrian level. *4th International Symposium on Computational Wind Engineering, 96*(10), 1667–1677. https://doi.org/10.1016/j.jweia.2008.02.055

Morakinyo, T. E., & Lam, Y. F. (2016). Simulation study on the impact of tree-configuration, planting pattern and wind condition on street-canyon's micro-climate and thermal comfort. *Building and Environment, 103*, 262–275. https://doi.org/10.1016/j.buildenv.2016.04.025

Morakinyo, T. E., Ouyang, W., Lau, K. K.-L., Ren, C., & Ng, E. (2020). Right tree, right place (urban canyon): Tree species selection approach for optimum urban heat mitigation—Development and evaluation. *Science of the Total Environment, 719*, 137461. https://doi.org/10.1016/j.scitotenv.2020.137461

Mortoja, M., & Yigitcanlar, T. (2020a). Local drivers of anthropogenic climate change: Quantifying the impact through a remote sensing approach in Brisbane. *Remote Sensing, 12*(14), 2270. https://doi.org/10.3390/rs12142270

Mortoja, M., Yigitcanlar, T., & Mayere, S. (2020b). What is the most suitable methodological approach to demarcate peri-urban areas? A systematic review of the literature. *Land Use Policy, 95*, 104601. https://doi.org/10.1016/j.landusepol.2020.104601

Nazarian, N., Acero, J. A., & Norford, L. (2019). Outdoor thermal comfort autonomy: Performance metrics for climate-conscious urban design. *Building and Environment, 155*, 145–160. https://doi.org/10.1016/j.buildenv.2019.03.028

Noel, M. M. (2012). A new gradient based particle swarm optimization algorithm for accurate computation of global minimum. *Applied Soft Computing, 12*(1), 353–359. https://doi.org/10.1016/j.asoc.2011.08.037

Ouyang, W., Sinsel, T., Simon, H., Morakinyo, T. E., Liu, H., & Ng, E. (2022). Evaluating the thermal-radiative performance of ENVI-met model for green infrastructure typologies: Experience from a subtropical climate. *Building and Environment, 207*, 108427. https://doi.org/10.1016/j.buildenv.2021.108427

Pace, R., De Fino, F., Rahman, M. A., Pauleit, S., Nowak, D. J., & Grote, R. (2021). A single tree model to consistently simulate cooling, shading, and pollution uptake of urban trees. *International Journal of Biometeorology, 65*(2), 277–289. https://doi.org/10.1007/s00484-020-02030-8

Park, C. Y., Yoon, E. J., Lee, D. K., & Thorne, J. H. (2020). Integrating four radiant heat load mitigation strategies is an efficient intervention to improve human health in urban environments. *Science of the Total Environment, 698*, 134259. https://doi.org/10.1016/j.scitotenv.2019.134259

Park, M., Hagishima, A., Tanimoto, J., & Narita, K. (2012). Effect of urban vegetation on outdoor thermal environment: Field measurement at a scale model site. *Building and Environment, 56*, 38–46. https://doi.org/10.1016/j.buildenv.2012.02.015

Park, S., Tuller, S. E., & Jo, M. (2014). Application of universal thermal climate index (UTCI) for microclimatic analysis in urban thermal environments. *Landscape and Urban Planning, 125*, 146–155. https://doi.org/10.1016/j.landurbplan.2014.02.014

Ren, X., Zhang, G., Chen, Z., & Zhu, J. (2023). The influence of wind-induced response in urban trees on the surrounding flow field. *Atmosphere, 14*(6), 1010. https://doi.org/10.3390/atmos14061010

Ren, Z., Zhao, H., Fu, Y., Xiao, L., & Dong, Y. (2022). Effects of urban street trees on human thermal comfort and physiological indices: A case study in Changchun city, China. *Journal of Forestry Research, 33*(3), 911–922. https://doi.org/10.1007/s11676-021-01361-5

Roudsari, M. S., & Pak, M. (2013). *Ladybug: A parametric environmental plugin for grasshopper to help designers create an environmentally-conscious design*. https://www.aivc.org/resource/ladybug-Parametric-Environmental-Plugin-Grasshopper-Help-Designers-Create-Environmentally

Sanusi, R., Johnstone, D., May, P., & Livesley, S. J. (2017). Microclimate benefits that different street tree species provide to sidewalk pedestrians relate to differences in plant area index. *Landscape and Urban Planning, 157,* 502–511. https://doi.org/10.1016/j.landurbplan.2016.08.010

Shaamala, A., Yigitcanlar, T., Nili, A., & Nyandega, D. (2024). Algorithmic green infrastructure optimisation: Review of artificial intelligence driven approaches for tackling climate change. *Sustainable Cities and Society, 101,* 105182. https://doi.org/10.1016/j.scs.2024.105182

Shivanna, K. R. (2022). Climate change and its impact on biodiversity and human welfare. *Proceedings of the Indian National Science Academy, 88*(2), 160–171. https://doi.org/10.1007/s43538-022-00073-6

Skutecki, R., Jalali, R., Dragańska, E., Cymes, I., Romaszko, J., & Glińska-Lewczuk, K. (2019). UTCI as a bio-meteorological tool in the assessment of cold-induced stress as a risk factor for hypertension. *Science of the Total Environment, 688,* 970–975. https://doi.org/10.1016/j.scitotenv.2019.06.280

Soltani, A., & Sharifi, E. (2017). Daily variation of urban heat island effect and its correlations to urban greenery: A case study of Adelaide. *Frontiers of Architectural Research, 6*(4), 529–538. https://doi.org/10.1016/j.foar.2017.08.001

Son, T., Weedon, Z., Yigitcanlar, T., Sanchez, T., Corchado, J., & Mehmood, R. (2023). Algorithmic urban planning for smart and sustainable development: Systematic review of the literature. *Sustainable Cities and Society, 94,* 104562. https://doi.org/10.1016/j.scs.2023.104562

Song, M., & Chen, D. (2018). An improved knowledge-informed NSGA-II for multi-objective land allocation (MOLA). *Geo-Spatial Information Science, 21*(4), 273–287. https://doi.org/10.1080/10095020.2018.1489576

Speak, A., Montagnani, L., Wellstein, C., & Zerbe, S. (2020). The influence of tree traits on urban ground surface shade cooling. *Landscape and Urban Planning, 197,* 103748. https://doi.org/10.1016/j.landurbplan.2020.103748

Stojakovic, V., Bajsanski, I., Savic, S., Milosevic, D., & Tepavcevic, B. (2020). The influence of changing location of trees in urban green spaces on insolation mitigation. *Urban Forestry & Urban Greening, 53,* 126721. https://doi.org/10.1016/j.ufug.2020.126721

Sudholt, D., & Thyssen, C. (2012). Running time analysis of ant colony optimization for shortest path problems. *Journal of Discrete Algorithms, 10,* 165–180.

Tams, L., Paton, E. N., & Kluge, B. (2023). Impact of shading on evapotranspiration and water stress of urban trees. *Ecohydrology, 16*(6), e2556. https://doi.org/10.1002/eco.2556

Tsoka, S., Tsikaloudaki, A., & Theodosiou, T. (2018). Analyzing the ENVI-met microclimate model's performance and assessing cool materials and urban vegetation applications – A review. *Sustainable Cities and Society, 43,* 55–76. https://doi.org/10.1016/j.scs.2018.08.009

Tuczek, M., Degirmenci, K., Desouza, K. C., Watson, R., Yigitcanlar, T., & Breitner, M. (2022). Mitigating urban heat with optimal distribution of vegetation and buildings. *Urban Climate, 44,* 101208. https://doi.org/10.1016/j.uclim.2022.101208

Upreti, M., Saikia, P., Shilky, Lal, P., & Kumar, A. (2024). Chapter 2 – Major challenges in the urbanizing world and role of earth observations for livable cities. In A. Kumar, P. K. Srivastava, P. Saikia, & R. K. Mall (Eds.), *Earth observation in urban monitoring* (pp. 23–52). Elsevier. https://doi.org/10.1016/B978-0-323-99164-3.00002-1

Vachhani, V. L., Dabhi, V. K., & Prajapati, H. B. (2016, 7-9 Jan. 2016). Improving NSGA-II for solving multi objective function optimization problems. In *2016 International conference on computer communication and informatics (ICCCI)* (pp. 1–6). https://doi.org/10.1109/ICCCI.2016.7479921

Vafaeinejad, A., & Mahmoudi Jam, S. (2021). Using particle swarm optimization algorithm and geospatial information system for potential evaluating of groundwater (case study: Mehran, Iran). *Arabian Journal of Geosciences, 14*(12), 1139. https://doi.org/10.1007/s12517-021-07475-8

Verma, S., Pant, M., & Snasel, V. (2021). A comprehensive review on NSGA-II for multi-objective combinatorial optimization problems. *IEEE Access, 9*, 57757–57791. https://doi.org/10.1109/ACCESS.2021.3070634

Versteeg, H. K., & Malalasekera, W. (2007). Turbulence and its modelling. *An Introduction to Computational Fluid Dynamics: The Finite Volume Method, 3*, 40–115.

Wallenberg, N., Lindberg, F., & Rayner, D. (2022). Locating trees to mitigate outdoor radiant load of humans in urban areas using a metaheuristic hill-climbing algorithm – Introducing TreePlanter v1.0. *Geoscientific Model Development, 15*(3), 1107–1128. https://doi.org/10.5194/gmd-15-1107-2022

Winbourne, J. B., Jones, T. S., Garvey, S. M., Harrison, J. L., Wang, L., Li, D., Templer, P. H., & Hutyra, L. R. (2020). Tree transpiration and urban temperatures: Current understanding, implications, and future research directions. *BioScience, 70*(7), 576–588. https://doi.org/10.1093/biosci/biaa055

Woodward, A., Hinwood, A., Bennett, D., Grear, B., Vardoulakis, S., Lalchandani, N., Lyne, K., & Williams, C. (2023). Trees, climate change, and health: An urban planning, greening and implementation perspective. *International Journal of Environmental Research and Public Health, 20*(18), 6798. https://doi.org/10.3390/ijerph20186798

Wu, Z., Shi, Y., Ren, L., & Hang, J. (2024). Scaled outdoor experiments to assess impacts of tree evapotranspiration and shading on microclimates and energy fluxes in 2D street canyons. *Sustainable Cities and Society, 108*, 105486. https://doi.org/10.1016/j.scs.2024.105486

Yang, A.-S., Juan, Y.-H., Wen, C.-Y., & Chang, C.-J. (2017). Numerical simulation of cooling effect of vegetation enhancement in a subtropical urban park. *Applied Energy, 192*, 178–200. https://doi.org/10.1016/j.apenergy.2017.01.079

Yang, J., Hu, X., Feng, H., & Marvin, S. (2021). Verifying an ENVI-met simulation of the thermal environment of Yanzhong Square Park in Shanghai. *Urban Forestry & Urban Greening, 66*, 127384. https://doi.org/10.1016/j.ufug.2021.127384

Yang, S., Wang, L. (L), Stathopoulos, T., & Marey, A. M. (2023). Urban microclimate and its impact on built environment – A review. *Building and Environment, 238*, 110334. https://doi.org/10.1016/j.buildenv.2023.110334

Yang, Y., Xu, Y., Duan, Y., Yang, Y., Zhang, S., Zhang, Y., & Xie, Y. (2023). How can trees protect us from air pollution and urban heat? Associations and pathways at the neighborhood scale. *Landscape and Urban Planning, 236*, 104779. https://doi.org/10.1016/j.landurbplan.2023.104779

Yigitcanlar, T. (2010). *Rethinking sustainable development: Urban management, engineering, and design*. IGI Global.

Yigitcanlar, T., Fabian, L., & Coiacetto, E. (2008). Challenges to urban transport sustainability and smart transport in a tourist city: The Gold Coast, Australia. *Open Transportation Journal, 2*, 29–46. https://doi.org/10.2174/1874447800802010029

Yin, Z., Liu, Z., Liu, X., Zheng, W., & Yin, L. (2023). Urban heat islands and their effects on thermal comfort in the US: New York and New Jersey. *Ecological Indicators, 154*, 110765. https://doi.org/10.1016/j.ecolind.2023.110765

Yoon, E. J., Kim, B., & Lee, D. K. (2019). Multi-objective planning model for urban greening based on optimization algorithms. *Urban Green Infrastructure – Connecting People and Nature for Sustainable Cities, 40*, 183–194. https://doi.org/10.1016/j.ufug.2019.01.004

Yu, Q., Ji, W., Pu, R., Landry, S., Acheampong, M., O' Neil-Dunne, J., Ren, Z., & Tanim, S. H. (2020). A preliminary exploration of the cooling effect of tree shade in urban landscapes. *International Journal of Applied Earth Observation and Geoinformation, 92*, 102161. https://doi.org/10.1016/j.jag.2020.102161

Zeng, F., Lei, C., Liu, J., Niu, J., & Gao, N. (2020). CFD simulation of the drag effect of urban trees: Source term modification method revisited at the tree scale. *Sustainable Cities and Society, 56*, 102079. https://doi.org/10.1016/j.scs.2020.102079

Zhang, B., & Brack, C. L. (2021). Urban forest responses to climate change: A case study in Canberra. *Urban Forestry & Urban Greening, 57*, 126910. https://doi.org/10.1016/j.ufug.2020.126910

Zhang, G., Zhang, C., & Zhang, H. (2018). Improved K-means algorithm based on density Canopy. *Knowledge-Based Systems, 145*, 289–297.

Zhang, J., & Gou, Z. (2021). Tree crowns and their associated summertime microclimatic adjustment and thermal comfort improvement in urban parks in a subtropical city of China. *Urban Forestry & Urban Greening, 59*, 126912. https://doi.org/10.1016/j.ufug.2020.126912

Zhang, Y., Zhang, Q., Farnoosh, A., Chen, S., & Li, Y. (2019). GIS-based multi-objective particle swarm optimization of charging stations for electric vehicles. *Energy, 169*, 844–853. https://doi.org/10.1016/j.energy.2018.12.062

Zhao, Q., Sailor, D. J., & Wentz, E. A. (2018). Impact of tree locations and arrangements on outdoor microclimates and human thermal comfort in an urban residential environment. *Urban Forestry & Urban Greening, 32*, 81–91. https://doi.org/10.1016/j.ufug.2018.03.022

Zhao, Q., Wentz, E. A., & Murray, A. T. (2017). Tree shade coverage optimization in an urban residential environment. *Building and Environment, 115*, 269–280. https://doi.org/10.1016/j.buildenv.2017.01.036

Zhao, Y., & Gong, P. (2024). Optimal site selection strategies for urban parks green spaces under the joint perspective of spatial equity and social equity. *Frontiers in Public Health, 12*(2296–2565). https://doi.org/10.3389/fpubh.2024.1310340

Zheng, Q., Feng, B.-W., Liu, Z.-Y., & Chang, H.-C. (2021). Application of improved particle swarm optimisation algorithm in hull form optimisation. *Journal of Marine Science and Engineering, 9*(9). https://doi.org/10.3390/jmse9090955

Zheng, S., He, C., Guldmann, J.-M., Xu, H., & Liu, X. (2023). Heat mitigation benefits of urban trees: A review of mechanisms, modeling, validation and simulation. *Forests, 14*(12), 2280. https://doi.org/10.3390/f14122280

7

AI-Driven Green Infrastructure Optimisation

7.1 Introduction and Background

7.1.1 Opening Remarks

Climate change is upon us, and we need effective, efficient, and immediate policy and action to tackle the catastrophic consequences (Arbolino et al., 2018; Mortoja & Yigitcanlar, 2020a; Sotto et al., 2019). Climate change, as highlighted by the Intergovernmental Panel on Climate Change (IPCC, 2014), threatens both natural and human systems, intensifying risks, such as heatwaves, floods, droughts, and biodiversity loss. To confront the multifaceted challenges of climate change, a spectrum of adaptable measures is essential. Mitigation strategies, for instance, range from advancing renewable energy technologies (Ang et al., 2022) and enhancing energy efficiency (Adua et al., 2021) to implementing policies for significant reductions in greenhouse gas emissions (Dubois et al., 2019). In addition, carbon capture and storage technologies have emerged as critical tools to balance our carbon footprints (Dziejarski et al., 2023).

Adaptation strategies are equally varied and crucial. These include fortifying infrastructure against the increasing frequency of extreme weather events (Markolf et al., 2019) and promoting water conservation methods to ensure sustainability in the face of shifting weather patterns (Mishra et al., 2021). Within the broad context of climate action, green infrastructure (GI) offers a unique convergence of mitigation and adaptation benefits (Anguelovski et al., 2022; Wang et al., 2023a). By integrating natural elements such as green roofs, parks, and urban wetlands into city planning, GI not only assists in carbon sequestration and energy conservation but also plays a vital role in enhancing urban resilience, such as mitigating the urban heat island effect and managing stormwater. Thus, GI stands as a testament to how nature-based solutions can be woven into the urban fabric to address the pressing concerns of climate change (Gaffin et al., 2012; Gill et al., 2007; Perveen et al., 2017).

Expanding on this, Choi et al. (2021) described GI as a complex network encompassing natural, semi-natural, and engineered elements, which are integrated within and around urban spaces at various scales. This infrastructure was intentionally designed and maintained to offer a multitude of ecosystem services and benefits. Similarly, the European Commission articulates a comparable definition, emphasising GI as "a strategically planned

DOI: 10.1201/9781003737056-9

network of natural and semi-natural areas with other environmental features, designed and managed to deliver a wide range of ecosystem services, while also enhancing biodiversity" (European Commission, 2013).

The concept of GI in Europe has become more closely associated with the principles of multifunctionality, climate change adaptation, and green growth, especially in recent decades. This intersection has given rise to a diverse research and policy agenda encompassing a wide range of subjects and catering to various audiences. This expanding scope reflects the growing recognition of GI's role in addressing complex environmental and socio-economic challenges (Jones et al., 2022; Ying et al., 2022).

In contrast to grey infrastructure, which refers to engineered constructions, GI implements natural processes to offer more sustainable and economical solutions for addressing climate change via adaptation and mitigation (Mortoja & Yigitcanlar, 2020b; Wang et al., 2022, 2023). Additionally, blue infrastructure serves the same purpose as GI but focuses on the water ecosystem, including urban ponds, rain gardens, and canals (De Silva et al., 2022; Yigitcanlar, 2010).

Moreover, GI has significant functions in addressing climate change, extending beyond its capacity to manage stormwater, sequester carbon, mitigate urban heat, and enhance biodiversity. These functions also include promoting ecological connectivity, improving air quality, reducing energy consumption, enhancing landscape aesthetics, and fostering social cohesion (Feng et al., 2022; Jiang et al., 2022; Kumar et al., 2022; Li et al., 2022a). Nonetheless, the promotion of GI as a tool to protect the environment is significantly challenging because of the difficulty of measuring environmental benefits (Akbari, 2002; Nordh et al., 2009; Schipperijn et al., 2010).

Furthermore, it is critical to realise that the benefits of GIs are unequal, and factors such as location, type, and size contribute to shaping their environmental impact. The "why" of optimisation is grounded in the need to enhance environmental impact and sustainability outcomes (Sadollah et al., 2020). The "when" relates to the strategic stages of urban planning and development, ensuring that GI is integrated effectively from the outset. As for the "how," it involves adopting a holistic approach that considers local ecological, social, and economic contexts (Hariram et al., 2023). This approach ensures that GI not only contributes to localised improvements but also aligns with broader environmental and sustainability goals. Such optimisation requires careful planning and implementation strategies tailored to specific urban settings and their unique challenges and opportunities (Marasinghe et al., 2024; Neema & Ohgai, 2013).

7.1.2 Literature Background

The initial introduction emphasised the essential role that GI plays in responding to the pressing issues posed by climate change. However, it is crucial to adopt innovative approaches that extend beyond traditional methods to fully realise the potential of GI. With cutting-edge advancements, artificial intelligence (AI) and machine learning (ML) have emerged as promising solutions for optimising GI in novel and unprecedented ways. To capitalise on these technologies, it is necessary to conduct a comprehensive review of the existing research. This literature review aims to explore the current state of AI and ML applications in GI optimisation, examining how these technologies are utilised to address various aspects of GI, such as their effectiveness, adaptability to diverse environments, and potential challenges.

The objective of optimising GI is to identify the best possible solution within the framework of specified goals. Scenario analysis is the most employed method in this context. In this approach, researchers have established various combined scenarios for GI deployment. Each scenario was rigorously evaluated to assess its performance against the set criteria (Wang et al., 2023). This process involves a detailed examination of the outcomes under different GI configurations, which enables researchers to identify the most effective solution.

Recently, propelled by advancements in AI technologies, algorithms, and models (Yigitcanlar et al., 2021), researchers have begun to optimise GI for different objectives. Examples include the genetic algorithm (GA) to reduce urban heat islands (Zhou et al., 2023), heuristic algorithm (HA) to preserve natural habitats (Li et al., 2022a), particle swarm optimisation (PSO) to improve energy efficiency (Kim et al., 2016), and non-dominated sorting GA (NSGA) to minimise flood damage (Martínez et al., 2021).

In addition, ML models were employed in GI optimisation studies. Examples include random forest (RF) to maximise carbon sinks (Liu et al., 2023a), artificial neural networks (ANN), and support vector machines (SVMs) to minimise traffic air pollution (Hashad et al., 2021). Nonetheless, these ML models are mainly predictive and exhibit strong performance within the defined parameter space; their effectiveness can vary when applied outside of the designated range (Kranjčić et al., 2019).

To optimise different aspects of GI, including location, size, and type accurately, effective indicators related to GI and objectives must be determined. Nonetheless, the absence of universal indicator standards led to the critical dependence on the local context during the indicator selection stage (Koc et al., 2018). For instance, air quality and carbon emissions indicators were only included in some studies that optimise GI for the objective of enhancing air quality (Jayasooriya et al., 2020). Furthermore, while certain studies added energy consumption indicators to the objective of improving energy efficiency, it was not considered in other studies or analyses.

Considering this, a more consolidated understanding of the GI optimisation process is required—one that integrates considerations of human vulnerability and prioritises community engagement. This involves guiding not just researchers, practitioners, and urban policymakers but also actively involving the wider public in the decision-making process. By doing so, GI optimisation can become a tool for fostering more resilient, equitable, and sustainable urban environments, in line with the principles of nature-based solutions (Seddon et al., 2020).

Many studies have contributed to developing a framework for GI optimisation. For instance, Wang et al. (2023) proposed a framework for GI zoning in the context of water management. In a related vein, Kim et al. (2016) formulated a GI optimisation framework to be considered for thermal comfort. Moreover, in the realm of ecosystems, Abdollahi et al. (2022) provided a framework for allocating landscape patterns. Nevertheless, these frameworks were highly localised, and utilising them elsewhere is a challenging endeavour (Liu et al., 2023).

Besides, previous studies predominantly concentrated on singular objectives within the spectrum of GI optimisation objectives (Hashad et al., 2021; Liu et al., 2023). This narrow focus limits their comprehensiveness in addressing the multifaceted aspects required for effective climate change mitigation. As a result, this study presents a comprehensive framework that standardises criteria for the GI optimisation process. Additionally, it presents a structured knowledge of the diverse elements of the optimisation approach within the context of addressing climate change challenges.

Furthermore, this study aims to enhance our understanding of AI-driven GI optimisation process to contribute to the efforts in tackling climate change. Moreover, it attempts to answer the following questions: (a) What aspects of GI need to be considered in AI-driven optimisation of GI for tackling climate change? and (b) How do we implement the process steps of GI optimisation regarding climate change mitigation? With the intention of addressing these research questions, we undertook a systematic literature review that reveals AI-driven GI optimisation objectives, models, indicators, and scales. After this introductory section, Section 7.2 outlines the methodology, Section 7.3 showcases the findings, Section 7.4 is the discussion, and lastly, Section 7.5 encompasses the conclusion of the study.

7.2 Research Design

A systemic literature review technique was employed by using the Preferred Reporting Items for Systemic Reviews and Meta-Analyses (PRISMA) protocol. This approach is extensively acknowledged with a transparent, consistent, and comprehensive systematic literature review framework design (Regona et al., 2022; Son et al., 2023). The protocol involves three stages, i.e., planning, review, and reporting. The first step of planning encompasses elements, such as the study objective, study question, keywords, as well as inclusion and exclusion criteria. As explained, the study goal was to present a review that helps to understand the GI optimisation process as a tool for tackling climate change.

It is important to note that the scope of this review has a particular focus on GI that excludes the review of the broader category of nature-based solutions (NBS). The rationale behind this is as follows. GI is typically defined as a network of natural and semi-natural spaces in urban areas (Choi et al.,

2021). On the contrary, NBS are broader, encompassing solutions inspired and supported by nature, which are cost-effective, simultaneously providing environmental, social, and economic benefits, and helping build resilience (Sowińska-Świerkosz & García, 2022). Having a specific focus on GI allows for an in-depth exploration of AI and optimisation techniques within the urban GI context, highlighting specific challenges and opportunities for technological innovation in urban environmental management. While acknowledging the relevance of NBS, our concentrated attention to GI aims to provide detailed insights into the intersection of AI, optimisation, and GI, in line with current climate adaptation/mitigation priorities.

In addition, a diverse set of search keywords was employed to encompass both established and emerging AI technologies to ensure the inclusivity of the collected documents from databases. A university library search engine was used to explore relevant online publications from Scopus and Web of Science (WoS) databases. In July 2023, the search query was carried out using the following Boolean operation: (title or abstract or keyword) contains ("green infrastructure" OR "green area*" OR "green spaces*" OR "landscape") AND ("artificial intelligence" OR "AI" OR "machine learning" OR "ML" OR "deep learning" OR "algorithm") AND ("optimisation" OR "optimization").

To maximise the number of relevant literature, the publication date was left unrestricted. In total, the research resulted in 6600 records. Primary inclusion and exclusion criteria were applied to limit the selected articles to English peer-reviewed articles published in a journal and full-text articles available online. Table 7.1 presents the inclusion and exclusion criteria. All references were imported to RStudio software to remove duplicate records using the (bibliometrix) library, and (n = 1223) duplicate records were detected and removed. The remaining list included 2800 documents (n = 2800).

During the screening phase, this study employed secondary inclusion and exclusion criteria to eliminate articles that were not aligned with the research aim or question from the remaining list of records (n = 2800). The number of records decreased to 223. Then, abstracts and keywords were assessed to determine their relevance to the research question, guiding the inclusion in the final analysis. This procedure resulted in the exclusion of 80 articles. Next, the number of 143 eligible articles was forwarded to the eligibility phase.

TABLE 7.1

Inclusion and Exclusion Criteria

Inclusionary Criteria	Exclusionary Criteria
Academic journal article Peer-reviewed English language Full-text article Available online Relevant to the research objective/question	Books, chapters, conference proceedings Articles published in languages other than English Irrelevant to the research objective/question

In the last phase, the shortlisted articles were thoroughly read, with particular attention to their research aim. This process led to the exclusion of 85 more articles that did not align with the research question and objectives. Following the second round of comprehensive full-text reading, the total number of studies was narrowed down to 58 articles (Figure 7.1). During the review of the articles that use ML models and algorithms for GI optimisation, we observed a recurring pattern in which studies commonly utilised five aspects that change according to the optimisation objective.

These aspects include the following: (a) Optimisation objective: which pertains to one or more categories of climate change challenges that are targeted to be tackled through the optimisation of GI; (b) Indicators: involving the

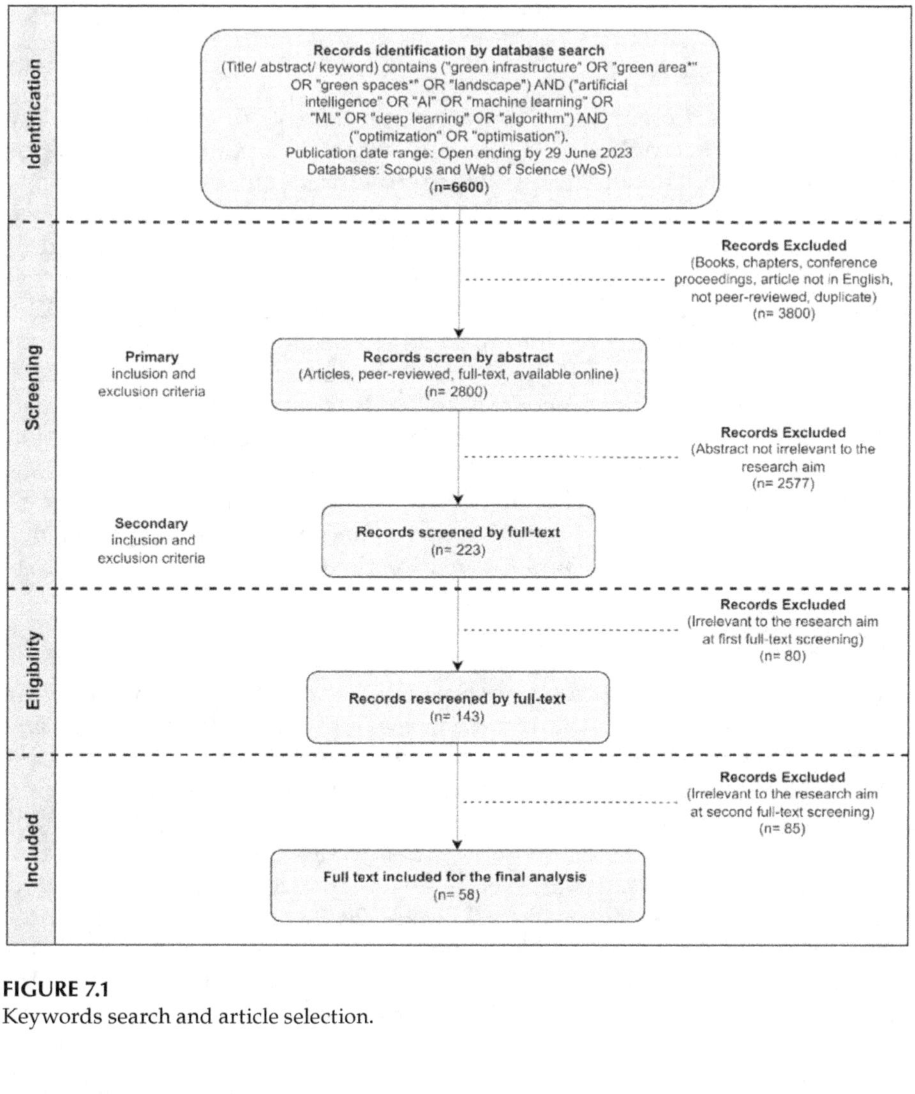

FIGURE 7.1
Keywords search and article selection.

environmental and built environment characteristics that are utilised to lead the optimisation process; (c) Models: where it defines the type of ML models or algorithms that are applied in the optimisation process; (d) Type of GI: denotes the GI category selected by a study to optimise it as a tool to mitigate climate change; (e) Scale: pertains to the size and coverage area of the optimisation within the urban area. These papers analysed these five aspects for the reviewed articles, and the compilation articles located in Table A7.1. This study proposes a systemic flowchart or framework that focuses on the GI optimisation process in the context of climate change adaptation and mitigation. This framework outlines a comprehensive approach to addressing the issue.

7.3 Analysis and Results

In this section, we present the comprehensive results of our systematic review that focused on the application of AI and ML in the optimisation of GI for climate change mitigation. The results are structured into several subsections to provide a coherent and detailed explanation of our findings. Initially, we offer a general observation of literature trends, highlighting the chronological evolution of research output up to 2023. This is followed by a geographical distribution analysis that sheds light on the regional focus of GI research. The subsequent subsections search into specific aspects of GI optimisation, including data and indicators used, the variety of optimisation models and algorithms employed, types of GI considered, and the scales of analysis applied in the studies. We conclude with a detailed discussion of the categorisation of GI optimisation objectives and introduce a novel framework for AI-driven GI optimisation in tackling climate change challenges. Each subsection synthesises the key findings and trends observed and provides a comprehensive overview of the current state and future directions in this field.

7.3.1 General Observations

The articles selected for review were organised chronologically by their year of publication to illustrate the trend in research output prior to 2023. The timeline begins in 2010, marked by the publication of the first two articles. Between 2010 and 2015, the field saw a relatively modest output, with only six research articles being published. However, a noticeable uptick in publication frequency occurred following the introduction of the EU strategy, culminating in a total of 23 articles by the year 2020. Notably, this period aligns with critical milestones in the Horizon 2020 and Horizon Europe programmes, which prioritised funding and research in areas relevant to GI and climate change adaptation. The surge in research output to 35 articles by 2023 can be partly attributed to the amplification of these programmes, alongside the introduction of significant policies such as the Water Infrastructure Improvement Act in the US and the Green Infrastructure Framework in England. These policy initiatives have not only provided the necessary financial support but also shaped the research agenda by highlighting priority areas such as sustainable urban development, climate resilience, and GI innovation. This policy-driven research trajectory underscores a growing convergence between academic interests and policy objectives in the field of GI, reflecting a more integrated approach to tackling climate change and environmental sustainability (Figure 7.2).

Additionally, research regions were classified due to the differences in geographical, demographical, and climate data that might influence research methods and results. Among the reviewed articles, Asia was the most frequent location for case studies, accounting for half of the total, with 27 articles. Europe followed with 17 studies, while North America was the setting for nine studies. South America and Oceania were comparatively less represented, serving as the focus for three and two studies, respectively. Notably, Africa was an exception, as no studies were conducted in this region, as depicted in Figure 7.3. Within these geographical contexts, the influence of China and the USA in GI research is significant, not simply due to the volume of studies but also because of the wide geographical, climatic, and landscape diversity within these nations. This diversity within each country means that studies conducted in different regions of China or the USA might yield varied insights, reflecting the distinct environmental and climatic conditions present across their large area. For instance, GI research in the arid regions of Western China or the southwestern USA would likely differ significantly from studies conducted in their more humid eastern regions.

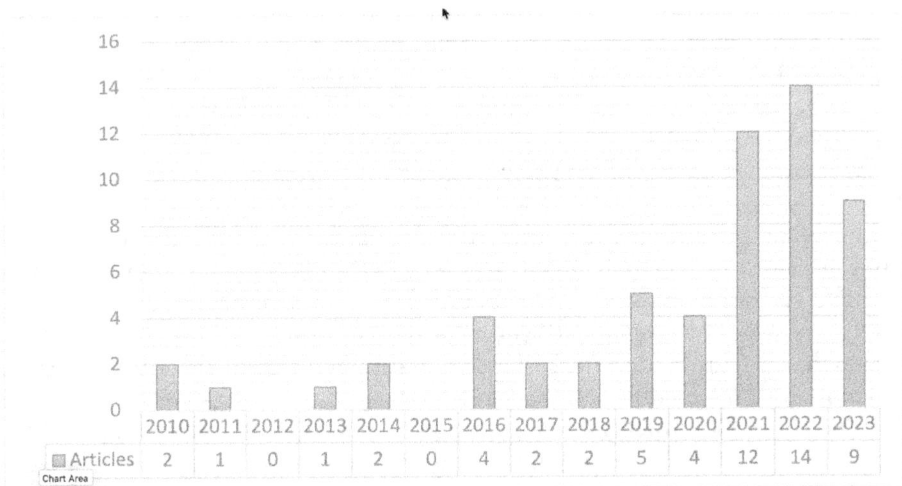

FIGURE 7.2
Publication distribution by year.

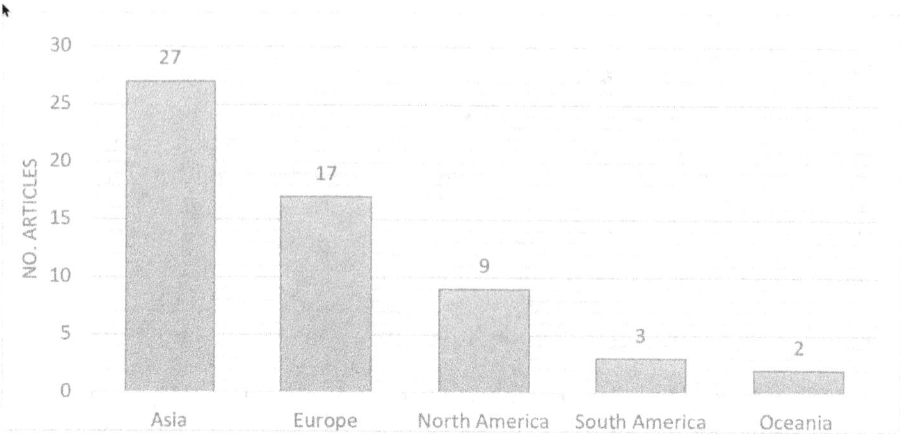

FIGURE 7.3
Geographic distribution of publications.

However, this distribution also underscores significant gaps in representation from certain regions, particularly Africa, suggesting a need for broader geographic inclusivity in future GI research. This would ensure a more comprehensive understanding of GI practices suitable for diverse climatic and geographical settings worldwide.

7.3.2 Data and Indicators

In the pursuit of optimising GI for climate change mitigation, a systematic approach to data collection is paramount. The initial step post-defining specific objectives entails a rigorous aggregation of data pertinent to the environmental and socio-economic facets of GI (Wang et al., 2021; Yoon et al., 2019). The identification and subsequent categorisation of optimisation indicators are crucial for reflecting the multidimensional objectives of GI optimisation. These indicators encapsulate not only the direct attributes of the GI but also the broader environmental impacts and socio-economic implications.

Consequently, the study has curated a suite of optimisation indicators, which are stratified based on their direct relevance to the GI, their alignment with the specified climate change objectives, and their role as general or supporting metrics. Each category serves a distinct function in the optimisation process: indicators related to GI provide a quantifiable measure of the infrastructure's physical and biological characteristics; those related to climate change objectives gauge the efficacy of GI in meeting environmental targets; and general/supporting indicators offer a contextual backdrop against which the effectiveness of GI is evaluated, integrating socio-economic and land use considerations.

Table 7.2 distils the indicators that have been implemented in selected articles, presenting a structured overview that underscores the prevalence

TABLE 7.2
Common Data and Indicators

Objective Categories	Relevance	Indicators	Number of Studies
Air quality	Related to GI	Green space map	5
		Green space type	1
		Canopy coverage of tree	3
		Tree type	2
		Tree height	1
		Tree price	1
		Leaf area density	1
	Related to the objective (air quality)	Air temperature	4
		Carbon emissions	5
		Carbon absorption of tree	1
		Wind speed	1
		Air pollution level	5
	Geospatial data	Land use map	5
		Digital elevation model	2
		Slope	2
		Population density	4
		Nighttime light data	1
		Distance to the road network	1
		Distance to the point of interest	1
		Proximity to river	1
Biodiversity and ecosystem security	Related to GI	Urban green spaces	5
		Trees type	1
		Vegetation density	1
	Related to the objective (biodiversity and ecosystem security)	Ecological protection control	1
		Urban water map	1
		Biophysical table	1
		Habitat map	1
		Resistance map	1
		Species distribution	1
		Soil type	4
		Bird diversity	1
		Flow path map	1
	Geospatial data	Land use map	7
		Land cover map	5
		DEM	5
		Temperature	4
		Population density	5
		Urban road system	3
		Nighttime light	1
		Points of interest	1

(Continued)

TABLE 7.2 *(Continued)*

Common Data and Indicators

Objective Categories	Relevance	Indicators	Number of Studies
Energy efficiency	Related to GI	Green space between buildings	3
		Tree type and volume	4
		Tree location	2
		Stomatal resistance	2
		Leaf area index	2
	Related to the objective (energy efficiency)	Energy demand	3
		Cooling and heating load	4
	Geospatial data	Building perimeter	5
		Sunshine distribution map	1
Public health	Related to GI	Urban green spaces	5
		Tree crown coverage area	1
		Leaf area index of vegetation	1
		Vegetation density	1
	Related to the objective (public health)	Land surface temperature	5
		Air pollution level	4
		Noise pollution level	5
		Climate health impact assessment	3
		Air temperature	3
		Water quality	2
	Geospatial data	Land use map	1
		Distance between road and trees	2
		Population density	2
		Building density	1
Heat island	Related to GI	Green space map	8
		Trees coverage	8
		Tree type and volume	8
	Related to the objective (heat island)	Wind speed	2
		Air temperature	3
		Humidity	3
		Solar radiation	4
		Land surface temperature	5
		Cooling effect map	2
	Geospatial data	Land use map	6
		Building dimensions	5
		Distance between tree and building	2
		Building density	1
		Points of interest	1
		Topographic map	1

TABLE 7.2 *(Continued)*

Common Data and Indicators

Objective Categories	Relevance	Indicators	Number of Studies
Water management	Related to GI	Urban green spaces	3
	Related to the objective (water management)	Soil type	8
		Catchment area	2
		Streams map	4
		Drainage system	6
		Precipitation map	2
		Flow Path map	2
		Flood hazard	3
		Outdoor water use	1
		Humidity	2
		Rainfall data	11
		Water quality	3
	Geospatial data	Land use map	13
		Land cover map	4
		Slope	4
		DEM	4

and significance of each indicator across various studies. This tabular compilation serves as a foundational reference for researchers and practitioners alike, guiding the strategic selection of indicators in GI optimisation efforts.

7.3.2.1 Air Quality

In the realm of air quality, a range of indicators were utilised to optimise the impact of GI, focusing on aspects like pollution levels, carbon emissions, and air temperature.

7.3.2.1.1 Indicators Related to the Objective

Data related to air quality as an objective of the GI optimisation reflects the air status, such as quality level, pollution level, and air temperature. The most frequently used indicators in the reported studies were air pollution levels and carbon emissions. The air pollution level indicator assesses various gases contributing to air pollution conditions, while the carbon emissions indicator measures only the carbon level in the air (Liu et al., 2023). Both indicators were employed by 50% of the reviewed studies (Choi & Lee, 2023; Daniele & Sciacca, 2021; Liu et al., 2023; Neema & Ohgai, 2010, 2013). The second most related indicator was the air temperature, with four studies of the air quality category where it is considered one of the most important indicators to assess air quality (Feng et al., 2022; Liu et al., 2023; Neema & Ohgai, 2013; Tao et al., 2023).

7.3.2.1.2 Indicators Related to GI

Green space map, which refers to the spatial representation showing the location of parks, gardens, tree cover, and other forms of vegetation, was the most used indicator regarding air quality objective with five related articles (Daniele & Sciacca, 2021; Neema & Ohgai, 2013; She et al., 2021), where most air quality studies focused on a large urban scale. Tree type and tree coverage indicators were the second most selected due to their essential role in air quality improvement (Choi & Lee, 2023; Hashad et al., 2021). Other indicators such as green space type, tree height, tree price, and leaf area density were applied by a few studies regarding the lack of studies considering GI in detail to optimise air quality (Choi & Lee, 2023; Hashad et al., 2021; Tao et al., 2023).

7.3.2.1.3 Geospatial Indicators

Geospatial indicators reflect those not directly associated with GI or air quality. The prevalence of land use map indicators can be attributed to their significance in gauging the land use influence on air quality and carbon emissions (Jia et al., 2019; Liu et al., 2023; Neema & Ohgai, 2010; Tao et al., 2023). The indicators utilised were the second most frequently used digital elevation model and slope because of their promoted resistance to ventilation (Liu et al., 2023; She et al., 2021; Tao et al., 2023). Further indicators used by different studies, including distance to certain facilities or road networks, were only applied depending on study area characteristics (Tao et al., 2023).

7.3.2.2 Biodiversity and Ecosystem Security

For biodiversity and ecosystem security, the chosen indicators reflect the health and diversity of natural habitats, encompassing factors such as soil type, habitat maps, and species diversity (Berry et al., 2015).

7.3.2.2.1 Indicators Related to the Objective

Indicators related to biodiversity and ecosystems are data demonstrating a variety of lives. Soil type has been widely used in four studies on biodiversity and ecosystem security (Abdollahi et al., 2022; Ahmadi-Mirghaed et al., 2020; Liu et al., 2023; Ou et al., 2019). This can be attributed to the fact that soil fertility and depth are significant factors in maintaining the ecological balance (Gavrilescu, 2021). The remaining indicators, such as habitat maps, bird diversity, flow path maps, and ecological protection control, were utilised based on the selected study area (Honeck et al., 2020; Liu et al., 2021; Yan & Tang, 2021).

7.3.2.2.2 Indicators Related to GI

Half of the reviewed articles that focused on optimising GI to enhance biodiversity and ecosystem security were implementing green space map indicators due to the large scope of the objective (Feng et al., 2022; Li et al., 2022a;

Liu et al., 2021, 2022; Yan & Tang, 2021). Additionally, green spaces are a valued part of the ecological system. Only one reviewed study specifically used tree type and vegetation density as an indicator related to GI to improve species diversity (Choi & Lee, 2023).

7.3.2.2.3 Geospatial Indicators

The top five most used geospatial indicators are land use and land cover maps, population density, digital elevation model, and temperature (Abdollahi et al., 2022; Abdollahi et al., 2022; Li et al., 2022b; Liu et al., 2022; Yan & Tang, 2021). Both land use and land cover maps were the widely utilised indicators with seven studies in consequence of their significance to reflect human activity and their effect on the ecological system, which improve GI optimisation processes (Abdollahi et al., 2022; Korosuo et al., 2014; Liu et al., 2022).

7.3.2.3 Energy Efficiency

In the context of energy efficiency, indicators were primarily concerned with measuring energy demand and the thermal performance of buildings, highlighting the role of GI in enhancing energy conservation.

7.3.2.3.1 Indicators Related to the Objective

Most reviewed articles have implemented energy demand and cooling and heating load, with almost 80% of the articles regarding energy efficiency (Calcerano & Martinelli, 2016; Mousavi et al., 2023; Murray et al., 2014). These two indicators express energy consumption to increase or decrease the temperature in buildings, which is the main objective of using GI to elevate thermal comfort to enhance energy efficiency.

7.3.2.3.2 Indicators Related to GI

The indicators related to tree type and tree volume, as well as the presence of green spaces between buildings, were frequently used by four and three studies, respectively (Calcerano & Martinelli, 2016; Hongbing et al., 2010; Kim et al., 2016; Rouhollahi et al., 2022). To improve energy efficiency, the optimisation process focused on trees and green spaces around buildings; consequently, these two indicators were most used. On the other hand, tree location, stomatal resistance, and leaf area index introduce more details about GI function that were assessed to achieve more accurate optimisation (Kim et al., 2016; Rouhollahi et al., 2022).

7.3.2.3.3 Geospatial Indicators

Building parameters often influence energy consumption and cooling and heating loads; for this reason, building parameter indicators was the most used indicator in terms of energy efficiency objectives (Calcerano & Martinelli, 2016; Hongbing et al., 2010; Rouhollahi et al., 2022). In addition, one study utilised the sunshine distribution map to study the areas that demand higher energy to reduce heat (Calcerano & Martinelli, 2016).

7.3.2.4 Public Health

Public health indicators in GI optimisation focused on elements like noise pollution, land surface temperature, and air quality, underscoring the direct impact of environmental conditions on human health.

7.3.2.4.1 Indicators Related to the Objective

Noise and high temperatures could affect people's physical and mental health. Noise pollution and land surface temperature indicators were frequently used by five studies individually (Feng et al., 2022; Neema & Ohgai, 2010, 2013; Trájer et al., 2022). In addition, air quality is an essential element of public health which was included in two indicators, including air pollution level and air temperature, by four and three studies, respectively (Feng et al., 2022; Neema & Ohgai, 2010, 2013). Climate health impact assessment is mandatory to achieve successful GI optimisation in terms of public health objectives. Three studies included climate health impact assessment indicators for the optimisation process (Huang et al., 2021; Trájer et al., 2022; Yan & Tang, 2021). The last indicator was water quality, with two studies as a primary element of public health where it dramatically affects people's health (Feng et al., 2022; Neema & Ohgai, 2013).

7.3.2.4.2 Indicators Related to GI

Optimising GI to improve public health is often applied on a large urban scale. Accordingly, five studies used urban green places to optimise GI regarding public health objectives (Feng et al., 2022; Neema & Ohgai, 2010; Trájer et al., 2022). Only one study was applied to the scale of trees using tree crown coverage, leaf area index, and vegetation density indicators to work as noise isolation on the roadside (Xu et al., 2022).

7.3.2.4.3 Geospatial Indicators

Two indicators were expressed in two different types of density, including population density and building density. These indicators are responsible for generating heat while being engaged in various activities that cause health effects (Huang et al., 2021). Also, the distance between the road and the trees was calculated by two studies to improve the noise isolation level (Abdollahi et al., 2022; Xu et al., 2022). Additionally, a land use map was used to identify areas with different levels of air quality and noise to optimise the location of GI (Neema & Ohgai, 2010).

7.3.2.5 Heat Islands

The indicators for assessing and mitigating urban heat islands primarily involve temperature metrics, solar radiation, and other factors influencing urban thermal dynamics.

7.3.2.5.1 Indicators Related to the Objective

The temperature factor, which is the main reason for urban heat, was utilised by eight studies in two different forms of indicators, including temperature of air and land (Jiang et al., 2022; Li et al., 2023; Yu et al., 2020; Wang et al., 2021; Zhang et al., 2017; Zhuang & Lu, 2021). The second most frequent indicator was solar radiation, which was employed by four studies to consider the shading influence while optimising the location of GI (Azcarate et al., 2021; Bajsanski et al., 2016; Calcerano & Martinelli, 2016; Zhou et al., 2023). A total of two studies attempted to utilise humidity since it effectively alleviates urban heat islands (Zhou et al., 2023; Zhuang & Lu, 2021). The last used indicators with two articles each were wind speed and cooling effect (Li et al., 2023; Zhou et al., 2023).

7.3.2.5.2 Indicators Related to GI

Three indicators were equally used in eight articles individually to optimise GI for reducing heat island. The first one was the green space map to implement the optimisation on a large urban scale (Feng et al., 2022; Wang et al., 2019; Yu et al., 2020). Meanwhile, in small-scale studies that optimise trees, two indicators were used equally: tree type and volume and tree coverage (Bajsanski et al., 2016; Hong et al., 2011; Zhao et al., 2017). In this scenario, the optimisation often focused on the scale of buildings.

7.3.2.5.3 Geospatial Indicators

In the context of a large urban scale, land use map was employed widely (n = 6) due to different activities' role in generating urban heat (Li et al., 2023; Yoon et al., 2019; Yu et al., 2020; Zhou et al., 2023; Zhuang & Lu, 2021). On the other hand, other studies (n = 5) used building dimension indicators (Calcerano & Martinelli, 2016; Hong et al., 2011; Hongbing et al., 2010; Li et al., 2023; Rouhollahi et al., 2022). The four indicators less considered to contribute were space between the tree and the building, building density, points of interest, and topographic map (Wang et al., 2019; Yu et al., 2020; Zhao et al., 2017).

7.3.2.6 Water Management

In water management, indicators such as rainfall data, soil type, and drainage systems are critical in evaluating and optimising the efficacy of GI in managing urban water resources.

7.3.2.6.1 Indicators Related to the Objective

Rainfall data indicator was the most used in GI optimisation with 11 studies (Lu et al., 2022; Liu et al., 2023; Tavakol-Davani et al., 2019; Torres et al., 2021; Zhu et al., 2023). Rainfall data is considered to contribute to evaluating the impact on urban areas and drainage systems. The second most used indicator was soil type (n = 8) due to its capacity to absorb and maintain water (Ahmadi-Mirghaed et al., 2020; Liu et al., 2016a; Martínez et al., 2018, 2021;

Wang et al., 2023). The drainage system was identified as one of the most used factors by six articles (Gao et al., 2022; Lu et al., 2022; Martínez et al., 2021; She et al., 2021; Wang et al., 2023; Zhu et al., 2023). Additionally, stream maps, flood hazard, water quality, precipitation maps, and catchment area indicators were utilised in various studies as a part of the water control system (Kumar et al., 2022; Liu et al., 2023; Zhou et al., 2022).

7.3.2.6.2 Indicators Related to GI

Optimisation of GI for improving water management is often done on a large scale; consequently, the only indicator utilised in relation to the GI was urban green space, as it functions as waters for rainfall and runoff. This indicator was applied by three articles utilising optimisation for water management objectives (Dong et al., 2021; Feng et al., 2022; Gao et al., 2022; Liu et al., 2016).

7.3.2.6.3 Other Indicators

Studies with the scope of water management usually illustrate the urban area in terms of green and grey infrastructure, which reflect land use and land cover maps (Martínez et al., 2021). For this reason, 13 studies implemented land use map indicators (Ahmadi-Mirghaed et al., 2020; Dong et al., 2021; Martínez et al., 2018, 2021; Wang et al., 2023; Zhou et al., 2022; Zhu et al., 2023), while four other studies used land cover maps for GI optimisation (Ahmadi-Mirghaed et al., 2020; Kumar et al., 2022; Liu et al., 2016; Wang et al., 2021). In addition, the digital elevation model and slope contribute significantly to the water movement direction and water pipe system orientation. Accordingly, these two indicators were used in four studies, respectively (Gao et al., 2022; Liu et al., 2023; She et al., 2021; Wang et al., 2023; Zhu et al., 2023).

7.3.3 Optimisation Models

This study gathered the optimisation methods from the reviewed articles and categorised them into five distinct classifications: (a) evolutionary algorithm (EA), (b) HA, (c) ML, (d) optimisation algorithm, and (e) other optimisation methods. Table 7.3 introduces a summary of frequently used models for GI optimisation.

7.3.3.1 Evolutionary Algorithms

EAs are a cluster of optimisation algorithms inspired by genetics and natural selection (Porta et al., 2013; Yoon et al., 2019). They are utilised to find a solution to complex problems when finding an optimal solution is difficult or expensive in terms of computation (Li & Parrott, 2016; Xiao et al., 2007). The fundamental idea behind EA is to mimic the natural evolution of the population and apply it as a potential solution (Xiao et al., 2002). Only one study used EA algorithm in its basic form to optimise tree location to reduce urban heat islands (Jia et al., 2019; Stojakovic et al., 2020). In addition, there are common types of

TABLE 7.3
Optimisation Algorithms and Models

Categories	Models	Objectives	Type of Infrastructure	Scale	Number of Studies
Evolutionary algorithms	Artificial immune system	Public health	Urban green space	City	1
	Annealing algorithm	Air quality, biodiversity, ecosystem security, public health, water management	Urban green space	City	3
	Genetic algorithm	Air quality, biodiversity, energy efficiency, heat islands, public health, water management	Urban green space, trees, residential green space	City, district, building	10
	Evolutionary algorithm	Heat islands	Trees	Street	1
	Non-dominated sorting genetic algorithm	Biodiversity, heat islands, water management, flood control, urban cooling, ecological impact	Urban green space, residential green space	City, district	12
	Particle swarm optimisation	Biodiversity, ecosystem security, energy efficiency, heat islands, water management	Natural green space	City	4
Heuristic algorithms	Heuristic algorithm	Biodiversity, ecosystem security	Urban green space	City	1
Machine learning	Artificial neural network	Air quality	Trees	Street	1
	Linear regression	Heat islands, air quality	Residential green space	Building	2
	Random forest	Air quality	Natural green space	City	2
	Support vector machine	Air quality	Residential green space	Building	1
Optimisation algorithms	Optimisation algorithm	Energy efficiency, heat islands, air quality, public health, biodiversity, water management, urban cooling	Urban green space, trees, residential green space	City, district, building, street	19
Other optimisation methods	Piecewise Linear model	Water management	Urban green space	City	1
	Taguchi method	Energy efficiency, heat islands	Residential green space	Building	1
	3D spatial optimisation model	Heat islands	Trees	Building	1

EAs, including GA, annealing algorithm (AA), PSO, artificial immune system (AIS), and non-dominated sorting GA (NSGA), and many more. Each algorithm has its strengths and weaknesses and is adapted to problems and solutions.

7.3.3.1.1 Genetic Algorithm

GA stands out as one of the most used algorithms that depend on Darwin's idea of natural selection (Konak et al., 2006). It was developed to select the optimal or near-optimal solution to problems based on the genetic selection concept when another traditional method may be inefficient (Marseguerra et al., 2006; Neema & Ohgai, 2010). GA is more effective for large-scale areas, non-linearity, and multi-objective; this prowess renders GA a valuable method in optimisation and AI fields (Yamachi et al., 2006). Many studies (n = 10, 20%) have used GA to optimise GI for various objectives related to climate change. These objectives include enhancing air quality (Jia et al., 2019), conserving biodiversity and ecosystem security (Feng et al., 2022), improving energy efficiency (Murray et al., 2014), public health (Neema & Ohgai, 2010), reducing heat islands (Zhuang & Lu, 2021), and water management improvement (Gao et al., 2022).

7.3.3.1.2 Non-Dominated Sorting Genetic Algorithm

The NSGA is a multi-objective algorithm that works under the GA principles (Liu et al., 2016). In the case where finding individual solutions is not sufficient, NSGA handles optimisation with multiple, incompatible objectives (Deb et al., 2002); it requires multi-objective optimisation to deal with a set of competing objectives concurrently (Verhagen et al., 2018). The solution is to find a trade-off between these goals, which is known as the Pareto set (Martínez et al., 2018). NSGA has been applied widely in GI optimisation (n = 12) within the scope of biodiversity and ecosystem security (Verhagen et al., 2018), reducing heat islands (Yoon et al., 2019), and water management (Kumar et al., 2022).

7.3.3.1.3 Annealing Algorithm

The AA is one of the optimisation algorithms inspired by the annealing process used in metallurgy (Kirkpatrick et al., 1983). It is employed for addressing intricate optimisation problems to find the optimal situation with challenging search fields (Cunha et al., 2016). Moreover, it is used initially to reduce defects and optimise the quality of materials (Oxley & Mays, 2014). In terms of optimisation, it is utilised as a powerful technique that counterpoises exploration and exploitation, which makes it efficient for finding optimal solutions to complicated problems (Andrade et al., 2013). Three studies have been implemented to optimise GI in the fields of air quality, biodiversity and ecosystem security (Korosuo et al., 2014), public health (Abdollahi et al., 2022), and water management (She et al., 2021).

7.3.3.1.4 Particle Swarm Optimisation

Social behaviours observed in fish or bird flock particles served as inspiration for the widely used algorithm known as PSO (Boeringer & Werner, 2004). The PSO algorithm works through the interactions between the movement of particles in multidimensional space (Liu et al., 2013). The position of each particle corresponds to the parameter of the possible solution of the optimisation (Boeringer & Werner, 2004; Masoomi et al., 2013). PSO is simple and easy to implement and can deal with high-dimensional problems, which makes it a preferable choice for the optimisation process (Ahmadi-Mirghaed et al., 2020). Three articles have utilised PSO in the optimisation of GI regarding biodiversity and ecosystem security (Kim et al., 2016), energy efficiency, reducing heat islands (Kim et al., 2016), and water management (Ahmadi-Mirghaed et al., 2020).

7.3.3.1.5 Artificial Immune System

Artificial immune system (AIS) is an algorithm inspired by the human immune system (Coello & Cortés, 2005). It is employed to identify optimal solutions for complex problems across various domains, such as pattern recognition, optimisation, and AI (Huang et al., 2013). AIS simulates the functioning of the human immune system to develop models to do specific tasks. Also, AIS offers distinctive qualities for solving complex optimisation problems. Nevertheless, like other optimisation models, the operation's success depends on selecting suitable designs and parameters (Coello & Cortés, 2005; Huang et al., 2021). AIS contributed to one article to optimise GI to improve public health (Huang et al., 2021).

7.3.3.2 Heuristic Algorithm

The HA is a method to solve complex problems using approximate strategies or rules of thumb to find quick solutions (Vallejo et al., 2015). Unlike detailed algorithms that present optimal solutions using intensive computation, HAs sacrifice the optimal for the efficient (Pukkala & Kurttila, 2005). HA is widely used for large-scale problems, especially when the optimal solution is challenging to compute within a reasonable time (Aerts & Heuvelink, 2002). This algorithm was applied to the objective of biodiversity and ecosystem security by GI optimisation in one article of the reviewed studies due to the importance of finding efficient optimisation for an ecosystem rather than an optimal one (Li et al., 2022a).

7.3.3.3 Machine Learning

ML models are commonly used to solve optimisation problems. The integration between ML and optimisation enables better exploration to find optimal solutions. Overall, ML for optimisation offers powerful tools to solve complex issues such as GI optimisation.

7.3.3.3.1 Linear Regression

Linear regression (LR) serves as an essential machine-learning technique in the realm of optimisation. It is widely used to reduce the gap between target values and predicted values (Maulud & Abdulazeez, 2020). Moreover, it is an interpretable optimisation model; however, for complex problems or variables with non-linear relationships, other ML models may be more suitable (Li et al., 2023). Nonetheless, the LR model remains a valuable technique mainly when relationship interpretability is essential. Two reviewed articles implemented LR to optimise GI, seeking to mitigate urban heat and enhance air quality (Hashad et al., 2021; Li et al., 2023).

7.3.3.3.2 Support Vector Machine

A widely recognised supervised model frequently employed for classification is the SVM (Drucker et al., 1997). However, it is also utilised for optimisation problems, especially in terms of constrained optimisation. SVM algorithm works to either maximise the margin or minimise the error (Bohra & Palivela, 2015; Hashad et al., 2021). An article used SVM to optimise GI to help to improve air quality. Nevertheless, SVM may not be the suitable model for optimisation goals practically with large data classes (Hashad et al., 2021).

7.3.3.3.3 Random Forest

RF is a widely recognised technique primarily utilised for prediction, yet it can also be used for optimisation processes offering valuable insights into features (Breiman, 2001). Implementing RF for optimisation tasks relies on the type of the problem, data characteristics, and objectives. Two studies optimised GI using FR in terms of the air quality objective. However, other algorithms might be more suitable like GA (Hashad et al., 2021; Liu et al., 2023).

7.3.3.4 Other Optimisation Models and Algorithms

There were other algorithms and techniques used to optimise GI for different objectives. The optimisation algorithm was used most frequently in the reviewed articles (n = 19) (Trájer et al., 2022; Xu et al., 2022; Yan & Tang, 2021). It indicates either the model was defined as an optimisation algorithm or the studies modified one of the basic algorithms for specific optimisation tasks. In addition, there were other techniques utilised in only one study to optimise GI, such as the Piecewise Linear model (Torres et al., 2021), Taguchi Method (Mousavi et al., 2023; Zhao et al., 2017), and 3D spatial optimisation model.

7.3.4 Green Infrastructure Types

A network of natural or semi-natural spaces, features, and systems is referred to as GI, and it is intended to offer environmental, economic, and social advantages within urban or rural settings (Vaeztavakoli et al., 2018). GI is a versatile approach that addresses difficulties and improves overall life

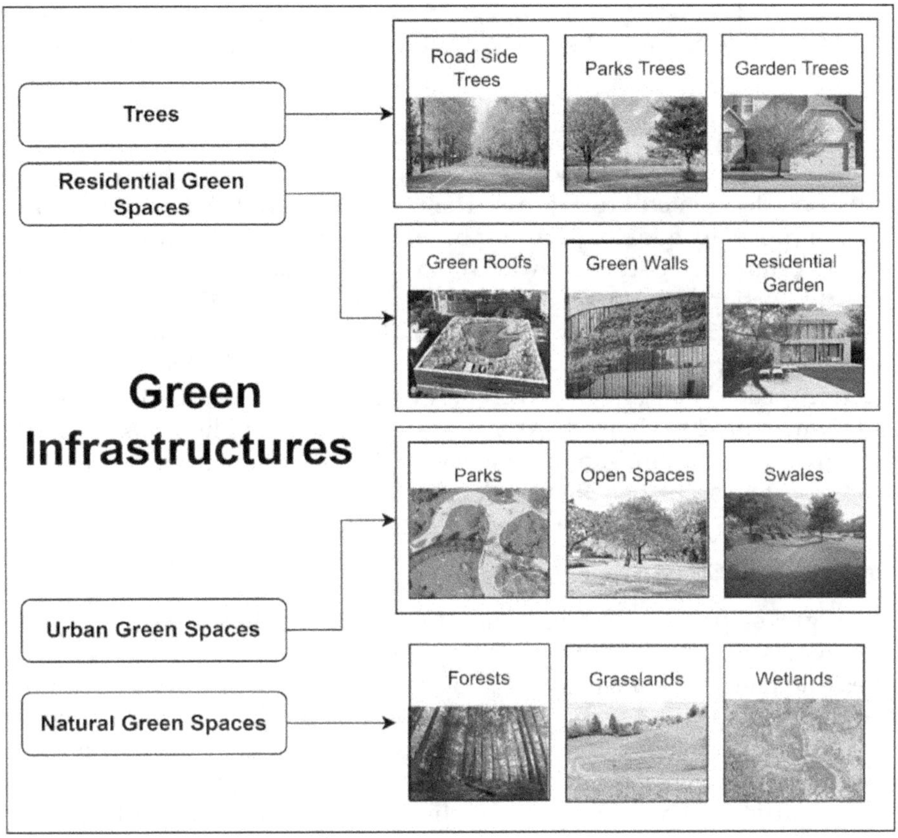

FIGURE 7.4
Green infrastructure types.

quality. This review grouped GI into four categories: (a) trees, (b) residential green spaces, (c) urban green areas, and (d) natural green spaces (Figure 7.4).

7.3.4.1 Trees

It indicates that the studies optimise the design or location of a tree, whether an individual or cluster, regardless of the application area, whether it is residential, a park, or roadside, if it utilises a tree for the optimisation tasks. In total, ten reviewed studies selected trees for the optimisation of different objectives. Most of them (n = 7) optimise trees to reduce urban heat (Bajsanski et al., 2016; Hong et al., 2011; Hongbing et al., 2010; Rouhollahi et al., 2022; Zhao et al., 2017). Two other studies optimised tree location and type to improve energy efficiency and air quality, respectively (Hashad et al., 2021; Hongbing et al., 2010; Rouhollahi et al., 2022). Remarkably, one study used the optimisation of trees to improve public health (Choi & Lee, 2023). Consequently, the scale/size of a tree for optimisation was implemented widely to reduce heat islands for its essential role of offering shade and reducing surrounding temperature.

7.3.4.2 Residential Green Spaces

Residential green spaces encompass elements such as green roofs, green walls, and designated green areas. They are designed to enhance residents' lives, improve air quality, protect biodiversity, and improve overall quality of life. A number of studies (n = 15) have used residential green spaces in the optimisation tasks (Kumar et al., 2022; Tavakol-Davani et al., 2019; Wang et al., 2022, 2023). Improving thermal comfort leads to minimising energy consumption; for this reason, five studies optimised residential green spaces to reduce heat islands and improve energy efficiency (Calcerano & Martinelli, 2016; Kim et al., 2016; Li et al., 2023; Mousavi et al., 2023; Zhuang & Lu, 2021). In a single study, optimisation of GI was applied to enhance the quality of air and ecosystem security (Murray et al., 2014).

7.3.4.3 Urban Green Spaces

Urban green spaces considered within this study included parks, gardens, open spaces, and swales within urban environments. Urban green spaces are an essential element of cities providing numerous benefits and contributing to improving various aspects of the city's well-being for humans and biodiversity. Efficiently managing urban green spaces is crucial for cities' sustainability. Urban green spaces were the most used in the optimisation models, with 33 articles. The top three objectives concern GI optimisation, water management (n = 13) (Liu et al., 2016; Lu et al., 2022; Martínez et al., 2018; Zhu et al., 2023), biodiversity and ecosystem security (n = 10) (Abdollahi et al., 2022; Li et al., 2022a; Verhagen et al., 2018; Yan & Tang, 2021), and reducing urban heat (n = 8) (Jiang et al., 2022; Wang et al., 2019; Yu et al., 2020; Zhang et al., 2017; Zhou et al., 2023). To improve water management, optimisation studies often use large-scale green areas because of their essential role in the water cycle. In addition, it provides a habitat for various plants and animals contributing to system security. It also contains a large number of trees that help together with green surfaces (e.g., shrubs, groundcovers, and grasses) to reduce heat island effects. Because of the interactions between urban green areas and people, it helps to improve public health, and this objective was utilised by six studies. Finally, five studies optimised urban green spaces to improve air quality (Jia et al., 2019; Daniele & Sciacca, 2021).

7.3.4.4 Natural Green Spaces

Natural green spaces refer to spaces that have been barely developed by humans, and it includes forests, grasslands, and wetlands on a city scale. Recognising the pivotal role natural green spaces play in protecting the ecological balance and biodiversity, three studies have optimised natural green spaces to preserve biodiversity and ecosystem security (Ahmadi-Mirghaed et al., 2020; Kim et al., 2016; Korosuo et al., 2014). Additionally, fewer studies (n = 1) were utilised in terms of air quality and water management objectives using natural green spaces; for this reason, there is a need for more natural green space optimisation studies for these objectives (Liu et al., 2023).

7.3.5 Analysis Scales

Defining the scale of the study is an essential step that shapes the research process. This step ensures the study is focused, relevant, and effective in addressing challenges and opportunities within the given constraints (Eizenberg et al., 2019). In the realm of GI optimisation studies, the extent of the analysis area can be adaptable, contingent upon the study's specific objectives. Each scale provides distinct insights into how urban green spaces impact the environment and public health. Here is a breakdown of four types of scales this review developed to categorise GI optimisation studies: (a) building, (b) street, (c) district, and (d) city (Table 7.4).

TABLE 7.4

GI Optimisation Analysis Scales

Categories	Scale	Number of Studies
Air quality	Building	1
	Street	1
	District	1
	City	6
Biodiversity and ecosystem security	Building	1
	Street	0
	District	1
	City	12
Energy efficiency	Building	4
	Street	0
	District	0
	City	0
Public health	Building	0
	Street	0
	District	1
	City	5
Heat island	Building	4
	Street	2
	District	6
	City	4
Water management	Building	0
	Street	0
	District	10
	City	7

7.3.5.1 Building Scale

At the building scale, studies often focus on areas surrounding individual buildings. Ten reviewed articles utilise the building scale. The top two most frequent objectives of building scale were reducing urban heat islands (n = 8) (Hong et al., 2011; Hongbing et al., 2010; Li et al., 2023; Mousavi et al., 2023; Zhao et al., 2017) and improving energy efficiency (n = 6) (Calcerano & Martinelli, 2016; Kim et al., 2016; Mousavi et al., 2023; Rouhollahi et al., 2022). Few studies used this scale to achieve other objectives, such as improving air quality (n = 2) (Choi & Lee, 2023; Murray et al., 2014) and preserving biodiversity and ecosystem security (n = 1) (Choi & Lee, 2023). The results indicate that employing a building scale primarily seeks to maximise the influence of green spaces on the microclimate, leading to enhanced energy efficiency.

7.3.5.2 Street Scale

In the analysis of street scale, studies may be along the whole extent of street. Optimisation studies of GI may utilise roadside trees, parking trees, or other types of GI. Limited studies (n = 4) have applied GI optimisation on a street scale, where three of them were to reduce urban heat islands and the other one was to improve air quality (Azcarate et al., 2021; Hashad et al., 2021; Stojakovic et al., 2020). The primary goal of optimising GI on the scale of streets is to assess the contribution of green spaces in alleviating urban heat islands and improving air quality.

7.3.5.3 District Scale

At the scale of the district, studies focused on large neighbourhoods within a city. Due to the connectivity of the district's scale to humans and the environment, GI optimisation studies employment for various objectives, including air quality (n = 2) (Jia et al., 2019; She et al., 2021), biodiversity and ecosystem security (n = 1) (Honeck et al., 2020), reducing heat islands (n = 6) (Yoon et al., 2019; Zhang et al., 2017; Zhuang & Lu, 2021), public health (n = 1) (Xu et al., 2022), and water management (n = 11) (Gao et al., 2022; Liu et al., 2016; Martínez et al., 2018; Wang et al., 2022; Zhou et al., 2022). As a result of a more considerable interaction between the district and the environment, we can notice the concentration on water management issues. Nevertheless, the other objectives need to be more considered in GI at the scale of the district.

7.3.5.4 City Scale

In the context of a city, studies adopt a comprehensive view of GI optimisation. The optimisation involved parks, forests, swales, and grasslands.

Consequently, we can observe the large number of studies (n = 12) that optimise GI related to biodiversity and ecosystem security objectives (Kim et al., 2016; Korosuo et al., 2014; Liu et al., 2021; Liu et al., 2022; Yan & Tang, 2021). Moreover, regarding the ability of GI in the city to offer additional services such as flooding mitigation, eight studies optimised GI to improve water management (Kumar et al., 2022; Martínez et al., 2021; Torres et al., 2021; Zhu et al., 2023). Finally, five studies each were utilised in terms of air quality (Daniele & Sciacca, 2021; Feng et al., 2022; Y. Liu et al., 2023; Tao et al., 2023), heat islands (Wang et al., 2019; Yu et al., 2020), and public health (Huang et al., 2021; Trájer et al., 2022).

7.3.6 Optimisation Objectives

This review categorised GI optimisation objectives into six climate change mitigation objectives involving the following: (a) air quality; (b) biodiversity and ecosystem security; (c) energy efficiency; (d) public health; (e) heat islands; (f) water management. Because of connections between these objectives, they were utilised individually (Liu et al., 2023; Zhou et al., 2023) or combined with two objectives (Abdollahi et al., 2022; Rouhollahi et al., 2022), four objectives (Neema & Ohgai, 2013), or five objectives (Feng et al., 2022). Although certain objective areas like heat island and water management were frequently addressed, there appears to be a gap in addressing the other objectives comprehensively (Figure 7.5).

7.3.6.1 Air Quality

GI provides important benefits in terms of improving air quality and mitigating air pollution impacts. In this review, air quality objective indicates studies that optimised GI to enhance the quality of air aspects, including minimising carbon emissions, maximising the number of carbon sinks, or minimising air temperature. This objective was applied by 11 reviewed articles (Daniele & Sciacca, 2021; Jia et al., 2019; Liu et al., 2023; Neema & Ohgai, 2010; Tao et al., 2023) that conclude the proper optimisation for GI is essential to maximise the benefits of GI to improve air quality.

7.3.6.2 Biodiversity and Ecosystem Security

To support biodiversity and provide a secure ecosystem, GI optimisation is essential where places such as parks, wetlands, and forests offer habitats for various animal species. Moreover, it provides crucial services for the ecosystem in terms of maintaining ecosystem security and human well-being. Many studies (n = 14) optimised different types of GI to preserve biodiversity and ecosystem security via various aspects such as maximising habitat quality, species diversity, and green space protection (Korosuo et al., 2014; Li et al., 2022a; Verhagen et al., 2018; Yan & Tang, 2021).

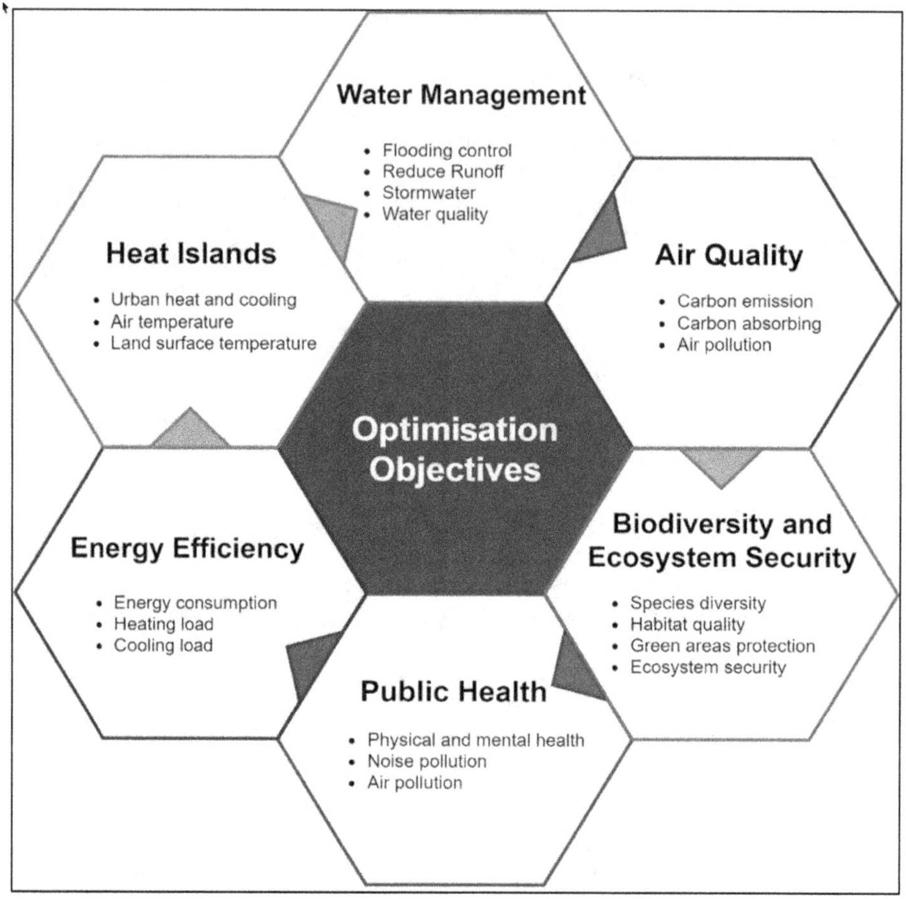

FIGURE 7.5
GI optimisation objectives categories.

7.3.6.3 Energy Efficiency

GI optimisation is important in reducing energy consumption in different energy demand sections. As a case in point, GI provides natural cooling, leading to a lower energy need for cooling and heating. Several studies (n = 6) optimised GI to improve energy efficiency by minimising energy demand for cooling and heating (Calcerano & Martinelli, 2016; Hongbing et al., 2010; Mousavi et al., 2023; Murray et al., 2014; Rouhollahi et al., 2022). Accordingly, GI optimisation with energy efficiency in mind can present positive impacts and create sustainable cities.

7.3.6.4 Public Health

GI optimisation that emphasises promoting public health and overall well-being became essential. GI helps to reduce stress, increase activity, and improve mental health. Furthermore, by enhancing air, water, and shaded GI safeguards during heatwaves and flooding. Seven reviewed studies used GI optimisation to improve different features of public health, such as minimising noise and air pollution and increasing body and mental health (Choi & Lee, 2023; Huang et al., 2021; Neema & Ohgai, 2010; Trájer et al., 2022; Xu et al., 2022).

7.3.6.5 Heat Islands

GI optimisation is a powerful tool that helps to mitigate urban heat islands. For example, trees provide shade, prevent solar radiation, and reduce temperature. Additionally, green spaces allow cool air to flow and enhance air humidity. Because of the interaction between heat and other climate change elements like air quality and energy efficiency, reducing heat islands was the most frequent objective employed by GI optimisation studies with 21 articles (Wang et al., 2019; Yoon et al., 2019; Yu et al., 2020; Zhang et al., 2017; Zhuang & Lu, 2021; Jiang et al., 2022).

7.3.6.6 Water Management

Water management was the second most selected objective by GI optimisation by 19 articles (Gao et al., 2022; Liu et al., 2016; Lu et al., 2022; Martínez et al., 2018; Zhu et al., 2023). Water management objectives indicate various sub-objectives, including water flooding, runoff, stormwater, and water quality. GI optimisation plays an essential role in effective water management. Moreover, GI absorbs rainwater, which reduces runoff, improves water quality, and minimises flood risk. Finally, implementing GI optimisation to achieve the goal of improving water management ensures water resilience against climate change.

7.3.7 An AI-Driven GI Optimisation Framework

In this section, this study presents a framework to enhance the understanding of the optimisation process within the realm of AI-driven GI (Figure 7.6). This framework or process flowchart facilitates understanding the steps and inputs contributing to GI optimisation. Moreover, it empowers researchers to conduct and evaluate their contributions or novel aspects of their work.

The proposed framework organises the AI-driven GI optimisation process seamlessly into seven steps:

- *Identify Environmental Issue:* This initial phase involves identifying a problem related to GI and climate change.

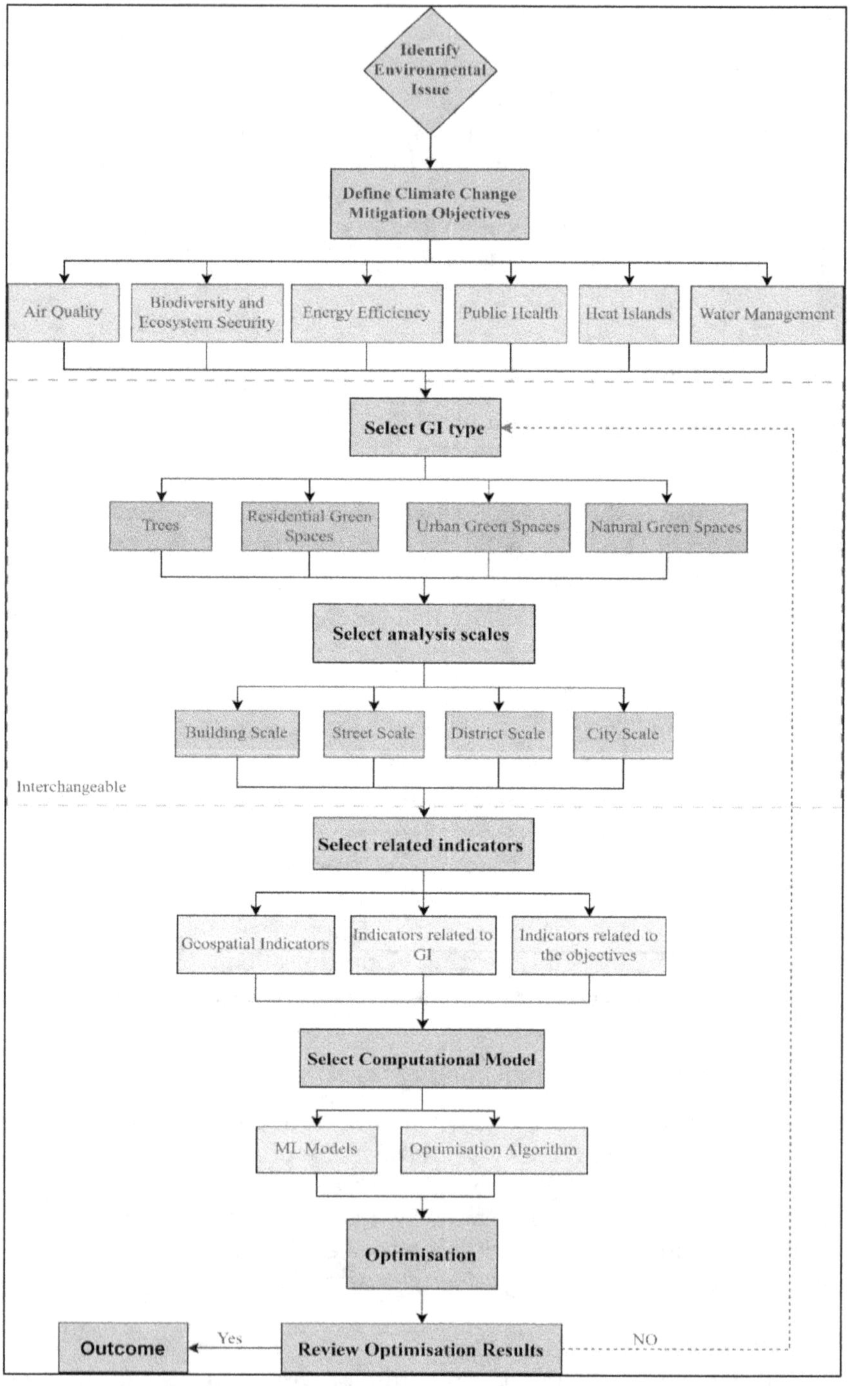

FIGURE 7.6
Framework of AI-driven GI optimisation for tackling climate change.

- *Objective Definition:* The subsequent step entails defining one or more objectives within the domain of climate change mitigation. These objectives are categorised into six primary domains: Air quality, biodiversity and ecosystem security, energy efficiency, public health, urban heat island, and water management.
- *GI Type Selection:* The type of GI is selected in alignment with the chosen objective. In this context, we categorise GI types into four distinct categories: Trees, residential green spaces, urban green spaces, and national green spaces.
- *Scale Analysis Selection:* The appropriate scale for a study is determined based on the type of objectives and the domain coverage. Scale analysis selection and the GI type analysis are reconcilable and interchangeable depending on the analysis circumstance.
- *Indicator Selection:* This process encompasses the formulation of indicators across three different levels defined by this review. These tiers include indicators interrelated with and influencing the chosen objectives, indicators related to GI characteristics, and other general/supporting indicators associated with the built environment.
- *Algorithm or Model Selection:* The study progresses to choose the most suitable ML model or optimisation algorithm responsible for driving the GI optimisation process.
- *Optimisation:* Finally, the selected model or algorithm is employed to execute the optimisation, utilising all chosen variables in the process.

By following the systematic process (Figure 7.6), researchers are equipped to methodically navigate the intricacy of GI optimisation regarding climate change.

7.4 Findings and Discussion

This study aimed to address two important questions: What aspects of GI need to be considered for tackling climate change? How do we implement the process steps of GI optimisation regarding climate change mitigation? The review answered these questions first by analysing different aspects of the optimisation process, starting with optimisation objectives intended to address climate change. Subsequently, other aspects included data and indicators, models and algorithms, GI types, and the scale of the optimisation area through a systematic review of the scholarly literature. Lastly, the study developed a framework for the process of GI optimisation. This framework serves to enhance understanding and facilitate practical implementation of the optimisation approach. This section discusses primary findings and other issues.

In terms of the selection of optimisation indicators, it is evident that their selection is primarily influenced by the optimisation objectives. Nevertheless, in the absence of standardised criteria for indicator selection, researchers have chosen these indicators to suit the local context. The findings indicate that the indicators typically fall into three categories: those related to the GI itself, those aligning with the optimisation objectives, and those pertinent to the specific place. Across diverse objectives, certain common indicators have emerged, suggesting a potential foundation for establishing standardised criteria for GI optimisation.

For instance, indicators such as carbon emissions and air pollution levels have been identified as critical elements for addressing air quality objectives (Daniele & Sciacca, 2021). Similarly, from the perspective of biodiversity and ecological security, soil type and characteristics have been the leading indicators, underscoring their importance in fostering healthier habitats (Abdollahi et al., 2022; Liu et al., 2023). Furthermore, land use patterns have been shown to significantly influence biodiversity preservation, thus warranting their inclusion in the optimisation process (Feng et al., 2022).

Regarding the objective of energy efficiency, reducing cooling and heating loads is the primary goal of GI optimisation (Mousavi et al., 2023). Consequently, indicators related to reducing heat, such as tree shading, solar radiation, and cooling energy demand, are important to include to improve energy efficiency (Calcerano & Martinelli, 2016; Murray et al., 2014; Rouhollahi et al., 2022). Regarding the objective of public health, four indicators were cornerstones for GI optimisation, including noise level, air and surface temperature, and air quality (Feng et al., 2022; Huang et al., 2021; Trájer et al., 2022). These indicators represent urban microclimate and exert a considerable impact on human well-being.

While the essential role of GI in enhancing physical and mental health is recognised, its integration in studies addressing climate change has been notably limited. This oversight can be attributed to the traditionally perceived minimal direct correlation between GI and climate change impacts. However, this perspective overlooks the significant social functions of GI, particularly in relation to the vulnerability of different population groups. Despite the essential role of GI in improving physical and mental health, it was absent in most studies; this can be attributed to minimal direct correlation with climate change (Huang et al., 2021; Neema & Ohgai, 2010; Trájer et al., 2022).

Due to their frequent use in GI optimisation tasks, the only semi-structured objectives regarding indicators are limited to heat islands and water management objectives. In the scope of heat islands, air temperature and surface temperature are the leading indicators that can be critical elements of the GI optimisation framework (Jiang et al., 2022; Zhang et al., 2017; Zhou et al., 2023). Therefore, they were present in most reviewed articles. On the other hand, rainfall data, drainage system, and soil type indicators reflect the water life cycle, which makes them fundamental indicators to be used for optimising GI in case of water management while the imperviousness of surfaces was overlooked in reviewed studies although critical to be considered water management (Kumar et al., 2022; Martínez et al., 2018; Torres et al., 2021; Wang et al., 2023; Zhu et al., 2023).

Focusing on the models used in GI optimisation, it is observed that a diverse array of optimisation algorithms and ML models have been employed, albeit with a scope for more advanced developments. In the realm of models, results indicate implementing various types of optimisation algorithms and ML models, yet more advanced models are required to enhance the optimisation outcomes. According to the reviewed studies, there were two main types of models: optimisation algorithms and ML models. Optimisation algorithms, including EAs and HAs, are powerful tools for searching and optimisation. These algorithms can tackle complex problems and discern suitable solutions. Nevertheless, reviewed articles are limited to a few old algorithms that need further improvements (Ahmadi-Mirghaed et al., 2020; Abdollahi et al., 2022; Li et al., 2022a; Neema & Ohgai, 2013). Additionally, indicators are crucial in directing the evolutionary process in optimisation algorithms (Dong et al., 2022). These indicators, encompassing a range of metrics or objectives, are the targets of optimisation by the algorithm (Audet et al., 2021).

Sometimes, indicators are not independent but form a chain or hierarchy in which one indicator directly influences another. This is particularly relevant in complex systems such as ecosystems and large-scale studies. Understanding these chains can help prioritise objectives and make more informed decisions about which indicators to optimise (Huang & Wang, 2021; Zhang et al., 2023). This aspect has gained significant importance in intricate systems, such as ecosystems, and in extensive research (Dong et al., 2022). A thorough understanding of these indicator chains is essential for establishing priorities among objectives and making informed decisions regarding the optimisation of specific indicators.

Notably, they are time-intensive due to the requirement of repeated calls for spatial analysis functions. Incorporating more advanced and evolving algorithms will undoubtedly enhance the optimisation of GI. On the other hand, ML models were utilised infrequently, and different models were implemented, for instance, ANNs, RFs, and SVMs. ML models have predominantly been predictive, yet they exhibit remarkable performance (Hashad et al., 2021; Liu et al., 2023). Hence, it is imperative to use more advanced ML models with dedicated optimisation capabilities to ensure the attainment of more accurate results.

The study findings also shed light on the dependency of GI type and scale on the specified objectives. For instance, studies focusing on water management objectives predominantly utilised city or district scales (Liu et al., 2016; Lu et al., 2022; Zhou et al., 2023), whereas energy efficiency studies were conducted exclusively at the building scale (Hongbing et al., 2010; Mousavi et al., 2023; Rouhollahi et al., 2022). The observed variation in the scale of studies pertaining to GI optimisation, based on different objectives such as water management or energy efficiency, underscores the necessity for a more comprehensive and unified framework. This framework should be designed to define the indicators clearly and systematically to select appropriate scales for GI optimisation.

This framework would not only streamline decision-making processes but also guarantee that the selected scale is optimally aligned with the distinct objectives of each GI project. Adopting this approach would enable more effective and focused execution of GI solutions, thereby amplifying their overall effectiveness in combating climate change challenges. Moreover, future studies should extend their focus to include various urban settings. This expansion is crucial for a more comprehensive understanding of GI implementation in diverse urban contexts.

Turning to the objectives of GI in the context of climate change, it was observed that no structured approach has addressed the objectives of GI regarding climate change. This study categorises GI optimisation objectives into six groups: air quality, biodiversity and ecological security, energy efficiency, public health, heat islands, and water management. Generally, missing standardised objectives led to a deficiency in the number of studies that focus on GI optimisation; however, most of the studies tend to focus on specific objectives (Zhu et al., 2023).

Concerning the objectives of heat islands and water management objectives, they were adequately addressed in more than half of the reviewed articles (Dong et al., 2021; Hong et al., 2011; Li et al., 2023; Tavakol-Davani et al., 2019); this advantage imparts a greater level of structure to these objectives compared to other objectives. Therefore, this study highlights the necessity for more comprehensive research on GI optimisation, given its critical role in climate change mitigation. Addressing this gap could involve exploring multi-objective models that maximise environmental benefits while addressing climate change.

From a planning and policy perspective, particularly in the European context, the study recognises the necessity of prioritising not only the systemic interconnection of GI elements but also their accessibility and maintenance of high environmental quality. This approach is pivotal for fostering urban resilience and enhancing climate change mitigation strategies. The Union's Green Deal serves as a relevant framework in this discussion, setting ambitious sustainability and climate neutrality targets that can be significantly supported through GI optimisation processes (Fetting, 2020).

Embedding GI optimisation within a coherent policy framework is essential to translate these objectives into actionable strategies (Breuer et al., 2019). This includes strategic long-term visions for sustainable development and operational instruments that guide the practical implementation of GI projects. The intersection of GI and urban planning emerges as a crucial point where policy directives and sustainability targets converge. For instance, integrating GI into urban landscapes can yield more effective results when strategically aligned with urban mobility plans, zoning regulations, and biodiversity conservation efforts (Voghera & Giudice, 2019).

The policy instruments that govern GI deployment, including funding mechanisms, governance structures, and regulatory requirements, are equally critical in implementing optimisation processes (Zhou et al., 2022). These not only provide the means for implementation but also establish the standards and benchmarks against which the success of GI initiatives can be measured. In the context of GI optimisation, policy instruments must be attuned to the overarching goals of enhancing biodiversity, reducing carbon emissions, and fostering sustainable urban ecosystems (Wakdok & Bleischwitz, 2021).

Thus, the optimisation process becomes a fulcrum for policy, planning, and practice, wherein the theoretical models of optimisation are intricately woven with the fabric of policy frameworks and implementation mechanisms. This synergetic approach ensures that GI optimisation is not an isolated academic exercise but a tangible component of the transition towards sustainable urban futures.

In sum, diverse studies have made significant strides in advancing GI optimisation frameworks (Abdollahi et al., 2022; Liu et al., 2023; Wang et al., 2023). Although useful, these contributions exhibit a notable level of localisation in the frameworks, which poses difficulties in their implementation in diverse situations (Liu et al., 2023). Moreover, many studies have focused on individual objectives within the broader spectrum of GI optimisation, limiting their scope to addressing the multifaceted aspects of climate change mitigation (Ahmadi-Mirghaed et al., 2020; Huang et al., 2021). Recognising these limitations, this study proposes a comprehensive framework. This framework not only standardises the criteria for GI optimisation but also provides a structured approach to manage its diverse components, making it particularly relevant for tackling the complex challenges of climate change.

In terms of incorporating policy considerations, it is evident that the effective implementation of GI optimisation relies heavily on supportive policy frameworks. Such frameworks should align with broader climate change adaptation and mitigation goals as outlined in international agreements and national policies. For instance, policies that incentivise sustainable urban planning and development can play a pivotal role in facilitating GI optimisation. Additionally, regulations that promote the integration of environmental considerations into urban and regional planning can help standardise GI optimisation criteria, ensuring their alignment with Sustainable Development Goals.

Nonetheless, this study acknowledged several limitations within its scope. A primary constraint is the limited number of studies that focus on GI optimisation in the context of climate change. This scarcity hinders the development of a comprehensive GI optimisation framework. Despite significant efforts to illustrate the GI optimisation process, there remains a need for more in-depth exploration of certain aspects, such as indicator selection and data suitability. Future research should aim to establish more standardised criteria for selecting effective and relevant datasets. These criteria should be in line with the objectives of climate change policies, further integrating policy considerations into the optimisation process.

Given the unique characteristics of each study area, there is also a need to differentiate between global- and local-level indicators, considering both generic and context-specific aspects. Future studies should carefully examine these factors. In doing so, they can contribute to a more nuanced understanding of how GI optimisation can be tailored to different environmental and policy contexts. Our team remains committed to furthering research in this field, with a particular emphasis on how policy frameworks can support and enhance GI optimisation efforts for climate change mitigation and adaptation.

7.5 Conclusion

As climate change persists with severe impacts on the ecosystem, it leads to increasing challenges for GI in addressing and mitigating its consequences. GI has emerged as a tool to increase sustainability and vitality—particularly the AI-driven approaches are found highly promising (Zhang & Chui, 2019). In addition, it plays an essential role in facing climate-related effects such as water flooding, carbon emission, heat islands, and loss of habitat. GI optimisation is important, as it enhances its efficiency and ensures it effectively contributes to climate change adaptation. Nevertheless, there needs to be more effectiveness of GI. There has been a need for a deep understanding of GI optimisation potential and limitations. Our study sought to address this gap by utilising a systematic literature review following the PRISMA protocol. The findings were analysed and discussed from six aspects, including optimisation objectives, objective categories, indicators, ML models and algorithms, GI types, and scale.

The study findings of the study highlighted that indicators can be categorised into three main groups: related to the GI, related to the optimisation objectives, and related to the location. In addition, the selection of optimisation indicators mainly relies on the chosen objective. However, the absence of standardised criteria for indicator selection has led researchers to adopt indicators based on local contexts. In the context of models, results indicate implementing various types of optimisation algorithms and ML models, yet more advanced models are required to enhance the optimisation outcomes. Additionally, there was a lack of covering climate change's different subjects, such as public health, energy efficiency, and air quality, which created a need for more studies in the field of GI optimisation.

AI-driven GI optimisation could aid efforts in climate change adaptation and mitigation in cities. This is to say, AI-driven GI optimisation represents a visionary approach to interacting between humans, the environment, and the climate. In the age of AI, advanced technology solutions offer invaluable opportunities to undertake GI optimisation in a much more timely and accurate manner. Prospective studies in this field, thus, are required to maximise the essential benefits of AI-driven GI optimisation in particularly tackling climate change—the greatest challenge to humanity.

In conclusion, our study emphasises that optimising GI transcends technical refinement. It demands thoughtful integration into the urban fabric, ensuring that GI not only contributes to environmental quality but also enhances social equity and accessibility. The true essence of GI's potential lies in its seamless integration with the urban ecosystem. As cities strive to become more resilient, the interconnectedness of GI with urban planning, policy frameworks, and community engagement becomes paramount. Future research and policy efforts should focus on harnessing GI's full potential in diverse urban contexts, ensuring that it serves as a cornerstone in our collective response to the complexities of climate change.

Acknowledgements

This chapter, with permission from the copyright holder, is a reproduced version of the following journal article: Shaamala, A., Yigitcanlar, T., Nili, A., & Nyandega, D., (2024). Strategic tree placement for urban cooling: a novel optimisation approach for desired microclimate outcomes. Urban Climate, 56(1), 102084.

Appendix

TABLE 7.A1

Salient Characteristics of Reviewed Articles

Author	Aim	Model	Objectives	Type of Infrastructure	Category	Scale
Li et al. (2022a)	• Find the optimal urban green space layout	HA	• Minimise the conversion of new urban green spaces	Urban green space	• Biodiversity and ecosystem security	City
Li et al. (2023)	• Optimise residential, public green spaces	LR	• Minimise urban heatwaves	Residential green space	• Heat island	Building
Hongbing et al. (2010)	• Optimise residential trees	OA	• Maximise tree shading provided	Trees	• Energy efficiency • Heat islands	Building
Hong et al. (2011)	• Optimise tree plant location	OA	• Maximise tree height	Trees	• Heat islands	Building
Bajsanski et al. (2016)	• Find the optimal location for trees in the parking	OA	• Maximum overshadowing of parking	Trees	• Heat islands	Street
Liu et al. (2023)	• Discover the optimal solution for determining the appropriate quantity of green spaces	RF	• Maximise the number of carbon sinks that meet land use constraints	Natural green space	• Air quality	City
Li et al. (2022b)	• Explore the optimal scale of landscape fragmentation	OA	• Maximise green area size	Urban green space	• Biodiversity and ecosystem security	City
Ahmadi-Mirghaed et al. (2020)	• Implement ecosystem services for land allocation	SWARM	• Maximise land suitability • Maximise water yield • Maximise habitat quality • Maximise aesthetic quality	Natural green space	• Biodiversity and ecosystem security • Water management	City

Zhou et al. (2023)	• Optimisation of green spaces planning	GA	• Minimise urban heat island	Urban green space	• Heat islands	District
Mousavi et al. (2023)	• Select optimal design of green roofs in buildings	Taguchi Method	• Minimise energy consumption • Minimise cooling load • Minimise heating load • Maximise comfort level	Residential green space	• Energy efficiency • Heat islands	Building
Rouhollahi et al. (2022)	• Optimise residential trees	OA	• Minimise energy consumption	Trees	• Energy efficiency • Heat islands	Building
Calcerano and Martinelli (2016)	• Select the optimal placement of trees around the buildings' first and second floors	GA	• Decrease in energy usage for summer cooling	Residential green space	• Energy efficiency • Heat islands	Building
Neema and Ohgai (2013)	• Select the optimal site for multitype green spaces	GA	• Maximise air quality • Minimise noise level • Maximise urban cooling • Maximise water quality	Urban green space	• Air quality • Public health • Heat islands • Water management	City
Abdollahi et al. (2022)	• Select the ecosystem service zones	AA	• Minimise noise pollution • Maximise recreational value. • Maximise aesthetic value	Urban green space	• Biodiversity and ecosystem security • Public health	City
Verhagen et al. (2018)	• Undertake green area optimisation	NSGA	• Maximise fruit yield • Maximise endangered species' habitat • Minimise losses in dairy farming	Urban green space	• Biodiversity and ecosystem security	City
Tavakol-Davani et al. (2019)	• Undertake GI location optimisation	NSGA	• Minimise runoff value • Minimise life cycle cost	Residential green space	• Water management	District

(Continued)

TABLE 7.A1 *(Continued)*
Salient Characteristics of Reviewed Articles

Author	Aim	Model	Objectives	Type of Infrastructure	Category	Scale
Korosuo et al. (2014)	• Optimise forest planning to assess the impact of various landscape fragmentation indices	AA	• Maximise Increase the total count of patches • Maximise the average size of each patch • Minimise the mean nearest neighbour distance • Maximise shape index • Minimise NPV of forestry • Minimise total green area	Natural green space	• Biodiversity and ecosystem security	City
Xu et al. (2022)	• Optimise urban green space on roadside	OA	• Minimise traffic noise	Trees	• Public health	District
Liu et al. (2022)	• Constructing regional GI	OA	• Minimise green area protection	Urban green space	• Biodiversity and ecosystem security	City
Martínez et al. (2021)	• Find the optimal number of GIs	NSGA	• Minimise flood damage	Urban green space	• Water management	City
Jiang et al. (2022)	• Optimise and explore green space patterns of the river-type city	OA	• Improve urban cooling intensity levels	Urban green space	• Heat islands	District
Yan and Tang (2021)	• Optimise green space planning	OA	• Maximise landscape ecosystem services	Urban green space	• Biodiversity and ecosystem security	City
Dong et al. (2021)	• Optimise GI location	NSGA	• Maximise water quality	Residential green space Urban green space	• Water management	District
Wang et al. (2019)	• Thermal environment optimisation based on urban green areas	OA	• Maximise urban thermal environment	Urban green space	• Heat islands	City

Daniele and Sciacca (2021)	• Find optimal green surface	OA	• Maximise carbon absorbing	Urban green space	• Air quality	City
Liu et al. (2021)	• Optimise ecosystem service	OA	• Maximise urban green spaces	Urban green space	• Biodiversity and ecosystem security	City
Wang et al. (2023)	• GI location optimisation	NSGA	• Maximise flood depth • Minimise runoff	Residential green space	• Water management	District
Trájer et al. (2022)	• Develop greening strategies	OA	• Minimise health impact	Urban green space	• Public health	City
Zhou et al. (2022)	• Optimise spatial distribution of GI	GA	• Maximise flood control • Maximise pollution control	Residential green space Urban green space	• Water management	District
Wang et al. (2022)	• GI location optimisation	NSGA	• Maximise flood risk reduction • Minimise life cycle cost • Minimise land occupied	Residential green space	• Water management	District
Torres et al. (2021)	• GI location optimisation	Piecewise Linear model	• Minimise runoff	Urban green space	• Water management	City
Tao et al. (2023)	• Optimise green spaces within urban areas	OA	• Maximise carbon sequestration	Urban green space	• Air quality	City
Choi and Lee (2023)	• Optimise tree plant location	OA	• Minimise tree cost • Maximise carbon absorption • Maximise species diversity • Maximise comfort level	Trees	• Air quality • Biodiversity and ecosystem security	Building
Zhang et al. (2017)	• Find the best location for new green spaces	OA	• Maximise cooling benefits	Urban green space	• Heat islands	District
Zhu et al. (2023)	• Select the optimal site for GI	NSGA	• Minimise runoff value • Minimise water pollution	Urban green space	• Water management	City

(Continued)

TABLE 7.A1 *(Continued)*

Salient Characteristics of Reviewed Articles

Author	Aim	Model	Objectives	Type of Infrastructure	Category	Scale
Gao et al. (2022)	• GI location optimisation	GA	• Minimise frequent urban flooding disasters	Residential green space Urban green space	• Water management	District
Hashad et al. (2021)	• Predict the spatial and size of roadside vegetation	ANN LR SVM XGBoost RF	• Minimise traffic air pollution	Trees	• Air quality	Street
Wang et al. (2021)	• Examines spatial vegetation effects on land surface temperature	OA	• Minimise land surface temperature • Minimise out-water use	Residential green space	• Heat islands • Water management	District
Liu et al. (2016)	• Optimal choice and positioning of GI	NSGA	• Minimise green spaces cost • Minimise runoff	Urban green space	• Water management	District
Yu et al. (2020)	• Optimise urban green spaces	GA	• Maximise urban cooling intensity levels	Urban green space	• Heat islands	City
Honeck et al. (2020)	• GI location optimisation	OA	• Minimise conflicts between environmental and other land use	Urban green space	• Biodiversity and ecosystem security	District
Kim et al. (2016)	• Optimise landscape pattern	SWARM	• Maximise economic development • Maximise ecological security • Maximise comprehensive benefits	Natural green space	• Biodiversity and ecosystem security	City

Feng et al. (2022)	• Identify the optimal location for green spaces	GA	• Increase the green cover rate to 40% • Improve microenvironment	Urban green space	• Air quality • Biodiversity and ecosystem security • Public health • Heat islands • Water management	City
Zhuang and Lu (2021)	• Optimise geographical arrangement of green roofs	GA	• Maximise the reduction of mean air temperature • Minimise heat exposure	Residential green space	• Heat islands	District
Huang et al. (2021)	• Optimise green space location	AIS	• Minimise climate change's effect on public health	Urban green space	• Public health	City
Stojakovic et al. (2020)	• Finding the best location for trees	EA	• Minimise urban heating	Trees	• Heat islands	Street
Yoon et al. (2019)	• Determines the location of greens based on effects	NSGA	• Maximise urban cooling • Minimise cost	Urban green space	• Heat islands	District
Martínez et al. (2018)	• Obtain the optimal configuration of GI	NSGA	• Minimise pollution load • Minimise runoff	Urban green space	• Water management	District
Liu et al. (2023)	• Optimise green and grey areas	NSGA	• Maximise runoff function • Minimise capital investment • Maximise ecological impact	RGS Urban green space	• Biodiversity and ecosystem security • Water management	City
Lu et al. (2022)	• Optimise urban green spaces	OA	• Minimise flood damage	Urban green space	• Water management	District
Neema and Ohgai (2010)	• Identify ideal locations for urban parks and open spaces	GA	• Minimise Air pollution • Minimise noise pollution	Urban green space	• Public health • Air quality	City
Kim et al. (2016)	• Identify the optimal green system	SWARM - Constriction Coefficient Hooke–Jeeves	• Minimise energy consumption • Maximise comfort level	Residential green space	• Energy efficiency • Heat islands	Building

(Continued)

TABLE 7.A1 *(Continued)*

Salient Characteristics of Reviewed Articles

Author	Aim	Model	Objectives	Type of Infrastructure	Category	Scale
Kumar et al. (2022)	• Optimise the location of green roofs	NSGA - SWARM	• Maximise flooding control	Residential green space	• Water management	City
Murray et al. (2014)	• Optimal design of the green roof	GA	• Minimise energy consumption • Minimise carbon emission	Residential green space	• Energy efficiency • Air quality	Building
Azcarate et al. (2021)	• Determine the best location of the tree	OA	• Maximise shading pedestrian zones	Trees	• Heat islands	Street
Zhao et al. (2017)	• Optimise tree placing	3D spatial optimisation model	• Maximise tree shade	Trees	• Heat islands	Building
Liu et al. (2023)	• Optimise GI layout	AA	• Maximise runoff reduction • Maximise pollution control • Maximise environmental benefits	Urban green space	• Air quality • Water management	District

References

Abdollahi, S., Ildoromi, A., Salmanmahini, A., & Fakheran, S. (2022). Optimisation of geographical space of ecosystem service areas and land-use planning, Iran. *Environmental Monitoring and Assessment, 194*(8), Article 527. https://doi.org/10.1007/s10661-022-10204-7

Adua, L., Zhang, K. X., & Clark, B. (2021). Seeking a handle on climate change: Examining the comparative effectiveness of energy efficiency improvement and renewable energy production in the United States. *Global Environmental Change, 70*, 102351. https://doi.org/10.1016/j.gloenvcha.2021.102351

Aerts, J., & Heuvelink, G. (2002). Using simulated annealing for resource allocation. *International Journal of Geographical Information Science, 16*(6), 571–587. https://doi.org/10.1080/13658810210138751

Ahmadi-Mirghaed, F., Mohammadzadeh, M., Salmanmahiny, A., & Mirkarimi, S. H. (2020). Decision scenarios using ecosystem services for land allocation optimisation across Gharehsoo watershed in Northern Iran. *Ecological Indicators, 117*, 106645. https://doi.org/10.1016/j.ecolind.2020.106645

Akbari, H. (2002). Shade trees reduce building energy use and CO2 emissions from power plants. *Environmental Pollution, 116*, S119–S126. https://doi.org/10.1016/S0269-7491(01)00264-0

Andrade, L., Mauri, G., & Mendonça, A. (2013). General multiobjective model and simulated annealing algorithm for waste-load allocation. *Journal of Water Resources Planning and Management, 139*(3), 339–344. https://doi.org/10.1061/(ASCE)WR.1943-5452.0000257

Ang, T.-Z., Salem, M., Kamarol, M., Das, H. S., Nazari, M. A., & Prabaharan, N. (2022). A comprehensive study of renewable energy sources: Classifications, challenges and suggestions. *Energy Strategy Reviews, 43*, 100939. https://doi.org/10.1016/j.esr.2022.100939

Anguelovski, I., Connolly, J. J. T., Cole, H., Garcia-Lamarca, M., Triguero-Mas, M., Baró, F., Martin, N., Conesa, D., Shokry, G., del Pulgar, C. P., Ramos, L. A., Matheney, A., Gallez, E., Oscilowicz, E., Máñez, J. L., Sarzo, B., Beltrán, M. A., & Minaya, J. M. (2022). Green gentrification in European and North American cities. *Nature Communications, 13*(1), 3816. https://doi.org/10.1038/s41467-022-31572-1

Arbolino, R., Carlucci, F., De Simone, L., Ioppolo, G., & Yigitcanlar, T. (2018). The policy diffusion of environmental performance in the European countries. *Ecological Indicators, 89*, 130–138. https://doi.org/10.1016/j.ecolind.2018.01.062

Audet, C., Bigeon, J., Cartier, D., Le Digabel, S., & Salomon, L. (2021). Performance indicators in multiobjective optimization. *European Journal of Operational Research, 292*(2), 397–422. https://doi.org/10.1016/j.ejor.2020.11.016

Azcarate, I., Acero, J., Garmendia, L., & Rojí, E. (2021). Tree layout methodology for shading pedestrian zones: Thermal comfort study in Bilbao (Northern Iberian Peninsula). *Sustainable Cities and Society, 72*, 102996. https://doi.org/10.1016/j.scs.2021.102996

Bajsanski, I., Stojakovic, V., & Jovanovic, M. (2016). Effect of tree location on mitigating parking lot insolation. *Computers, Environment and Urban Systems, 56*, 59–67. https://doi.org/10.1016/j.compenvurbsys.2015.11.006

Berry, E. M., Dernini, S., Burlingame, B., Meybeck, A., & Conforti, P. (2015). Food security and sustainability: Can one exist without the other? *Public Health Nutrition, 18*(13), 2293–2302. https://doi.org/10.1017/S136898001500021X

Boeringer, D. W., & Werner, D. H. (2004). Particle swarm optimisation versus genetic algorithms for phased array synthesis. *IEEE Transactions on Antennas and Propagation, 52*(3), 771–779. https://doi.org/10.1109/TAP.2004.825102

Bohra, P., & Palivela, H. (2015, 10-12 Dec. 2015). Understanding and formulation of various kernel techniques for support vector machines. 2015 IEEE International Conference on Computational Intelligence and Computing Research (ICCIC).

Breiman, L. (2001). Random forests. *Machine Learning, 45*(1), 5–32. https://doi.org/10.1023/A:1010933404324

Breuer, A., Janetschek, H., & Malerba, D. (2019). Translating sustainable development goal (SDG) interdependencies into policy advice. *Sustainability, 11*(7).

Calcerano, F., & Martinelli, L. (2016). Numerical optimisation through dynamic simulation of the position of trees around a stand-alone building to reduce cooling energy consumption. *Energy and Buildings, 112*, 234–243. https://doi.org/10.1016/j.enbuild.2015.12.023

Choi, C., Berry, P., & Smith, A. (2021). The climate benefits, co-benefits, and trade-offs of green infrastructure: A systematic literature review. *Journal of Environmental Management, 291*, 112583. https://doi.org/10.1016/j.jenvman.2021.112583

Choi, J., & Lee, G. (2023). Optimisation of tree planting for urban residential green spaces. *Landscape and Ecological Engineering, 19*(1), 107–121. https://doi.org/10.1007/s11355-022-00524-8

Coello, C., & Cortés, N. C. (2005). Solving multiobjective optimisation problems using an artificial immune system. *Genetic Programming and Evolvable Machines, 6*(2), 163–190. https://doi.org/10.1007/s10710-005-6164-x

Cunha, M. C., Zeferino, J. A., Simões, N. E., & Saldarriaga, J. G. (2016). Optimal location and sizing of storage units in a drainage system. *Environmental Modelling & Software, 83*, 155–166. https://doi.org/10.1016/j.envsoft.2016.05.015

Daniele, P., & Sciacca, D. (2021). An optimisation model for the management of green areas. *International Transactions in Operational Research, 28*(6), 3094–3116. https://doi.org/10.1111/itor.12987

Deb, K., Pratap, A., Agarwal, S., & Meyarivan, T. (2002). A fast and elitist multiobjective genetic algorithm: NSGA-II. *IEEE Transactions on Evolutionary Computation, 6*(2), 182–197. https://doi.org/10.1109/4235.996017

Dong, J., Gong, W., Ming, F., & Wang, L. (2022). A two-stage evolutionary algorithm based on three indicators for constrained multi-objective optimization. *Expert Systems with Applications, 195*, 116499. https://doi.org/10.1016/j.eswa.2022.116499

Dong, X., Yuan, P., Song, Y., & Yi, W. (2021). Optimizing green-gray infrastructure for non-point source pollution control under future uncertainties. *International Journal of Environmental Research and Public Health, 18*(14), 7586. https://doi.org/10.3390/ijerph18147586

Drucker, H., Burges, C. J., Kaufman, L., Smola, A., & Vapnik, V. (1996). Support vector regression machines. *Advances in Neural Information Processing Systems, 9*, 1996.

Dubois, G., Sovacool, B., Aall, C., Nilsson, M., Barbier, C., Herrmann, A., Bruyère, S., Andersson, C., Skold, B., Nadaud, F., Dorner, F., Moberg, K. R., Ceron, J. P., Fischer, H., Amelung, D., Baltruszewicz, M., Fischer, J., Benevise, F., Louis, V.

R., & Sauerborn, R. (2019). It starts at home? Climate policies targeting household consumption and behavioral decisions are key to low-carbon futures. *Energy Research & Social Science, 52,* 144–158. https://doi.org/10.1016/j.erss.2019.02.001

Dziejarski, B., Krzyżyńska, R., & Andersson, K. (2023). Current status of carbon capture, utilization, and storage technologies in the global economy: A survey of technical assessment. *Fuel, 342,* 127776. https://doi.org/10.1016/j.fuel.2023.127776

Eizenberg, E., Sasson, O., & Shilon, M. (2019). Urban morphology and qualitative topology: Open green spaces in high-rise residential developments. *Urban Planning, 4*(4), 73–85.

European Commission. (2013). Building a Green Infrastructure for Europe. https://doi.org/10.2779/54125.

Feng, L., Mi, X., & Yuan, D. (2022). Optimal planning of urban greening system in response to urban microenvironments in a high-density city using genetic algorithm: A case study of Tianjin. *Sustainable Cities and Society, 87,* 104244. https://doi.org/10.1016/j.scs.2022.104244

Fetting, C. (2020). "The European Green Deal", ESDN Report, December 2020, ESDN Office, Vienna.

Gaffin, S. R., Rosenzweig, C., & Kong, A. Y. (2012). Adapting to climate change through urban green infrastructure. *Nature Climate Change, 2*(10), 704–704. https://doi.org/10.1038/nclimate1685

Gao, Z., Zhang, Q. H., Xie, Y. D., Wang, Q., Dzakpasu, M., Xiong, J. Q., & Wang, X. C. (2022). A novel multi-objective optimisation framework for urban green-gray infrastructure implementation under impacts of climate change. *Science of the Total Environment, 825,* 153954. https://doi.org/10.1016/j.scitotenv.2022.153954

Gavrilescu, M. (2021). Water, soil, and plants interactions in a threatened environment. *Water, 13*(19), 2746.

Gill, S. E., Handley, J. F., Ennos, A. R., & Pauleit, S. (2007). Adapting cities for climate change: The role of the green infrastructure. *Built Environment, 33*(1), 115–133. https://doi.org/10.2148/benv.33.1.115

Hariram, N. P., Mekha, K. B., Suganthan, V., & Sudhakar, K. (2023). Sustainalism: An integrated socio-economic-environmental model to address sustainable development and sustainability. *Sustainability, 15*(13).

Hashad, K., Gu, J., Yang, B., Rong, M., Chen, E., Ma, X., & Zhang, K. M. (2021). Designing roadside green infrastructure to mitigate traffic-Voghera related air pollution using machine learning. *Science of the Total Environment, 773,* 144760. https://doi.org/10.1016/j.scitotenv.2020.144760

Honeck, E., Moilanen, A., Guinaudeau, B., Wyler, N., Schlaepfer, M. A., Martin, P., Sanguet, A., Urbina, L., Von Arx, B., Massy, J., Fischer, C., & Lehmann, A. (2020). Implementing green infrastructure for the spatial planning of peri-urban areas in Geneva, Switzerland. *Sustainability, 12*(4), 1387. https://doi.org/10.3390/su12041387

Hong, B., Lin, B., Hu, L., & Li, S. (2011). Optimal tree design for sunshine and ventilation in residential district using geometrical models and numerical simulation. *Building Simulation, 4*(4), 351–363. https://doi.org/10.1007/s12273-011-0056-1

Hongbing, W., Jun, Q., Yonghong, H., & Li, D. (2010). Optimal tree design for daylighting in residential buildings. *Building and Environment, 45*(12), 2594–2606. https://doi.org/10.1016/j.buildenv.2010.05.019

Huang, H., Yang, H., Chen, Y., Chen, T., Bai, L., & Peng, Z. R. (2021). Urban green space optimisation based on a climate health risk appraisal – A case study of Beijing city, China. *Urban Forestry and Urban Greening, 62*, 127154. https://doi.org/10.1016/j.ufug.2021.127154

Huang, K., Liu, X., Li, X., Liang, J., & He, S. (2013). An improved artificial immune system for seeking the Pareto front of land-use allocation problem in large areas. *International Journal of Geographical Information Science, 27*(5), 922–946. https://doi.org/10.1080/13658816.2012.730147

Huang, Z., & Wang, F. (2021, 7-8 Nov. 2021). A Review on Indicator-based Multi-objective Evolutionary Algorithms. 2021 IEEE 7th International Conference on Cloud Computing and Intelligent Systems (CCIS).

IPCC (2014). *Climate change 2014: Synthesis report. Contribution of working groups I, II and III to the fifth assessment report of the intergovernmental panel on climate change.* IPCC.

Jayasooriya, V. M., Ng, A. W., Muthukumaran, S., & Perera, C. B. (2020). Optimization of green infrastructure practices in industrial areas for runoff management: A review on issues, challenges and opportunities. *Water, 12*(4), 1024. https://doi.org/10.3390/w12041024

Jia, P., Fu, S., Li, Z., & He, H. (2019). Low-carbon optimisation of spatial pattern in Shenfu New District based on genetic algorithm. 2019 International Conference on Modeling, Simulation, Optimisation and Algorithm, ICMSOA 2019.

Jiang, Y., Li, X., & Huang, J. (2022). Zoning optimisation method of a riverfront greenspace service function oriented to the cooling effect: A case study in Shanghai. *International Journal of Environmental Research and Public Health, 19*(23), 16191. https://doi.org/10.3390/ijerph192316191

Jones, L., Anderson, S., Læssøe, J., Banzhaf, E., Jensen, A., Bird, D. N., Miller, J., Hutchins, M. G., Yang, J., Garrett, J., Taylor, T., Wheeler, B. W., Lovell, R., Fletcher, D., Qu, Y., Vieno, M., & Zandersen, M. (2022). A typology for urban Green Infrastructure to guide multifunctional planning of nature-based solutions. *Nature-Based Solutions, 2*, 100041. https://doi.org/10.1016/j.nbsj.2022.100041

Kim, J., Hong, T., Jeong, J., Koo, C., & Jeong, K. (2016). An optimisation model for selecting the optimal green systems by considering the thermal comfort and energy consumption. *Applied Energy, 169*, 682–695. https://doi.org/10.1016/j.apenergy.2016.02.032

Kirkpatrick, S., Gelatt, C. D., & Vecchi, M. P. (1983). Optimisation by simulated annealing. *Science, 220*(4598), 671–680. https://doi.org/doi:10.1126/science.220.4598.671

Koc, C., Osmond, P., & Peters, A. (2018). Evaluating the cooling effects of green infrastructure: A systematic review of methods, indicators and data sources. *Solar Energy, 166*, 486–508. https://doi.org/10.1016/j.solener.2018.03.008

Konak, A., Coit, D. W., & Smith, A. E. (2006). Multi-objective optimisation using genetic algorithms: A tutorial. *Reliability Engineering & System Safety, 91*(9), 992–1007. https://doi.org/10.1016/j.ress.2005.11.018

Korosuo, A., Heinonen, T., Ohman, K., Holmström, H., & Eriksson, L. O. (2014). Spatial optimisation in forest planning using different fragmentation measures. *Forest Science, 60*(4), 764–775. https://doi.org/10.5849/forsci.13-074

Kranjčić, N., Medak, D., Župan, R., & Rezo, M. (2019). Machine learning methods for classification of the green infrastructure in city areas. *ISPRS International Journal of Geo-Information, 8*(10), 463. https://doi.org/10.3390/ijgi8100463

Kumar, S., Guntu, R. K., Agarwal, A., Villuri, V. G. K., Pasupuleti, S., Kaushal, D. R., Gosian, A. K., & Bronstert, A. (2022). Multi-objective optimisation for stormwater management by green-roofs and infiltration trenches to reduce urban flooding in central Delhi. *Journal of Hydrology, 606*, 127455. https://doi.org/10.1016/j.jhydrol.2022.127455

Li, Q., Jin, T., Peng, Q., Lin, J., Zhang, D., Huang, J., & Liu, B. (2022b). Identifying the extent of the spatial expression of landscape fragmentation based on scale effect analysis in Southwest China. *Ecological Indicators, 141*, 109120. https://doi.org/10.1016/j.ecolind.2022.109120

Li, X., Li, X., & Ma, X. (2022a). Spatial optimisation for urban green space (UGS) planning support using a heuristic approach. *Applied Geography, 138*, 102622. https://doi.org/10.1016/j.apgeog.2021.102622

Li, X., & Parrott, L. (2016). An improved Genetic Algorithm for spatial optimisation of multi-objective and multi-site land use allocation. *Computers, Environment and Urban Systems, 59*, 184–194. https://doi.org/10.1016/j.compenvurbsys.2016.07.002

Li, Y., Ouyang, W., Yin, S., Tan, Z., & Ren, C. (2023). Microclimate and its influencing factors in residential public spaces during heat waves: An empirical study in Hong Kong. *Building and Environment, 236*, 110225. https://doi.org/10.1016/j.buildenv.2023.110225

Liu, Q., Tian, Y., Yin, K., Zhang, F., Huang, H., & Chen, F. (2021). Landscape pattern theoretical optimisation of urban green space based on ecosystem service supply and demand. *ISPRS International Journal of Geo-Information, 10*(4), 263. https://doi.org/10.3390/ijgi10040263

Liu, W., Xu, H., Zhang, X., & Jiang, W. (2022). Green infrastructure network identification at a regional scale: The case of Nanjing Metropolitan Area, China. *Forests, 13*(5), 735. https://doi.org/10.3390/f13050735

Liu, X., Ou, J., Li, X., & Ai, B. (2013). Combining system dynamics and hybrid particle swarm optimisation for land use allocation. *Ecological Modelling, 257*, 11–24. https://doi.org/10.1016/j.ecolmodel.2013.02.027

Liu, Y., Cibin, R., Bralts, V. F., Chaubey, I., Bowling, L. C., & Engel, B. A. (2016a). Optimal selection and placement of BMPs and LID practices with a rainfall-runoff model. *Environmental Modelling & Software, 80*, 281–296. https://doi.org/10.1016/j.envsoft.2016.03.005

Liu, Y., Theller, L. O., Pijanowski, B. C., & Engel, B. A. (2016b). Optimal selection and placement of green infrastructure to reduce impacts of land use change and climate change on hydrology and water quality: An application to the Trail Creek Watershed, Indiana. *Science of the Total Environment, 553*, 149–163. https://doi.org/10.1016/j.scitotenv.2016.02.116

Liu, Y., Xia, C., Ou, X., Lv, Y., Ai, X., Pan, R., Zhang, Y., Shi, M., & Zheng, X. (2023a). Quantitative structure and spatial pattern optimisation of urban green space from the perspective of carbon balance: A case study in Beijing, China. *Ecological Indicators, 148*, 110034. https://doi.org/10.1016/j.ecolind.2023.110034

Liu, Z., Han, Z., Shi, X., Liao, X., Leng, L., & Jia, H. (2023b). Multi-objective optimisation methodology for green-gray coupled runoff control infrastructure adapting spatial heterogeneity of natural endowment and urban development. *Water Research, 233*, 119759. https://doi.org/10.1016/j.watres.2023.119759

Lu, W., Xia, W., & Shoemaker, C. A. (2022). Surrogate global optimisation for identifying cost-effective green infrastructure for urban flood control with a computationally expensive inundation model. *Water Resources Research, 58*(4), e2021WR030928. https://doi.org/10.1029/2021WR030928

Marasinghe, R., Yigitcanlar, T., Mayere, S., Washington, T., & Limb, M. (2024). Computer vision applications for urban planning: A systematic review of opportunities and constraints. *Sustainable Cities and Society, 100*, 105047. https://doi.org/10.1016/j.scs.2023.105047

Markolf, S. A., Hoehne, C., Fraser, A., Chester, M. V., & Underwood, B. S. (2019). Transportation resilience to climate change and extreme weather events – Beyond risk and robustness. *Transport Policy, 74*, 174–186. https://doi.org/10.1016/j.tranpol.2018.11.003

Marseguerra, M., Zio, E., & Martorell, S. (2006). Basics of genetic algorithms optimisation for RAMS applications. *Reliability Engineering & System Safety, 91*(9), 977–991. https://doi.org/10.1016/j.ress.2005.11.046

Martínez, C., Sanchez, A., Galindo, R., Mulugeta, A., Vojinovic, Z., & Galvis, A. (2018). Configuring green infrastructure for urban runoff and pollutant reduction using an optimal number of units. *Water, 10*(11), 1528. https://doi.org/10.3390/w10111528

Martínez, C., Vojinovic, Z., & Sanchez, A. (2021). Multi-objective model-based assessment of green-grey infrastructures for urban flood mitigation. *Hydrology, 8*(3), 110. https://doi.org/10.3390/hydrology8030110

Masoomi, Z., Mesgari, M. S., & Hamrah, M. (2013). Allocation of urban land uses by multi-objective particle swarm optimisation algorithm. *International Journal of Geographical Information Science, 27*(3), 542–566. https://doi.org/10.1080/13658816.2012.698016

Maulud, D., & Abdulazeez, A. M. (2020). A review on linear regression comprehensive in machine learning. *Journal of Applied Science and Technology Trends, 1*(4), 140–147. https://doi.org/10.38094/jastt1457

Mishra, B. K., Kumar, P., Saraswat, C., Chakraborty, S., & Gautam, A. (2021). Water security in a changing environment: Concept, challenges and solutions. *Water*, 13(4).

Mortoja, M., & Yigitcanlar, T. (2020a). How does peri-urbanization trigger climate change vulnerabilities? An investigation of the Dhaka megacity in Bangladesh. *Remote Sensing, 12*(23), 3938. https://doi.org/10.3390/rs12233938

Mortoja, M., & Yigitcanlar, T. (2020b). Local drivers of anthropogenic climate change: Quantifying the impact through a remote sensing approach in Brisbane. *Remote Sensing, 12*(14), 2270. https://doi.org/10.3390/rs12142270

Mousavi, S., Gheibi, M., Wacławek, S., & Behzadian, K. (2023). A novel smart framework for optimal design of green roofs in buildings conforming with energy conservation and thermal comfort. *Energy and Buildings, 291*, 113111. https://doi.org/10.1016/j.enbuild.2023.113111

Murray, S. N., Walsh, B. P., Kelliher, D., & O'Sullivan, D. T. J. (2014). Multi-variable optimisation of thermal energy efficiency retrofitting of buildings using static modelling and genetic algorithms – A case study. *Building and Environment, 75*, 98–107. https://doi.org/10.1016/j.buildenv.2014.01.011

Neema, M. N., & Ohgai, A. (2010). Multi-objective location modeling of urban parks and open spaces: Continuous optimisation. *Computers, Environment and Urban Systems, 34*(5), 359–376. https://doi.org/10.1016/j.compenvurbsys.2010.03.001

Neema, M. N., & Ohgai, A. (2013). Multitype green-space modeling for urban planning using GA and GIS. *Environment and Planning B: Planning and Design, 40*(3), 447–473. https://doi.org/10.1068/b38003

Nordh, H., Hartig, T., Hagerhall, C. M., & Fry, G. (2009). Components of small urban parks that predict the possibility for restoration. *Urban Forestry & Urban Greening, 8*(4), 225–235. https://doi.org/10.1016/j.ufug.2009.06.003

Ou, D., Yao, X., Xia, J., Gao, X., Wang, C., Chen, W., Li, Q., Hu, Z., & Yang, J. (2019). Development of a composite model for simulating landscape pattern optimisation allocation: A case study in the Longquanyi District of Chengdu City, Sichuan Province, China. *Sustainability, 11*(9), 2678. https://doi.org/10.3390/su11092678

Oxley, R. L., & Mays, L. W. (2014). Optimisation – Simulation model for detention basin system design. *Water Resources Management, 28*(4), 1157–1171. https://doi.org/10.1007/s11269-014-0552-z

Perveen, S., Kamruzzaman, M., & Yigitcanlar, T. (2017). Developing policy scenarios for sustainable urban growth management: A Delphi approach. *Sustainability, 9*(10), 1787. https://doi.org/10.3390/su9101787

Pimentel da Silva, L., da Fonseca, M. N., de Moura, E. N., & de Souza, F. T. (2022). Ecosystems services and green infrastructure for respiratory health protection: A data science approach for Paraná, Brazil. *Sustainability, 14*(3), 1835.

Porta, J., Parapar, J., Doallo, R., Rivera, F. F., Santé, I., & Crecente, R. (2013). High performance genetic algorithm for land use planning. *Computers, Environment and Urban Systems, 37*, 45–58. https://doi.org/10.1016/j.compenvurbsys.2012.05.003

Pukkala, T., & Kurttila, M. (2005). Examining the performance of six heuristic optimisation techniques in different forest planning problems (Vol. 39). https://doi.org/10.14214/sf.396

Regona, M., Yigitcanlar, T., Xia, B., & Li, R. (2022). Opportunities and adoption challenges of AI in the construction industry: A PRISMA review. *Journal of Open Innovation: Technology, Market, and Complexity, 8*(1), 45. https://doi.org/10.3390/joitmc8010045

Rouhollahi, M., Whaley, D., Byrne, J., & Boland, J. (2022). Potential residential tree arrangement to optimise dwelling energy efficiency. *Energy and Buildings, 261*, 111962. https://doi.org/10.1016/j.enbuild.2022.111962

Sadollah, A., Nasir, M., & Geem, Z. W. (2020). Sustainability and optimization: From conceptual fundamentals to applications. *Sustainability, 12*(5).

Schipperijn, J., Stigsdotter, U. K., Randrup, T. B., & Troelsen, J. (2010). Influences on the use of urban green space – A case study in Odense, Denmark. *Urban Forestry & Urban Greening, 9*(1), 25–32. https://doi.org/10.1016/j.ufug.2009.09.002

Seddon, N., Chausson, A., Berry, P., Girardin, C., Smith, A., & Turner, B. (2020). Understanding the value and limits of nature-based solutions to climate change and other global challenges. *Philosophical Transactions of the Royal Society B, 375*(1794), 20190120. https://doi.org/10.1098/rstb.2019.0120

She, L., Wei, M., & You, X. Y. (2021). Multi-objective layout optimisation for sponge city by annealing algorithm and its environmental benefits analysis. *Sustainable Cities and Society, 66*, 102706. https://doi.org/10.1016/j.scs.2021.102706

Son, T., Weedon, Z., Yigitcanlar, T., Sanchez, T., Corchado, J. M., & Mehmood, R. (2023). Algorithmic urban planning for smart and sustainable development: Systematic review of the literature. *Sustainable Cities and Society, 94*, 104562. https://doi.org/10.1016/j.scs.2023.104562

Sotto, D., Philippi, A., Yigitcanlar, T., & Kamruzzaman, M. (2019). Aligning urban policy with climate action in the global south: Are Brazilian cities considering climate emergency in local planning practice? *Energies, 12*(18), 3418. https://doi.org/10.3390/en12183418

Sowińska-Świerkosz, B., & Garcia, J. (2022). What are nature-based solutions (NBS)? Setting core ideas for concept clarification. *Nature-Based Solutions, 2,* 100009. https://doi.org/10.1016/j.nbsj.2022.100009

Stojakovic, V., Bajsanski, I., Savic, S., Milosevic, D., & Tepavcevic, B. (2020). The influence of changing location of trees in urban green spaces on insolation mitigation. *Urban Forestry and Urban Greening, 53,* 126721. https://doi.org/10.1016/j.ufug.2020.126721

Tao, P., Lin, Y., Wang, X., Li, J., Ma, C., Wang, Z., Dong, X., Yao, P., & Shao, M. (2023). Optimisation of green spaces in plain urban areas to enhance carbon sequestration. *Land, 12*(6), 1218. https://doi.org/10.3390/land12061218

Tavakol-Davani, H. E., Tavakol-Davani, H., Burian, S. J., McPherson, B. J., & Barber, M. E. (2019). Green infrastructure optimisation to achieve pre-development conditions of a semiarid urban catchment. *Environmental Science: Water Research and Technology, 5*(6), 1157–1171. https://doi.org/10.1039/c8ew00789f

Torres, M. N., Fontecha, J. E., Walteros, J. L., Zhu, Z., Ahmed, Z., Rodríguez, J. P., & Rabideau, A. J. (2021). City-Scale optimal location planning of Green Infrastructure using piece-wise linear interpolation and exact optimisation methods. *Journal of Hydrology, 601,* 126540. https://doi.org/10.1016/j.jhydrol.2021.126540

Trájer, A. J., Sebestyén, V., Domokos, E., & Abonyi, J. (2022). Indicators for climate change-driven urban health impact assessment. *Journal of Environmental Management, 323,* 116165. https://doi.org/10.1016/j.jenvman.2022.116165

Vaeztavakoli, A., Lak, A., & Yigitcanlar, T. (2018). Blue and green spaces as therapeutic landscapes: Health effects of urban water canal areas of Isfahan. *Sustainability, 10*(11), 4010. https://doi.org/10.3390/su10114010

Vallejo, M., Rieser, V., & Corne, D. W. (2015). Genetic algorithm evaluation of green search allocation policies in multilevel complex urban scenarios. *Journal of Computational Science, 9,* 57–63. https://doi.org/10.1016/j.jocs.2015.04.004

Verhagen, W., van der Zanden, E. H., Strauch, M., van Teeffelen, A. J. A., & Verburg, P. H. (2018). Optimizing the allocation of agri-environment measures to navigate the trade-offs between ecosystem services, biodiversity and agricultural production. *Environmental Science and Policy, 84,* 186–196. https://doi.org/10.1016/j.envsci.2018.03.013

Voghera, A., & Giudice, B. (2019). Evaluating and planning green infrastructure: A strategic perspective for sustainability and resilience. *Sustainability, 11*(10).

Wakdok, S. S., & Bleischwitz, R. (2021). Climate change, security, and the resource nexus: Case study of Northern Nigeria and Lake Chad. *Sustainability, 13*(19).

Wang, C., Turner, V. K., Wentz, E. A., Zhao, Q., & Myint, S. W. (2021). Optimisation of residential green space for environmental sustainability and property appreciation in metropolitan Phoenix, Arizona. *Science of the Total Environment, 763,* 144605. https://doi.org/10.1016/j.scitotenv.2020.144605

Wang, G. Q., Zheng, B. H., Yu, H., & Peng, X. (2019). Green space layout optimisation based on microclimate environment features. *International Journal of Sustainable Development and Planning, 14*(1), 9–19. https://doi.org/10.2495/SDP-V14-N1-9-19

Wang, J., Liu, J., Mei, C., Wang, H., & Lu, J. (2022). A multi-objective optimisation model for synergistic effect analysis of integrated green-gray-blue drainage system in urban inundation control. *Journal of Hydrology, 609*, 127725. https://doi.org/10.1016/j.jhydrol.2022.127725

Wang, J., Liu, J., Yang, Z., Mei, C., Wang, H., & Zhang, D. (2023a). Green infrastructure optimisation considering spatial functional zoning in urban stormwater management. *Journal of Environmental Management, 344*, 118407. https://doi.org/10.1016/j.jenvman.2023.118407

Wang, M., Zhong, X., Sun, C., Chen, T., Su, J., & Li, J. (2023b). Comprehensive performance of green infrastructure through a life-cycle perspective: A review. *Sustainability, 15*(14).

Xiao, N., Bennett, D. A., & Armstrong, M. P. (2002). Using evolutionary algorithms to generate alternatives for multiobjective site-search problems. *Environment and Planning A, 34*(4), 639–656. https://doi.org/10.1068/a34109

Xiao, N., Bennett, D. A., & Armstrong, M. P. (2007). Interactive evolutionary approaches to multiobjective spatial decision making: A synthetic review. *Computers, Environment and Urban Systems, 31*(3), 232–252. https://doi.org/10.1016/j.compenvurbsys.2006.08.001

Xu, C., Han, B., Lu, F., & Wu, T. (2022). Assessing the traffic noise reduction effect of roadside green space using LiDAR point cloud data in Shenzhen, China. *Forests, 13*(5), 765. https://doi.org/10.3390/f13050765

Yamachi, H., Tsujimura, Y., Kambayashi, Y., & Yamamoto, H. (2006). Multi-objective genetic algorithm for solving N-version program design problem. *Reliability Engineering & System Safety, 91*(9), 1083–1094. https://doi.org/10.1016/j.ress.2005.11.045

Yan, S., & Tang, J. (2021). Optimisation of green space planning to improve ecosystem services efficiency: The case of Chongqing urban areas. *International Journal of Environmental Research and Public Health, 18*(16), 8441. https://doi.org/10.3390/ijerph18168441

Yigitcanlar, T. (2010). *Sustainable urban and regional infrastructure development: Technologies, applications and management.* IGI Global.

Yigitcanlar, T., Mehmood, R., & Corchado, J. (2021). Green artificial intelligence: Towards an efficient, sustainable and equitable technology for smart cities and futures. *Sustainability, 13*(16), 8952. https://doi.org/10.3390/su13168952

Ying, J., Zhang, X., Zhang, Y., & Bilan, S. (2022). Green infrastructure: Systematic literature review. *Economic Research-Ekonomska Istraživanja, 35*(1), 343–366. https://doi.org/10.1080/1331677X.202

Yoon, E. J., Kim, B., & Lee, D. K. (2019). Multi-objective planning model for urban greening based on optimisation algorithms. *Urban Forestry & Urban Greening, 40*, 183–194. https://doi.org/10.1016/j.ufug.2019.01.004

Yu, Y., Zhang, W., Fu, P., Huang, W., Li, K., & Cao, K. (2020). The spatial optimisation and evaluation of the economic, ecological, and social value of urban green space in Shenzhen. *Sustainability, 12*(5), 1-18, 1844. https://doi.org/10.3390/su12051844

Zhang, C., Tang, L., & Zhang, J. (2023). Identifying critical indicators in performance evaluation of green supply chains using hybrid multiple-criteria decision-making. *Sustainability, 15*(7).

Zhang, K., & Chui, T. F. M. (2019). A review on implementing infiltration-based green infrastructure in shallow groundwater environments: Challenges, approaches, and progress. *Journal of Hydrology, 579*, 124089. https://doi.org/10.1016/j.jhydrol.2019.124089

Zhang, Y., Murray, A. T., & Turner, B. L. II (2017). Optimizing green space locations to reduce daytime and nighttime urban heat island effects in Phoenix, Arizona. *Landscape and Urban Planning, 165,* 162–171. https://doi.org/10.1016/j.landurbplan.2017.04.009

Zhao, Q., Wentz, E. A., & Murray, A. T. (2017). Tree shade coverage optimisation in an urban residential environment. *Building and Environment, 115,* 269–280. https://doi.org/10.1016/j.buildenv.2017.01.036

Zhou, H., Wang, Q., Zhu, N., Li, Y., Li, J., Zhou, L., Pei, Y., & Zhang, S. (2023). Optimisation methods of urban green space layout on tropical islands to control heat island effects. *Energies, 16*(1), 368. https://doi.org/10.3390/en16010368

Zhou, Q., Luo, J., Qin, Z., Su, J., & Ren, Y. (2022). Conceptual planning approach of low impact developments for combined water quality-quantity control at an urban scale: A case study in Southern China. *Journal of Flood Risk Management, 15*(1), e12760. https://doi.org/10.1111/jfr3.12760

Zhu, Y., Xu, C., Liu, Z., Yin, D., Jia, H., & Guan, Y. (2023). Spatial layout optimization of green infrastructure based on life-cycle multi-objective optimization algorithm and SWMM model. *Resources, Conservation and Recycling, 191,* 106906.

Zhuang, Q., & Lu, Z. (2021). Optimisation of roof greening spatial planning to cool down the summer of the city. *Sustainable Cities and Society, 74,* 103221. https://doi.org/10.1016/j.scs.2021.103221

Part III

Policy, Planning, and Global Perspectives

This final part of the book situates technological and design innovations within the wider policy and planning landscape. It examines urban climate policy challenges, particularly in the Global South, and explores how transit-oriented development and urban heat management intersect. The chapters also discuss the integration of policy and technology to develop proactive responses to urban heat risks. Emphasising the importance of governance, regulatory frameworks, and context-sensitive planning, this part demonstrates that climate resilience requires coordinated strategies linking technology, green infrastructure, and policy. It provides global insights for decision-makers seeking equitable, effective, and scalable solutions to urban heat challenges.

DOI: 10.1201/9781003737056-10

8

Urban Climate Policy in the Global South

8.1 Introduction and Background

The Anthropocene, a geological era of human domination on Earth's resources, has generated many complex problems that threaten the well-being and existence of many species, including humankind (Van Loon et al., 2016; Yigitcanlar et al., 2019a, 2019b, 2019c). The single most important problem we face today, in the Anthropocene age, is climate change or more correctly the climate emergency or crisis (Scranton, 2015; Yigitcanlar et al., 2019). The impacts, just to name a few, include more frequent and extreme heat episodes, rising seas and increased coastal flooding, longer and more damaging wildfire seasons, more destructive hurricanes, costly and growing health impacts, heavier precipitation and flooding, destruction of marine ecosystems, more severe droughts in some areas, increased pressure on groundwater supplies, melting ice cap, disruptions to food supplies, plant and animal range shifts, climate refugees, and changing seasons (Burke et al., 2015; Parmesan & Yohe, 2003; Pecl et al., 2017).

During the last few years, various solutions have been put forward to combat the problem, particularly targeting the unsustainable urban development and cleaner production (Aldieri et al., 2019; Jakob & Steckel, 2016; Yigitcanlar, 2009). These include adopting new development paradigms to make cities more sustainable, resilient, and smarter—e.g., smart city and smart urbanism movements (Aldieri et al., 2019; Yigitcanlar et al., 2018a, 2018b)—and incentivising green technology and cleaner industrial development (Arbolino et al., 2017; Ingrao et al., 2018; Rottz et al., 2019). However, the solutions targeting more sustainable urbanisation and industrialisation with limited emissions have not found wider application grounds across the globe, and also existing initiatives were not efficient enough to make a significant difference (Arbolino et al., 2018; Yigitcanlar et al., 2019).

There have been international and national platforms and initiatives to intervene with the status quo and also provide possible directions and solutions to combat the climate change problem—more correctly, the factors causing it. For example, as nation signatories of the United Nations' Framework Convention on Climate Change (UNFCCC) and to the Paris Agreement (Zapf et al., 2019), many countries have made the commitment, expressed in their nationally determined contribution (NDC), to reduce their greenhouse gas (GHG) emissions (Yigitcanlar & Kamruzzaman, 2018). However, as most of the world population is urban, compliance with NDC depends on the

DOI: 10.1201/9781003737056-11

engagement of local governments—i.e., municipalities (Manowska & Nowrot, 2019). Furthermore, it is required for cities to take firm action on climate adaptation and resilience promotion, as the worst effects of global warming—floods, droughts, landslides, epidemics—are experienced by the population at the local level, deeply affecting public health and quality of life, especially among the most vulnerable social sectors (Brasil. Sumário Executivo, 2017).

Engagement of cities across the world in tackling climate change, autonomously of their respective nation-states, is a relatively recent phenomenon (Broto, 2017; Hale, 2016; Octaviano et al., 2016), with significant and positive impacts on the latest accomplishments of international climate negotiations. Despite the lack of a legal status under international law, cities have intensified their actions in international relations through the construction of networks, such as Local Governments for Sustainability (ICLEI), United Cities and Local Governments (UCLG), 40 Cities for Climate, and Compact of Mayors (C40 Group), active in sustainable urban development, and climate change. As a result, cities have been identified as key actors for environmental protection and sustainable development—firstly participating in UNFCCC Conferences as "observant organizations," and from the 2010 Conference of the Parties (COP-16) onward as "governmental actors."

The final declaration of the United Nations Conference on Sustainable Development (Rio +20), "The Future We Want" (United Nations, General Assembly, 2012), has reaffirmed the strategic role of cities in promoting sustainable development. The Rio +20 declaration also set out the commitment to formulate "sustainable development goals" (SDGs), replacing the "millennium development goals" (MDGs). These SDGs were issued in September 2015 (United Nations, General Assembly, 2015), a set of 17 main goals, broken down into 169 targets and monitored by corresponding indicators, to be pursued by all nations, regardless of their developmental stage.

The sustainable development goal (SDG) no. 13 (SDG-13) refers specifically to "climate action," expressing the global commitment to take urgent action on climate change and its impacts within the UNFCCC framework. SDG Target 13.B fosters capacity building, promotion for productive climate change-related planning and management in local communities in the least developed countries, as well as women, youth, and vulnerable minorities. SDG-11, on the other hand, expresses the commitment to make cities and human settlements safe, resilient, and sustainable. Among the foreseen axes of action, Target 11.B establishes the goal to substantially increase, by 2020, the number of cities and human settlements adopting and implementing integrated policies and plans towards inclusion, resource efficiency, mitigation, and adaptation to climate change and resilience to disasters, in line with the Sendai Framework for Disaster Risk Reduction (2015–2030)—the first major international agreement of the post-2015 development agenda.

Both SDG-11 and -13 strongly connect to strategic goals and actions established by the 2016 "new urban agenda" (NUA), approved in the city of Quito (Ecuador) (United Nations, General Assembly, 2016). NUA expressly envisions cities and human settlements that adopt and implement disaster risk reduction and management, reduce vulnerability, build resilience and responsiveness to natural and human-made hazards, and promote mitigation and adaptation to climate change. By doing so, NUA highlights the mandatory connection existing between SDG-13 (action against global climate change), SDG-11 (sustainable cities and communities), and SDG-3 (health and well-being).

Not only is SDG-3, health and well-being, indispensable for the quality of life in cities, but it is also a central component in the construction of urban resilience. In this sense, Target 3.D commits the nation-states to strengthen the capacity of all countries, and also local governments, for early warning, reduction, and management of these risks, considering the increasing national and global risks associated with extreme climatic events.

It is important to stress that the most vulnerable social segments, which already have the worst housing and sanitation conditions and the worst levels of access to essential services, including health, are also the most exposed to the adverse effects of extreme weather events and other shocks (Jacobi & Sulaiman, 2016; Nobre, 2001; Pessini & Sganzerla, 2016). Hence, the commitment expressed by Target 1.E, deployed from SDG-1, eradicating poverty, to build the resilience of the poor and those in a vulnerable situation and to reduce their exposure and vulnerability to extreme events, shocks, and disasters. In an increasingly urban world, building the resilience of the poorest and most vulnerable is one of the significant challenges to be faced by urban planners and managers.

Against this backdrop, the aim of this study is to examine how cities, and particularly cities in the global south (an emerging term used by the World Bank to refer to countries located in Asia, Africa, Latin America (including Brazil), and the Caribbean and considered to have low- and middle-income levels (Lahsen, 2009)), are considering the climate emergency in their local planning mechanisms. The study concentrates on Brazil as the case study context, and places five major Brazilian cities—i.e., Manaus, Salvador, Goiânia, São Paulo, and Curitiba—under the microscope to address the research aim. The methodology of the investigation involves undertaking a literature review, collecting public policy data, and conducting a quantitative analysis. The chapter is structured as follows: after this introduction section, Section 8.2 presents materials and methods. Section 8.3 discloses the results. Section 8.4 discusses the findings, and Section 8.5 concludes the chapter.

8.2 Research Design

8.2.1 Case Study

The study selects Brazil as a relevant country context from the global south—with vast land, large population, pristine natural assets, and a big appetite for economic growth—to explore how well local urban policy is aligned with climate action (Engle & Lemos, 2010). The research is developed as a multiple case study (Yin, 2003). It involved literature review, data collection in digital repositories, and the qualitative analysis of the data using a content analysis software—i.e., NVivo (Owen, 2014). As it would not be possible to study all 5,570 Brazilian municipalities, in order to cover the different realities and main biomes of Brazilian territory, five major Brazilian cities were targeted to be selected for the analysis at the ratio of one city per each region of the country—i.e., North, Northeast, Midwest, Southeast, and South.

The case study selection process obeyed the following criteria: (a) population above 500,000 people (the Brazilian Institute of Geography and Statistics (IBGE) classifies Brazilian cities in three categories: small (less than 100,000 inhabitants); medium (more than 100,000 inhabitants and less than 500,000 inhabitants); and large (more than 500,000 inhabitants). IBGE's methodology does not divide large cities into any other additional categories. Large cities in Brazil are more typically "urban" than medium and small cities, which is why this investigation opted to select only cities with more than 500,000 inhabitants); (b) Local environmental bodies and councils, according to 1981 National Environmental Policy (Brasil, 1981); (c) Existence of a Master Plan, approved accordingly to 2001 Statute of the City (Brasil, 2001); (d) City's participation in at least one of the following international city networks, selected for their outstanding performance in the field of climate change: Local Governments for Sustainability (ICLEI), Mercociudades, as representative of the Communities and United Local Governments in South America (UCLG), and C40 Group, and; (e) Availability of data and information pertinent to municipal management and planning, accessible through the Internet. Based on these criteria (Table 8.1), the following cities were selected as the object of the case study: Goiânia (Central-West Region), Salvador (Northeast Region), Manaus (North Region), São Paulo (Southeast Region) (in Brazil, all municipalities with more than 500,000 inhabitants must follow the same institutional framework as cities with a population of one million, 2 million or 12 million people. Therefore, the fact that São Paulo is significantly larger than the other selected cities does not distort or affect the results of the proposed investigation), and Curitiba (Southern Region) (Brasília, capital city of the nation, was excluded from the selection process because, as the seat of the Federal District, it presents unique institutional characteristics, distinct from the other Brazilian municipalities). As seen in Table 8.1, despite listed similarities, the case cities are to different sizes and characteristics, thus a one-on-one comparison would not be a correct approach (Yigitcanlar et al., 2012). Here, the purpose is to capture the general trends related to climate policy in Brazilian cities. Hence, having cities with different attributes is a valid approach to address the research question (Yigitcanlar, 2014).

TABLE 8.1

Case Study Cities' Salient Characteristics

City	Population	Environmental Body and Council	Master Plan in Accordance with the Statute of the City	City's Networks	Data Available on the Internet
Goiânia State of Goiás Midwest Region	1,466,105	Yes	2007	ICLEI UCLG	Yes
Salvador State of Bahia Northeast Region	2,953,986	Yes	2016	C40 ICLEI UCLG	Yes
Manaus State of Amazonas North Region	2,130,264	Yes	2014	ICLEI	Yes
São Paulo State of São Paulo Southwest Region	12,106,920	Yes	2014	C40 ICLEI UCLG	Yes
Curitiba State of Paraná South Region	1,908,359	Yes	2015	C40 ICLEI UCLG	Yes

8.2.2 Methodology and Data

Assuming that an effective climate policy at any level must take as its starting point the findings of GHG emissions inventories, we firstly collected the inventories drawn up by the surveyed cities, in order to compare their findings with Emission Inventories edited by the respective states and 2010 and 2015 national emissions inventories issued by the Brazilian government. The policy document search was organised according to four areas, or climate action axis: (a) Mitigation; (b) Adaptation; (c) Resilience as defined by the IPCC (Intergovernmental Panel on Climate Change—IPCC, 2018); and (d) Cooperation (cooperation viewed as decentralised cooperation, or city's diplomacy, comprehending agreements celebrated between cities and any other international actors, such as other cities, other countries, international organisations, as well as actions developed within the context of city's international networks). Table 8.2 presents local policies, plans, programmes, and projects, classified according to the four climate action axes of mitigation, adaptation, resilience, and cooperation.

The documents, compiled by the researchers in the first quarter of 2018 in official digital data bases (In Brazil, due to national law provisions, all federative entities, municipalities included, must provide free access to all public interest information. In compliance, the union (federation), states and most of the large municipalities have made their legislation (laws and

TABLE 8.2

Local Policies, Plans, Programmes, and Projects

Mitigation	Adaptation	Resilience	Cooperation
GHG emissions inventories	GHG emissions inventories	GHG emissions inventories	City networks
Master plan	Master plan	Master plan	Master plan
Land use regulations	Land use regulations	Land use regulations	Decentralised cooperation
Housing local plan	Housing local plan	Housing local plan	Decentralised cooperation
Mobility local plan	Mobility local plan	Mobility local plan	Decentralised cooperation
Sanitation local plan	Sanitation local plan	Sanitation local plan	Decentralised cooperation
Waste management plan	Waste management plan	Waste management plan	Decentralised cooperation
Drainage local plan	Drainage local plan	Drainage local plan	Decentralised cooperation
Public health local plan	Public health local plan	Public health local plan	Decentralised cooperation
Climate mitigation plan	Climate adaptation plan	Disasters management plan	Decentralised cooperation

regulations) available online, in official digital databases), and organised by city and topic and listed in Table 8.3. Compiled documents were processed with NVivo, firstly sorted out in cases, each case corresponding to one of the investigated cities as well as to the nation. The word search tool was used to find references to the four axes of climate action—i.e., adaptation, cooperation, mitigation, and resilience—in the texts of the compiled documents. The word search tool was also used to locate references, in the local documents, to national regulations related to climate change—e.g., National Emissions Inventory, 2009 National Climate Change Policy (PNMC), 2016 National Adaptation Plan (PNA), Brazil's NDC, 2013 Sectoral Health Plan for Climate Change Mitigation and Adaptation, and 2012 National Protection and Civil Defense Policy (PNPDEC). All policy documents were in the Portuguese language. Analyses were conducted in Portuguese, and the results were translated into English.

We have used the AND, OR, NEAR, (*), REQUIRED, FUZZY, and (") operators in at least five successive rounds of word search queries in order to obtain more precise results, avoiding redundancies, ambiguities, and false-positive results (NVivo, as qualitative analysis tool, allows the election of objective criteria and comparison standards for data contained in large masses of text). The obtained results present a satisfactory degree of precision for the proposed analysis, but not, of course, absolute precision, as despite the efforts to reduce the investigated topics to expressions as accurate and straightforward

TABLE 8.3

Compiled Policy Documents Organised by City and Topic

Topic	Type	City				
		Curitiba	**Goiânia**	**Manaus**	**Salvador**	**São Paulo**
Climate	Emissions Inventory	2013 Emissions Inventory—3rd Edition	2016 Emissions Inventory—2nd Edition		2013 Emissions Inventory—1st Edition	2009 Emissions Inventory—2nd Edition
	Plan	2009 Climate Action Plan	2011 Sustainable Goiânia	The Card of Amazonia	Coastal Management Plan	2009 Climate Change Municipal Policy
	Legislation/ Report	2016 Curitiba Strategic Actions: Climate and Resilience	2016 Local Policy on the Prevention, Reduction and Compensation of CO_2 and Other Vehicular Emissions	2011 Environmental Code (proxy to a local climate plan)	2015 Sustainable Development and Innovation Incentives Municipal Program	
	Report	2014 Curitiba's Environmental and Socio-Economic Vulnerabilities Evaluation			2015 Environmental and Sustainable Development Municipal Policy	
Urban Planning	Plan	2015 Master Plan	2007 Master Plan	2014 Master Plan	2016 Master Plan	2014 Master Plan
Land Use Regulations	Legislation	2000 Land Use Regulations	1994 Urban Zoning	2014 Land Use Legislation	2016 Land Use Regulations	2016 Zoning and Land Use Regulations
			2008 Environmental Zoning	2014 Urban Perimeter Legislation	2017 Building Code	2017 Building Code
			2008 Building Code	2014 Building Code		

(*Continued*)

TABLE 8.3 *(Continued)*

Compiled Policy Documents Organised by City and Topic

		City				
Topic	**Type**	**Curitiba**	**Goiânia**	**Manaus**	**Salvador**	**São Paulo**
Housing	Plan	Social Housing Municipal Plan	Provisions on Social Housing		2008–2015 Municipal Housing Plan	Municipal Housing Plan
Mobility	Legislation/Plan	2014–2017 Urban Mobility Plan	Urban Mobility Plan	2015 Urban Mobility Plan	2018 Sustainable Urban Mobility Municipal Plan	2015 Urban Mobility Plan 2001 Municipal Transportation System
Sanitation	Plan	2017 Sanitation Municipal Plan 2017 Water Distribution Municipal Plan 2017 Sewage Municipal Plan	Municipal Sanitation Plan	2014 Municipal Sanitation Plan	2010–2011 Municipal Sanitation Plan	2010–2030 Municipal Sanitation Plan
Waste Management	Plan	2017 Waste Management Municipal Plan	2014 Waste Management Local Plan	2010 Waste Management Municipal Plan	2012 Waste Management Municipal Plan	2014 Waste Management Municipal Plan
Drainage	Plan	2017 Urban Drainage Plan	2014 Drainage Regulations			2012 Drainage and Rainwater Management Manual
Disasters Management	Legislation/Plan	Municipal Civil Defense System	1998/2010 Municipal Civil Defense System	2014–2017 Municipal Public Health Plan	2015 Rain Contingency Plan 2017 Urban Arborization Plan 2013 Civil Defense System	2006 Municipal Civil Defense System

Public Health	Plan	2018–2021 Municipal Public Health Plan	2014/2017 Municipal Public Health Plan	2014–2017 Municipal Public Health Plan	2018–2021 Municipal Public Health Plan	2014–2017 Municipal Public Health Plan
International Cooperation	Legislation/ Platform/ Agreement	Participation in city's networks: ICLEI, CGLU, C40; International Affairs Secretariat reports; Nazca platform	Participation in Cities Networks: ICLEI, CGLU, UBERLAC, Sustainable and Emergent Cities Initiative; International Affairs Municipal Advisory reports; Nazca Platform; Vulnerability Assessment (UNFCCC Adaptation Platform)	Participation in City's Networks: ICLEI; Nazca Platform; Municipality's reports	Participation in City's Networks: ICLEI, CGLU, C40; Nazca Platform; Decentralised cooperation agreements	Participation in City's Networks: ICLEI, CGLU, C40; International Affairs Secretariat Reports Twin-Cities Agreement Regulations

as possible, some inconsistencies have persisted, mostly due to redundancies and ambiguities typical of the Portuguese language). The references obtained in the word search rounds were grouped in four nodules, corresponding to each of the climate action axes: Adaptation, Cooperation, Mitigation, and Resilience, as well as a fifth node, devoted to the identification of possible connections between public health and other local policies (Table 8.4). Lastly, several rounds of queries were performed, crossing the evidence compiled in each node with the cases, thus obtaining results that are further discussed in order to offer some insights.

8.3 Analysis and Results

8.3.1 National, State, and Local GHG Emissions

The Brazilian government, through the Ministry of Communications, Technology, Innovation, and Science (MCTIC), has been inventorying national GHG emissions since 1991. The 3rd National Inventory, published in 2016, with data up to 2010, indicates that the economic sectors responsible for the majority of the national emissions are "Agriculture", with 33% and "Energy", with 29%—e.g., Brazil is world's biggest ethanol producer (Pacheco & Silva, 2019)—followed closely by "Land Use and Forests," with 27% of the total emissions. "Industrial Processes" and "Waste Treatment," respectively, account for 7% and 4% of the national emissions (Figure 8.1). The estimated total emissions for 2010, calculated by the IPCC 2006 methodology (GWP-SAR), was 1,364,197 GtCO2e (Ministério da Ciência, 2018).

Data from the National Emission Register System (SIRENE) compiled in preparation for the 4th National Inventory indicates slightly higher emissions for the year 2015: 1,368,000 GtCO2e in total, but still well below the emission limit set in Brazil's NDC of 2,068,000 GtCO2e for 2020. The "Energy" sector, with 33% of emissions, seems to have surpassed the "Agriculture" sector, with 31% of emissions, followed by the "Land Use" sector, with 24%, "Industrial Processes" with 7% and "Waste Treatment," with 5% (Figure 8.1).

The evolution of the national emissions shows that, between 1990 and 2015, the total emissions line follows the same path as the "Land Use" emissions line, so that the emissions peaks observed in 1995 and 2005 correspond precisely to the "Land Use" high points, observed due to the increase in deforestation in the same periods (Figure 8.2), especially in the Amazon region (the Amazon, the lungs of the planet, cannot be recovered once it is gone). The August 2019 bushfires blazing in the Amazon are part of a larger deforestation crisis, accelerated by President Jair Bolsonaro's policies. According to an article published on 28 August 2019 in New York Times, hundreds of government workers on the front lines of enforcing Brazil's environmental

TABLE 8.4

NVivo Codebook

Node	Expressions
Climate change	"mudança climática" OR "mudanças climáticas" OR "mudança do clima" OR "mudanças do clima" OR "alteração do clima" OR "alterações do clima" OR "alteração climática" OR "alterações climáticas"
Adaptation	Adaptaç*
Risk	Risco*
Vulnerability	Vulnerabilidade OR vulnerave*
Water	Água OR "recursos hídricos" OR permeabilidade OR drenagem
International agreements	"acordo internacional" OR "acordos internacionais" OR "cooperação internacional" OR "cooperação descentralizada" OR "relações internacionais" OR "internacionalização"
International financing	"financiamento internacional" OR "financiamentos internacionais"
International organisations	"organização internacional" OR "organizações internacionais" OR "agência internacional" OR "agências internacionais" OR "ONU" OR "HABITAT" OR "Banco Mundial" OR "Banco Interamericano de Desenvolvimento" OR "BID" OR "World Bank" OR "Organização Mundial da Saúde"
International networks	"rede internacional" OR "redes internacionais" Or "ICLEI" OR "CGLU" OR "C40"
NDC	Ndc OR indc
PNA	PNA OR "Plano Nacional de Adaptação" OR "Plano Nacional de Adaptação à Mudança do Clima"
PNMC	"política nacional sobre mudanças do clima" OR "política nacional sobre mudança do clima" OR "política nacional de mudanças climáticas" OR "política nacional de mudança do clima" OR "PNMC"
Global warming	"aquecimento global"
Greenhouse effect	"efeito estufa"
Carbon emissions	"emissão de carbono" OR "emissão de gases" OR "emissões"
Energy	energia
Mitigation	Mitig*
Pollution	poluição
Resilience	Resiliência OR resiliente*
PNPDEC	"política nacional de proteção e defesa civil" OR PNPDEC OR "Sistema Nacional de Proteção e Defesa Civil" OR SINPDEC OR "Conselho Nacional de Proteção e Defesa Civil" OR CONPDEC
Disaster	Desastre* OR enchente* OR escorregamento* OR seca* OR INCÊNDIO*
Extreme events	"evento extremo" OR "eventos extremos" OR "evento climático extremo" OR "eventos climáticos extremos" OR evento*+extremo*
Resilience	Resiliência OR resiliente*
Sectoral public health plan for climate change mitigation and adaptation	"plano sectorial da saúde para mitigação e adaptação à mudança do clima" OR "PSMC-Saúde" OR "plano setorial da saúde para mitigação adaptação às mudanças climáticas" OR "plano sectorial da saúde para mitigação e adaptação à mudança climática"
Health	saúde

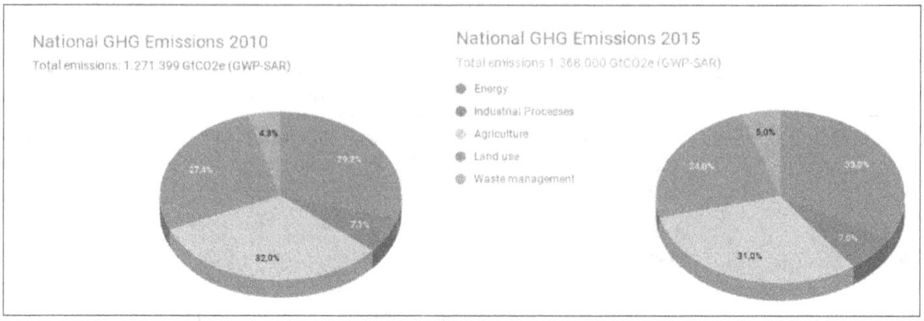

FIGURE 8.1
National GHG emissions in 2010 and 2015 (Ministério da Ciência, 2018).

laws signed an open letter warning that their work has been hampered by President Jair Bolsonaro, contributing to a rise in deforestation and the fires sweeping through the Amazon. One of the reasons for such heightened reaction is that the Amazon rainforest fires that have been blazing out of control in Brazil for weeks could have far-reaching effects on environmental and public health—e.g., air quality, weather patterns, loss of biodiversity deforestation, diseases, and mental health (Azevedo, 2017; Caviglia-Harris, 2018; Malhi et al., 2008).

The contribution of the "Land Use" and "Agriculture" sectors is comparatively higher than the other sectors', coherently with the fact that, although

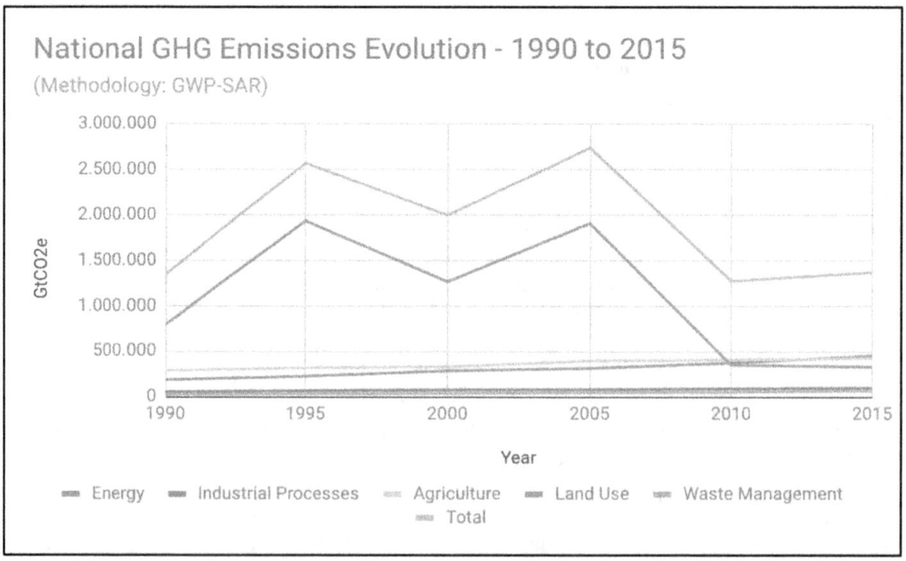

FIGURE 8.2
National-level GHG emissions (1990–2015) (Ministério da Ciência, 2018).

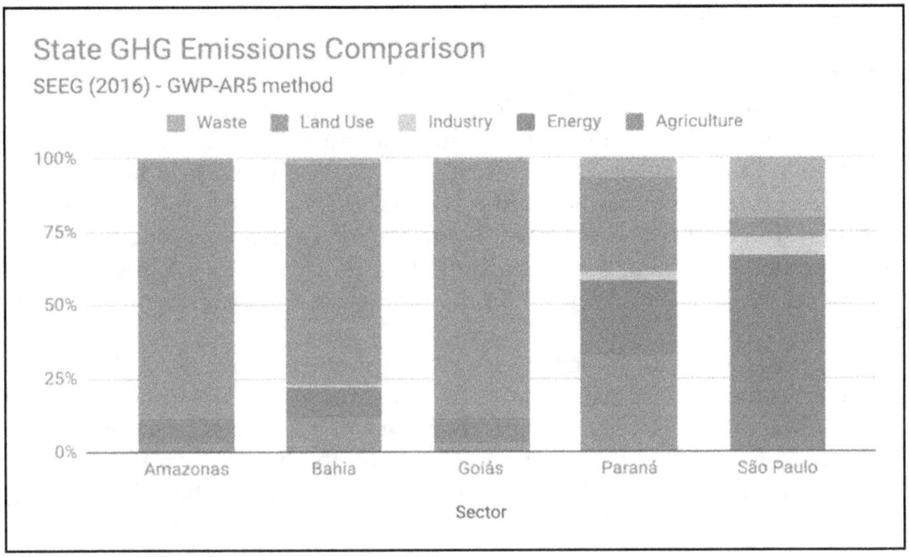

FIGURE 8.3
State-level GHG emissions (SEEG. Plataforma SEEG Brasil).

85% of the Brazilian population lives in cities, only 0.63% of the total 8,156 million km² of national territory correspond to urban areas (Farias et al., 2017). Therefore, at the national level, deforestation control is decisive to the fulfilment of the mitigation targets established in Brazil's NDC. The examination of the inventories of the states of Amazonas, Bahia, Goiânia, Paraná, and São Paulo, where the five cities investigated in this study are located, also corroborates the significant weight of the "Land Use" and "Agriculture Sectors" to the fulfilment of Brazil's NDC mitigation targets (Figure 8.3).

According to the SEEG GHG inventories (Amazonas and Goiás did not issue official GHG inventories. To fill in this gap, this study has used the SEEG platform emissions estimates, issued by the Climate Observatory according to the IPCC guidelines (De Azevedo et al., 2018)), the states that are closest to the national emissions framework are Paraná and Goiás. In both, the "Agriculture" sector is responsible for most of the emissions (33% in Paraná and 54% in Goiás), followed by the "Land Use" sector (32% in Paraná and 28% in Goiás). The "Industry" sector has practically no contribution to emissions in Goiás, while in Paraná, it accounts for 3% of emissions. Coherently, the "Energy" sector accounts for 25% in Paraná, while in Goiás, it corresponds to 13% of the emissions.

The states closest to the national emissions framework are Paraná and Goiás. In both states, the "Agriculture" sector is responsible for most of the emissions (33% in Paraná and 54% in Goiás), followed by the "Land Use" sector (32% in Paraná and 28% in Goiás). The "Industry" sector has practically no contribution to emissions in Goiás, while in Paraná, it accounts for 3% of emissions. Coherently, the "Energy" sector accounts for 25% in Paraná, while in Goiás, it corresponds to 13% of the emissions.

In the states of Bahia and Amazonas, the "Land Use" sector is mainly responsible for most of the emissions, 75% in Bahia and 88% in Amazonas. Nevertheless, there are some significant differences between the two states' inventories: in Bahia, the "Agriculture" and "Energy" sectors have higher weight than in Amazonas, accounting respectively for 12% and 10% of total emissions, in comparison to only 3% and 8% in Amazonas. The "Industry" sector practically does not contribute to the emissions of the state of Amazonas, while in the State of Bahia, industrial emissions represent only 1% of the total.

In comparison to the national emissions framework, the most discrepant inventory is the state of São Paulo's, the most urbanised and industrialised state in the country. In São Paulo's inventory, the "Energy" sector alone is responsible for 53% of the emissions. Although the "Agriculture" sector comes in second place, with 20% of the State's emissions, the other sectors, "Energy," "Industry," and "Waste," more closely related to urban activities, account for 75% of the total carbon emitted in the State.

About the five investigated cities, it is essential to highlight that all of them have issued legal provisions determining the edition of local emissions inventories, as Insofar, Curitiba, Goiânia, Salvador, and São Paulo have all issued local emissions inventories, in compliance with municipal regulations. Manaus 2014 Master Plan foresees the implementation of a local vehicle and industrial emissions control programme, as the city—located in the heart of the Amazon Forest—has an estimated fleet of 713,000 motor vehicles (Souza, 2017) and about 500 industries installed in the Industrial Zone (Zona Franca), operating mainly in the segments of electronics, two-wheeled vehicles, and chemicals (Suframa, 2019). Regardless, the city has not yet issued an emissions inventory.

A study published by Nobre et al. (2016) indicates that the Amazon Forest can undergo irreversible changes due to the continuous loss of biodiversity and deforestation. According to the authors, either a temperature increases above 4°C or deforestation exceeding 40% of the forest area would be sufficient to lead to the definitive s*avanisation* of the region, and in the last 60 years, the temperature in the Amazon has increased by about 1°C, with deforestation of about 20% of the total area of tropical forest. The Amazonas emissions inventory published in the SEEG platform, in turn, indicates that 89% of total emissions from the state—where the city of Manaus stands—result from changes in land use and 3% from agricultural activities. Following a downward trend in emissions observed between 2003 and 2009, the survey of the SEEG platform indicates that emissions from the state of Amazonas have been increasing since 2014.

A study published by Liu et al. (2016) indicates that pollution from the city of Manaus, more specifically, nitrogen oxides produced by automotive vehicles and industries, significantly interferes with the climate and the ecosystem equilibrium in the Amazon Forest by changing isoprene oxidation, thus interfering with the formation of rain in the region. In view of this situation, the absence of an official emissions inventory for the city of Manaus is hugely worrying, especially considering the significant weight that the Amazon Forest preservation represents for the country's (if not the world's) climate balance as a whole and the fulfilment of the goals assumed by the Brazilian government in its NDC.

As for the other surveyed cities, Curitiba, Salvador, and Goiânia have all issued emissions inventories using the Global Protocol for Community-Scale GHG Emission Inventories (GPC), a method developed specifically for cities by the World Resources Institute (WRI), the C40 Group and ICLEI, based on parameters signed by the Intergovernmental Panel on Climate Change (IPCC). Only the city of São Paulo has applied the IPCC 2006 method (the same used by the Federal Union in the National Inventory). There are no significant incompatibilities between the two methods, thus allowing some comparison between the four local inventories (Figure 8.4).

Considering the three cities that used the GPC method, the sector that occupies the first place in all three inventories is the "Transportation" sector: 78% of emissions in Salvador, 73% in Curitiba, and 58.47% in Goiânia. In Salvador and Goiânia, the second place in emissions is occupied by the "Stationary Energy" sector, with 20.58% in Goiânia and 18% in Salvador. In Curitiba, the "Stationary Energy" sector occupies the third place, with 12%, just behind the "Waste" sector, with 15% of total emissions. In Salvador and Goiânia,

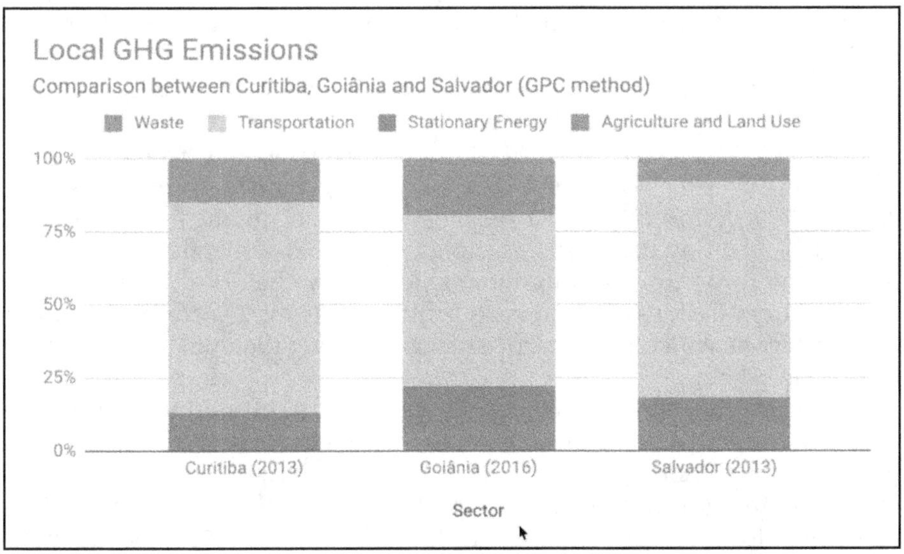

FIGURE 8.4
Local-level GHG emissions (Abreu, 2017; Prefeitura de Curitiba, 2016; Prefeitura de Salvador, 2016).

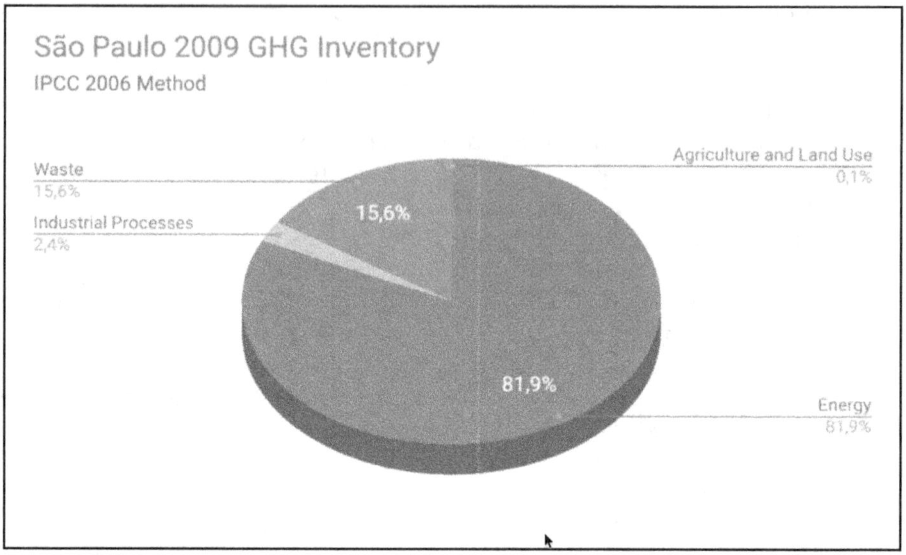

FIGURE 8.5
São Paulo's 2009 GHG inventory (Instituto Ekos Brasil & Geolock Consultoria e Engenharia Ambiental, 2013).

the "Waste" sector also has critical participation, occupying the third place, respectively, with 8% and 9.55% of the emissions. Among the three cities, only the city of Goiânia recorded some contribution to the "Agriculture and Land Use" sector, with 1.4% of total emissions (Figure 8.5).

São Paulo's emissions inventory, prepared by the IPCC 2006 method, has computed emissions in four sectors: "Agriculture, Forestry and other Soil Use" (AFOLU), "Energy," "Industrial Processes and Product Usage" (IPPU), and "Waste." The first place, accounting for 82% of the emissions, was the "Energy" sector, covering mostly the fuel combustion emissions (99.7%) and in a tiny part fugitive energy (0.3%). Secondly, the "Waste" sector accounted for 16% of emissions, followed by the "IPPU" sector, with about 2% of emissions. The emissions of the "AFOLU" sector were practically irrelevant, which is not surprising considering that most of São Paulo's territory is urban.

Considering that "Energy" emissions are a product of energy consumption in transportation and buildings, it is possible to conclude that the emissions distribution in the city of São Paulo does not differ from the pattern observed in the other three cities, where "Transportation" and "Stationary Energy" are responsible for more than 75% of all three cities' emissions. Another noteworthy observation, concerning decentralised cooperation, lies in the fact that the majority of the official emissions inventories had the technical support of an international actor, as shown in Table 8.5.

TABLE 8.5

State- and City-Level GHG Inventories[a]

Entity	Edition	Year	Method	Source	Technical Support
Manaus	No official emissions inventory				
Amazonas	No official emissions inventory				
Salvador	1st	2013	GPC	Prefeitura De Salvador (2016)	ICLEI WRI Brazil
Bahia	1st	2008	IPCC 2006	Governo Do Estado Da Bahia (2010)	None
Goiânia	2nd	2016	GPC	Abreu (2017)	Inter-American Development Bank
Goiás	No official emissions inventory				
São Paulo	2nd	2009	IPCC 2006	Instituto Ekos Brasil and Geolock Consultoria e Engenharia Ambiental (2013)	World Bank
Estado de São Paulo	1st	2008	IPCC 2006	Governo Do Estado De São Paulo (2011)	British Embassy
Curitiba	3rd	2013	GPC	Prefeitura De Curitiba (2016)	ICLEI
Paraná	1st	2012	IPCC 2006	Governo Do Estado Do Paraná (2015)	None

[a] During the study, the City of Goiânia had not officially published the results of its emissions inventory. Request for access to the contents of the inventory based on the right of information, made at the Transparency Portal of the City of Goiânia, was not met.

The comparison between the local, state, and national inventories allows for some additional considerations. Firstly, the main connection point between the national inventory and the state and municipal inventories is the "Waste" sector, which highlights the importance of synergies and coordination between the National, State and Municipal Sanitation and Waste Management Policies in the country.

Secondly, the main connection point between state and municipal inventories is the "Energy" sector, which in cities is strongly related to the high energy consumption by both the "Transport"—mostly, fossil fuels—and the "Stationary Energy" sectors—hydropower and thermoelectric energy. This points out to the need not only to structure state and municipal policies focused on energy efficiency and sustainability but also to coordinate these policies, as well as the resulting plans, programmes and projects (according to the 1988 Brazilian Constitution, the Union and the Federated states, not the municipalities, are in charge of all energy sources regulation, production and distribution. Therefore, municipality actions in this field are restricted mostly to energy consumption). This also highlights the strategic importance of Urban Mobility and Public Transport Plans, in their various modalities, including mobility on foot and by bicycle, to emissions control at both the state and the local level.

Finally, the two main connection points between the national and state inventories, respectively, the "Agriculture" and "Land Use" sectors, present very little weight as emission sources in the local inventories. This circumstance, however, does not authorise the discarding of these sectors as focal points of public policies, also at the local level, especially considering the environmental services provided by the environmental protection areas and the preventive role to environmental degradation that can be played by sustainable agriculture in the urban fringes of large cities. The comparison between national, state, and local inventories indicates that cities can contribute to Brazil's NDC more in terms of climate adaptation and resilience than in emissions reduction. Nevertheless, mitigation actions at the local level are also of critical importance, as they contribute to air pollution control, microclimate regulation, and the promotion of public health and quality of life.

8.3.2 Local Climate Plans

Starting in 2009, when the National Climate Change Policy (PNMC) was published, all investigated cities, except for Manaus, have issued their own climate plans. The cities of Curitiba and São Paulo edited their climate plans in the same year of 2009, contemplating Mitigation, Adaptation, and Resilience actions. Goiânia, in turn, edited its Sustainable Goiânia Plan in 2011, within the framework of the Inter-American Development Bank's Emerging and Sustainable Cities Platform (IDB). Goiânia's Plan is broader than a strict climate plan but contemplates in-depth climate actions. The city of Salvador, in a similar way, incorporated climate actions into two regulations issued in 2015, one approving the Municipal Program for Sustainable Development and Incentives for Innovation and another approving the Environmental and Sustainable Development Policy of the Municipality. For the city of Manaus, its 2011 Environmental Code has been used in this study as a proxy of "climate plan," since it also contemplates, indirectly, some local climate actions. Table 8.6 and Figure 8.6 present the qualitative analysis results of climate plans.

In four of the five local climate plans, except Goiania's, Mitigation was the most referenced climate axis, which is a consequence of the investigated

TABLE 8.6

Climate Action References in Local Climate Plans

Climate Plan	Adaptation	Cooperation	Mitigation	Resilience
Curitiba (2009)	3	1	7	0
Goiânia (2011)	63	15	46	26
Manaus (2011)	35	1	45	0
Salvador (2015)	28	4	57	2
São Paulo (2009)	23	1	99	7
Total	152	22	254	35

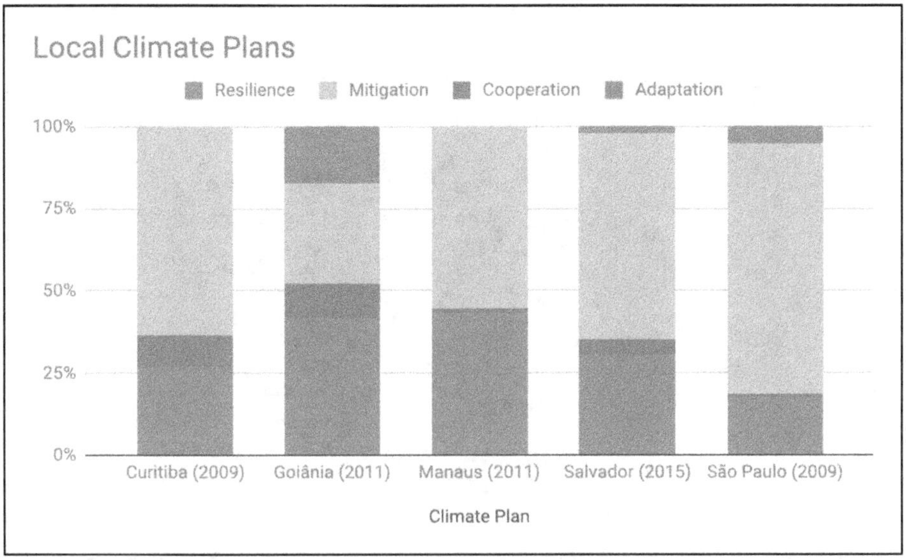

FIGURE 8.6
Climate action references in local climate plans.

cities' concern with the consumption of fossil fuels and sustainable waste management. In the case of São Paulo, more than 75% of the references relate to Mitigation.

This finding is consistent with the fact that combating air pollution, mainly particulate matter produced by motor vehicles, is a critical challenge for the city of São Paulo (World Health Organization—WHO, 2016). According to Saldiva (2018), the reduction of air pollution in the city of São Paulo to the levels recommended by the World Health Organization (WHO) could bring about an increase in the life expectancy of the population of about three years. Thus, alternatives to fossil fuels in urban mobility are a crucial issue in São Paulo's climate strategy. Besides, in these four investigated cities, adaptation was the second most mentioned climate action axis. Goiânia's Climate Plan presented a higher number of references to Resilience in comparison to the other cities' Climate Plans. In all cities, the cooperation axis was the least mentioned. Considering the total number of references observed in the five Climate Plans, Mitigation obtained 54% of the references, Adaptation, 32.8%, followed by references to Resilience, with 7.6%, and Cooperation, with only 4.8% (Figure 8.7).

8.3.3 City Master Plans

In Brazil, according to the 1988 Brazilian Constitution (Brasil), whereas environmental protection is a competence shared by all the federated entities, urban planning, and urban management are, mostly, local or municipal

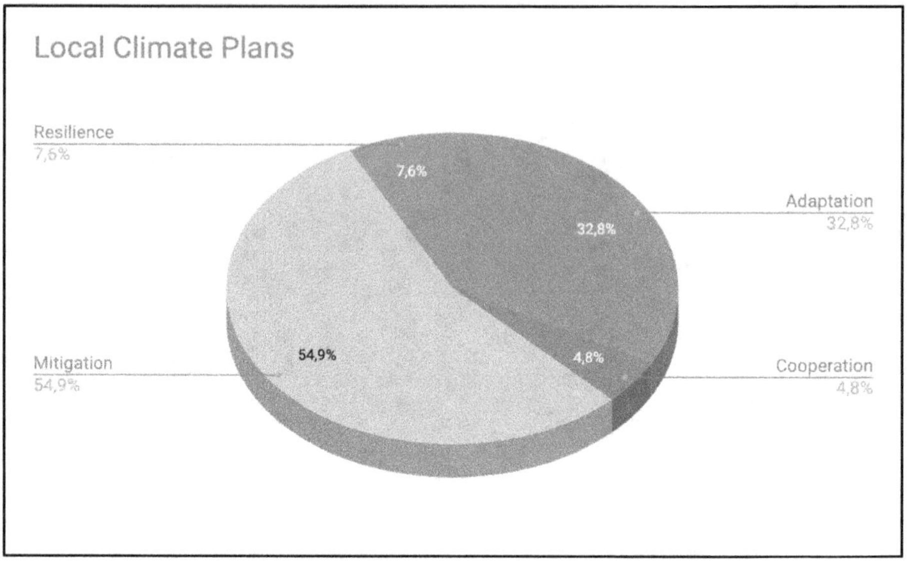

FIGURE 8.7
Climate action references in local climate plans.

issues. Municipalities with a minimum population of 20,000 inhabitants are hence obligated to edit their Master Plans, not only to set their essential territorial planning guidelines but also to coordinate and integrate land use, environmental protection, and sectoral policies and regulations at the local level. Hence, Master Plans should present at least some reference to climate action amongst their provisions. All five investigated cities have issued Master Plans following the guidelines and minimum requirements set by the 2001 Statute of the City, Brazil's national law on urban planning and management (Table 8.7).

All of the investigated Master Plans contain references to Adaptation, Mitigation, and Resilience actions, except for Manaus, which did not show results for Resilience actions. Furthermore, only Curitiba's Master Plan has

TABLE 8.7
Climate Action References in Master Plans

Master Plan	Adaptation	Cooperation	Mitigation	Resilience
Curitiba (2015)	44	1	22	9
Goiânia (2007)	59	0	10	7
Manaus (2014)	42	0	3	0
Salvador (2016)	143	0	63	45
São Paulo (2014)	253	0	62	42
Total	541	1	160	253

presented one reference to decentralised cooperation actions. All five Master Plans have shown a higher number of references to Adaptation actions, in the first place, and to Mitigation actions, in the second place. Taking the sum of all the Master Plans' references to climate action, the most mentioned axis was Adaptation, with 67.2% of the references, followed by Mitigation, with 19.9%, Resilience, with 12.8% and Cooperation, with 0.1% (Figure 8.8).

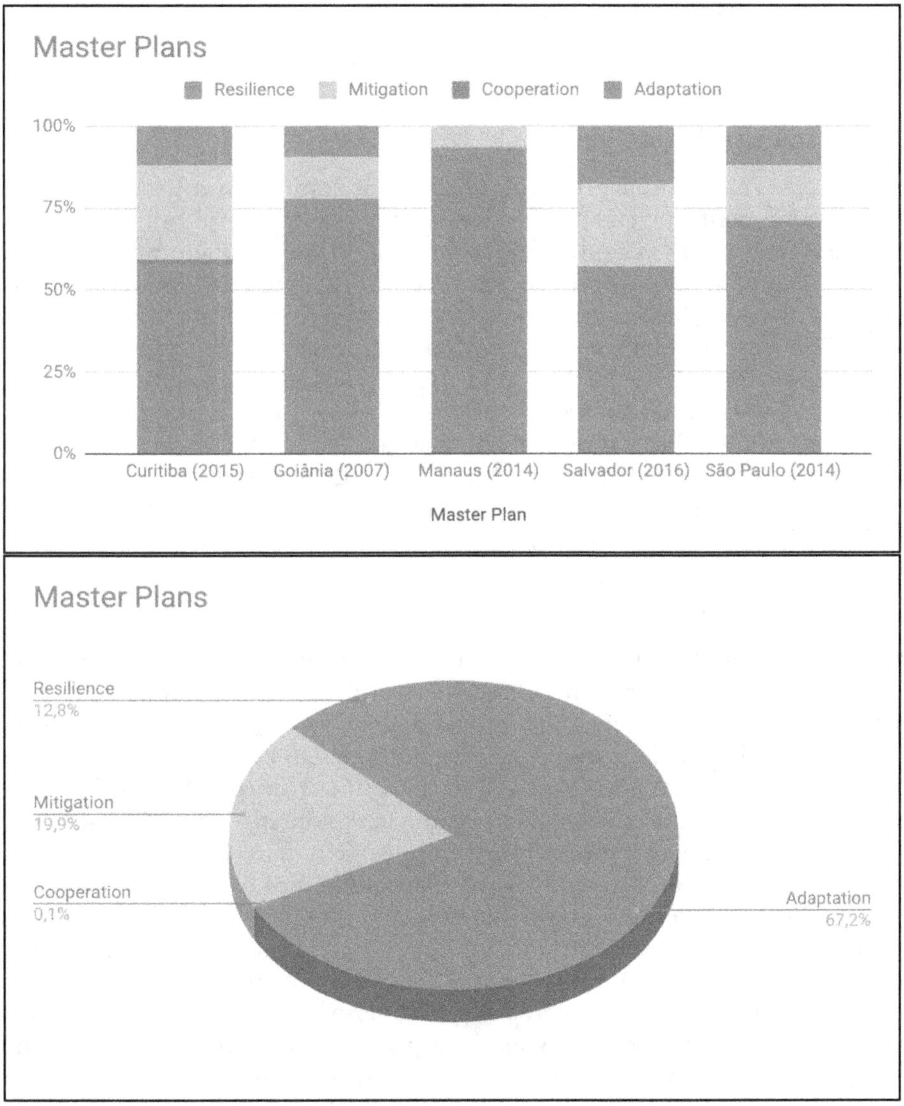

FIGURE 8.8
Climate action references in master plans.

TABLE 8.8

Climate Action References in City Land Use Regulations

City	Adaptation	Cooperation	Mitigation	Resilience
Curitiba	137	2	23	31
Goiânia	358	13	61	552
Manaus	22	0	7	4
Salvador	70	0	23	10
São Paulo	64	0	25	8
Total	651	15	139	605

8.3.4 Land Use Regulations

For the sake of analysis, all regulations about zoning, land use, parcelling, and development, such as zoning laws, land use laws, and building codes, have been gathered under the unique topic "land use regulations" (Table 8.8).

In Manaus, Salvador, and São Paulo, the most referenced climate action axis was, firstly, Adaptation, followed by Mitigation and Resilience. The cities of Goiânia and Curitiba were the only ones to present references to decentralised cooperation actions in their land use regulations. In the city of Goiânia, the most mentioned axis was Resilience, followed by Adaptation and Mitigation. In the city of Curitiba, the most mentioned axis was Adaptation, followed by Resilience and Mitigation (Figure 8.9). Considering the overall sum of references, the Adaptation axis was the most mentioned, with 46.2%, followed by Resilience, with 42.9%, Mitigation, with 9.9% and Cooperation with only 1.1% (Figure 8.9).

8.3.5 Sectoral Plans

According to the 1988 Brazilian Constitution, municipalities have the power to provide for all essential public services at the local level, following both national and state directives. Thus, the data survey involved the following public policies: housing, mobility, sanitation, solid waste, urban drainage, disasters, and health. The goal was to detect the incorporation of local climate actions also in these fields. According to the Brazilian 1988 Constitution and the 2001 Statute of the City, local public policies must be coordinated and integrated, following the city's Master Plan guidelines.

According to national regulations, the edition of local plans on sanitation, waste management, and urban mobility is mandatory for municipalities (Brasil; Brasil; Brasil). Drainage plans are not mandatory; however, due to extreme weather events, an increasing number of cities have recently issued, if not proper plans, at least specific regulations on urban drainage and rainwater management. Concerning housing, even though local Housing Plans are not mandatory, in order to access National Social Housing Fund resources, municipalities must edit Social Housing Plans, focused on housing provision for low-income families (Brasil).

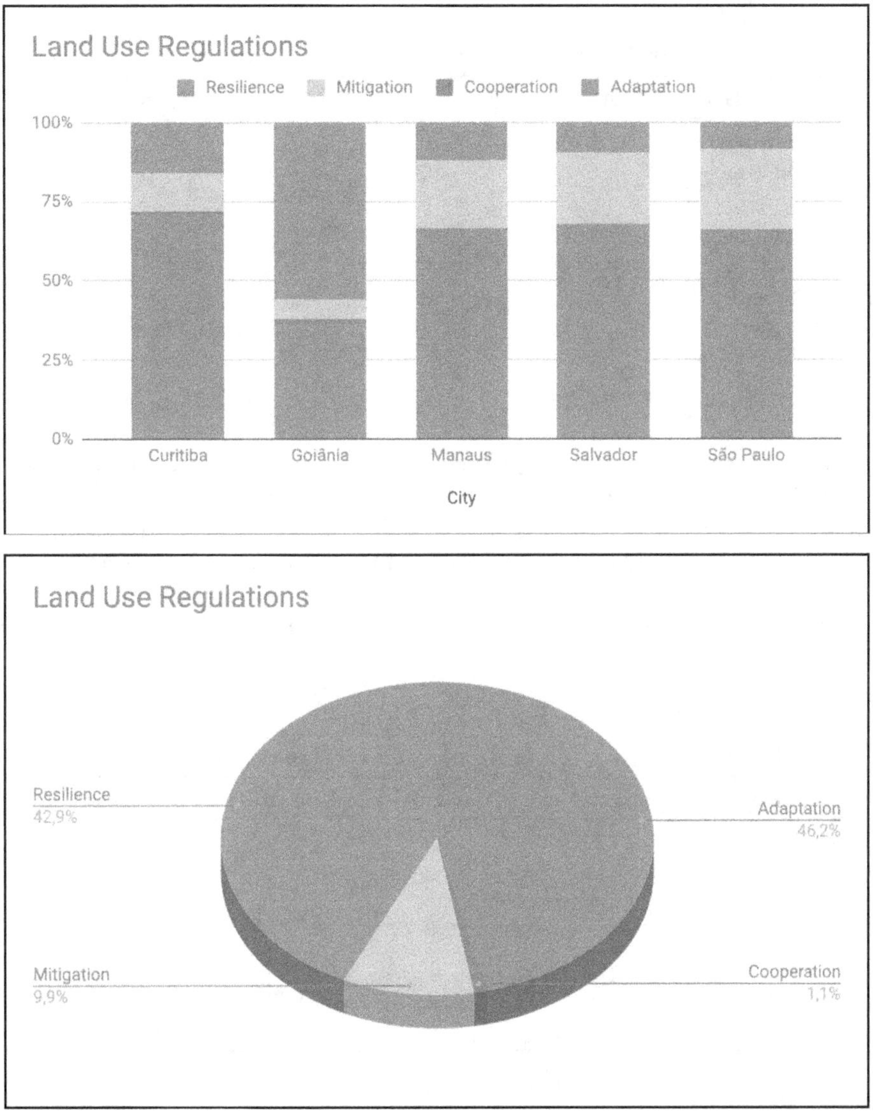

FIGURE 8.9
Climate action references in city land use regulations.

Regarding disasters, the National System of Civil Protection and Defense, regulated by a 2012 National Law (Brasil), expressly integrates the Municipalities as Members, with a set of local attributions, amongst them the powers to incorporate protection and civil defence actions in municipal planning, to map disaster risk areas and to declare emergency situations and public calamity at the local level. Thus, it is incumbent on the Municipalities,

TABLE 8.9

Climate Action References in Curitiba's Sectoral Plans

CURITIBA	Adaptation	Cooperation	Mitigation	Resilience
Disasters	9	0	0	24
Drainage	531	1	35	179
Health	2	0	0	2
Housing	112	10	23	4
Mobility	1	0	2	0
Sanitation	1318	18	37	64
Waste	59	4	37	5

if not to draft typical local disaster prevention and mitigation plans, at least to structure their respective civil defence at the local level.

Lastly, concerning Public Health, the 1988 Brazilian Constitution has expressly granted Municipalities the powers to provide, with the technical and financial cooperation of the union and the states, health care services for the population at the local level. The Municipalities are thus included in the Unified Health System (SUS), and must, therefore, prepare and periodically update their respective Local Health Plans (Brasil). Tables 8.9–8.13 and Figures 8.10–8.14 present the analysis results obtained for each investigated city.

TABLE 8.10

Climate Action References in Goiânia's Sectoral Plans

Goiânia	Adaptation	Cooperation	Mitigation	Resilience
Disasters	20	0	0	2
Drainage	57	0	3	4
Health	44	0	1	6
Housing	0	0	0	0
Mobility	0	0	0	0
Sanitation	0	0	0	0
Waste	76	8	86	23

TABLE 8.11

Climate Action References in Manaus' Sectoral Plans

Manaus	Adaptation	Cooperation	Mitigation	Resilience
Disasters	39	6	0	2
Drainage	0	0	0	0
Health	39	6	0	2
Housing	0	0	0	0
Mobility	16	0	131	107
Sanitation	553	3	30	12
Waste	40	0	117	3

TABLE 8.12

Climate Action References in Salvador's Sectoral Plans

Salvador	Adaptation	Cooperation	Mitigation	Resilience
Disasters	99	0	2	118
Drainage	0	0	0	0
Health	195	13	6	52
Housing	0	0	0	0
Mobility	36	9	62	11
Sanitation	18	0	1	0
Waste	1	0	0	0

TABLE 8.13

Climate Action References in São Paulo's Sectoral Plans

São Paulo	Adaptation	Cooperation	Mitigation	Resilience
Disasters	25	0	0	41
Drainage	1666	4	202	425
Health	271	0	3	0
Housing	103	0	8	12
Mobility	43	7	228	4
Sanitation	721	10	202	93
Waste	84	6	153	48

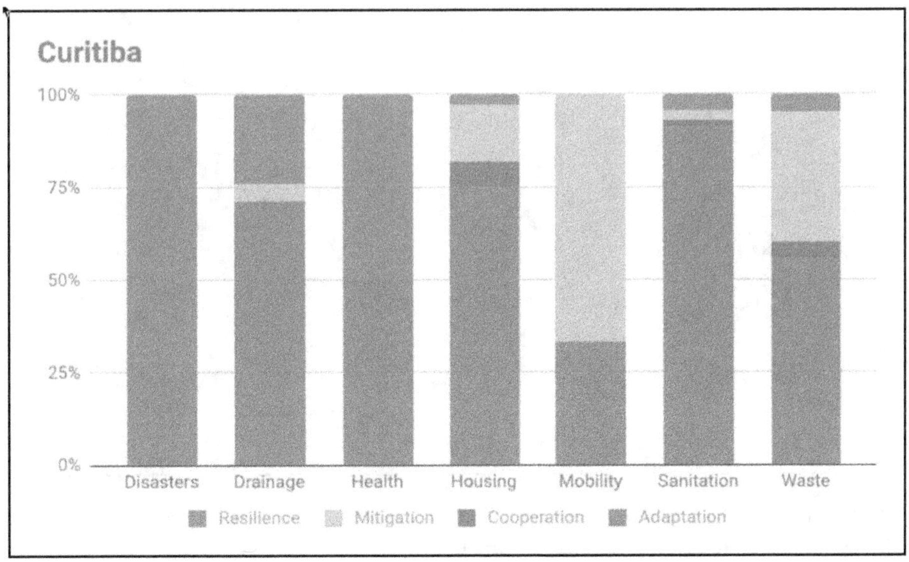

FIGURE 8.10
Climate action references in Curitiba's sectoral plans.

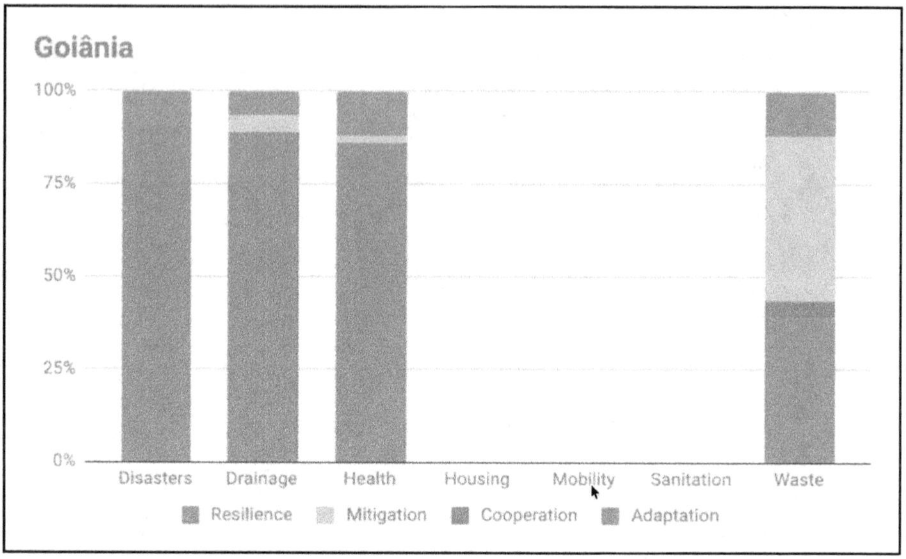

FIGURE 8.11
Climate action references in Goiânia's sectoral plans.

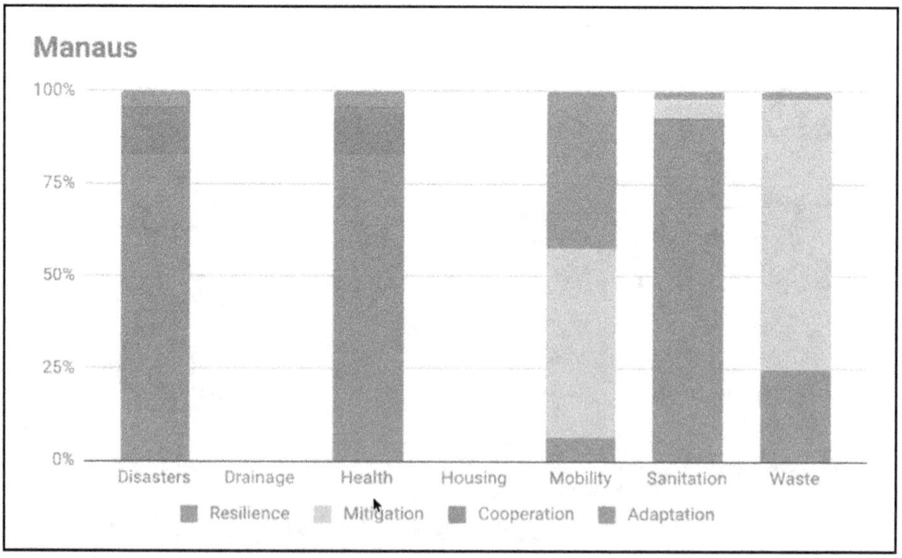

FIGURE 8.12
Climate action references in Manaus' sectoral plans.

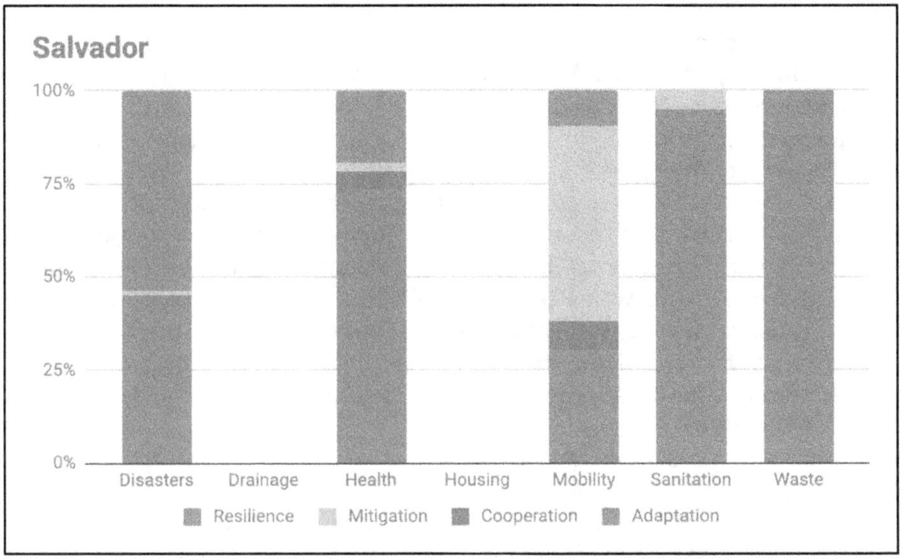

FIGURE 8.13
Climate action references in Salvador's sectoral plans.

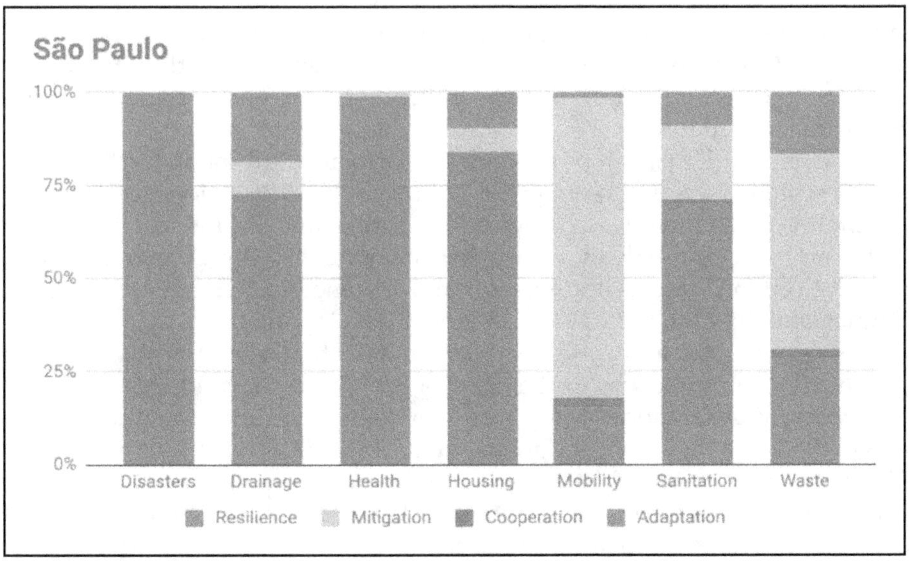

FIGURE 8.14
Climate action references in São Paulo's sectoral plans.

Among the five investigated cities, only São Paulo and Curitiba presented references to climate actions in all sectoral policies: Disasters, Drainage, Health, Housing, Mobility, Sanitation, and Waste management. Goiânia did not present references to climate actions in Housing, Mobility, and Sanitation. Manaus and Salvador, in turn, did not present references to climate actions in Drainage and Housing.

Firstly, concerning Disaster Management, the most referenced climate action axis in the cities of Curitiba, Salvador—the latter, the Brazilian city with the most significant number of inhabitants in the risk area (IBGE, 2018)—and São Paulo was Resilience, followed by Adaptation. In Goiânia, the axis most commonly referred to was Adaptation. Unlike the other investigated cities, the Municipality of Manaus dealt with the prevention and remediation of disasters in its Municipal Health Plan and not, as expected, in civil defence regulations.

In what concerns Drainage, the most referenced climate action axis in the cities of Curitiba, Goiânia, and São Paulo was Adaptation, followed by Resilience, in second place, and Mitigation, in third place. Only the cities of São Paulo and Curitiba presented references to cooperation actions in urban drainage. Manaus and Salvador did not present any results in this field, as the document survey did not locate any documents specifically dedicated to urban drainage for both cities.

Only São Paulo and Curitiba presented references to climate action in Housing policy documents. Adaptation was the axis with the highest number of references, followed, in the case of São Paulo, by Resilience, in the second place, and Mitigation, in the third place; in the case of Curitiba, Mitigation occupied the second place, followed respectively by Cooperation, in third, and Resilience, in fourth.

In the field of Urban Mobility, except for Goiânia, all the investigated cities presented a higher number of references to Mitigation, followed by Adaptation in São Paulo, Salvador and Curitiba, and Resilience in Manaus. Only São Paulo and Salvador presented references to the Cooperation axis. The city of Goiânia did not present any references to climate action in its Municipal Urban Mobility Plan, still under development.

Concerning Sanitation, the axis of climate action with the most significant number of references in the cities of São Paulo, Salvador, Manaus, and Curitiba, was Adaptation. Mitigation was the second most mentioned axis in São Paulo, Salvador, and Manaus; in Curitiba, it was the Resilience axis. There are references to Cooperation only in the Sanitation Plans of Curitiba and São Paulo. The Municipal Sanitation Plan of Goiânia (in preparation) did not present references to any climate action axis.

For Waste Management, Adaptation was the most referenced axis in both Curitiba and Salvador Waste Management Plans. In the cities of São Paulo, Manaus, and Goiânia, Mitigation was the most referenced axis, followed by Adaptation and Resilience. Only the cities of Curitiba, Goiânia and São Paulo presented references to the Cooperation axis in their Local Waste Management Plans.

Finally, concerning Public Health, Adaptation was the most referenced climate action axis in all five Local Health Plans, followed by Mitigation, in São Paulo, and by Resilience, in Curitiba, Goiânia, and Salvador. References to Coordination actions were found only in the Health Plans of Salvador and Manaus.

A word search focused on the Municipal Public Health Plans pointed out that only Salvador's contains references to the expression "climate change," detected in two contexts: the articulation of health and climate agendas and the prevention of arboviruses such as yellow fever. Furthermore, based on the assumption that health promotion is a necessary component of sustainable urban development, another specific Word Search was carried out, aimed at locating references to "health" in all the investigated local climate plans, master plans, land use regulations, and other sectoral plans (Table 8.14). The results confirm the initial hypothesis that "health" is an integral dimension of urban planning in the five investigated cities, including concerning climate change issues.

Figure 8.15 illustrates that there are references to health in the master plans and the land use regulations of all the investigated cities. All local climate plans, except Curitiba's, also make references to health. Finally, there are references to health in most of the sectoral plans of the investigated cities, especially sanitation and waste management plans. Overall results are sorted by sectoral policy and climate action axis, in turn, the following findings are presented (Table 8.15).

TABLE 8.14

References to Health in City Sectoral Plans

Documents	Curitiba	Goiania	Manaus	Salvador	São Paulo
Climate plan	0	5	23	34	8
Master plan	43	46	9	69	33
Land use regulations	62	37	7	11	25
Disasters	1	0	948[a]	21	3
Drainage	6	0	0	0	31
Housing	84	1	0	0	6
Sanitation	279	0	18	0	343
Waste	110	124	82	0	224

[a] Disaster management is a component of the Manaus Municipal Public Health Plan, hence returns the highest number of results.

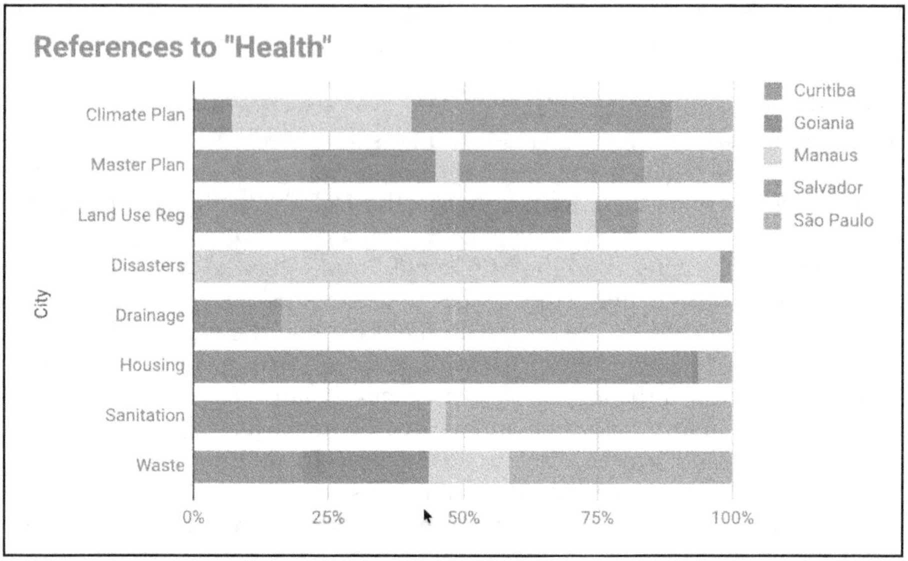

FIGURE 8.15
References to health in city sectoral plans.

Adaptation was the most referenced axis in the majority of the sectoral policies documents: disasters, drainage, housing, sanitation, and public health. The mobility and waste management document presented more references to mitigations. Resilience was the second most referred axis in drainage, disaster, and mobility. Considering the total number of references, the adaptation axis was the most referenced, with 71.2%, followed by resilience, with 14.3%, mitigation, with 13.3% and cooperation, with 1.2% (Figure 8.16).

TABLE 8.15

Climate Action References in Sectoral Policies

Sectoral Policy	Adaptation	Cooperation	Mitigation	Resilience
Disasters	192	6	2	187
Drainage	2254	5	24	608
Housing	215	10	31	16
Mobility	96	16	423	122
Sanitation	1437	31	270	169
Public health	561	19	10	62
Waste	260	18	394	79
Total	5015	105	1370	1243

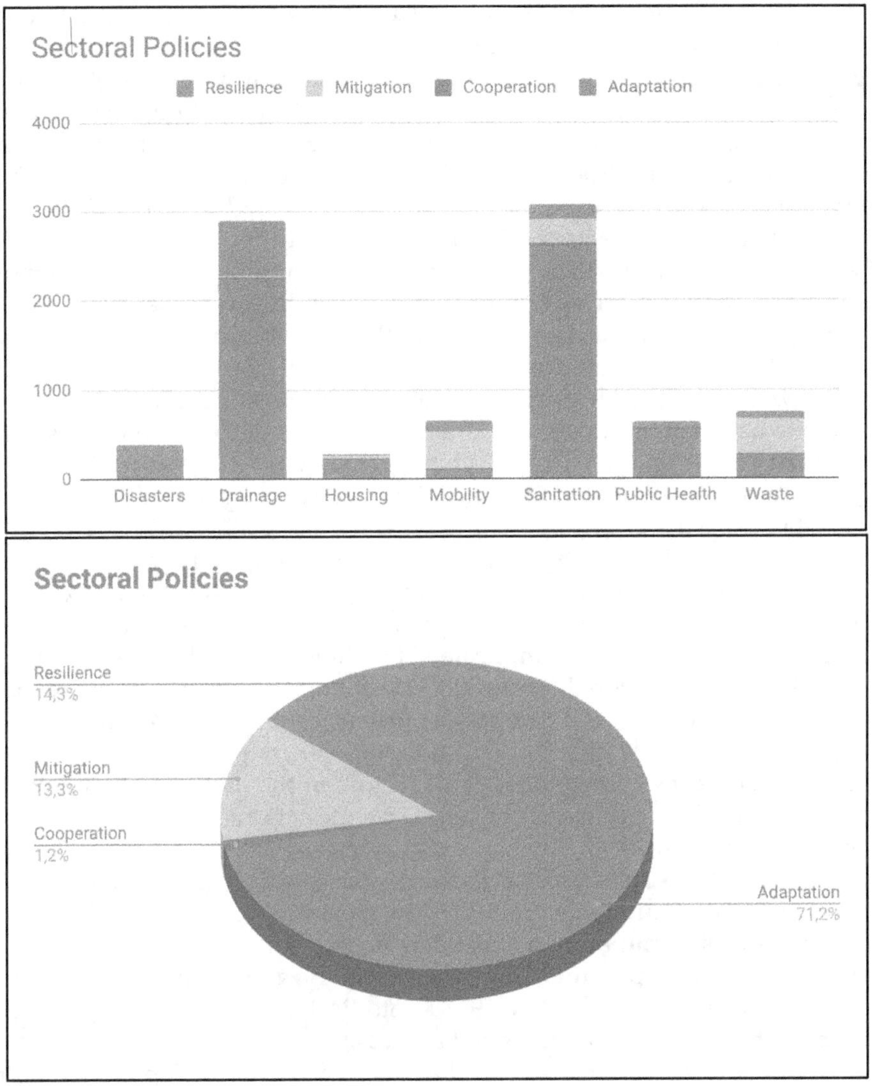

FIGURE 8.16
Climate action references in sectoral policies.

8.4 Findings and Discussion

8.4.1 Institutional Frameworks and Climate Action in Brazil

The 1992 United Nations Conference on Environment and Development (UNCED) fostered the establishment of the 1992 United Nations Framework

Convention on Climate Change (UNFCCC). As its host, Brazil has been, for some time, one of the Global South's main protagonists in climate change negotiations (the situation has changed dramatically with the election of Brazilian President Jair Bolsonaro in October 2018, a far-right, open climate sceptic, anti-environmentalist politician. Just before taking office, the new Presidential Cabinet withdrew Brazil's proposal to host the 25th UNFCCC Conference of the Parties (COP 25), to be held in December 2019, and since the very first days of the federal mandate, the appointed Minister of Environment, Ricardo Salles, has been making significant cuts on climate programmes and initiatives at the national level. In this context, it is essential to highlight that this chapter portrays the institutional framework before President Bolsonaro's inauguration, as comprised in the years 2017 and 2018. In other words, this study discloses a much more positive standing of Brazil in climate action than it actually is today (Lahsen, 2009). Nevertheless, the country has only issued a national climate change policy (PNMC) long after the adoption of the 1997 Kyoto Protocol, through the approval of a national law in 2009 (Brasil).

In resonance with the Brazilian Federation's legal framework, PNMC expressly contemplates amongst its primary goals the promotion of climate adaptation at all levels—national, regional, and local—with the participation and collaboration of all implicated stakeholders. Furthermore, all measures taken at the national level must consider and integrate regional and local actions taken by both public and private entities. PNMC also fosters integration between mitigation and adaptation strategies at all federative levels, as well as the engagement of all sectors in the implementation of climate policies, plans, programmes, and strategic actions.

Under the 1992 UNFCCC, all states signatory to the 2015 Paris Agreement, regardless of their level of economic development, are obliged to comply with the voluntary commitments embodied in their respective NDC. The NDC mechanism was formulated to effectively engage and commit the UNFCCC member states in the pursuit of the 1.5°C target. As a result, it can be affirmed that, unlike its predecessor—the 1997 Kyoto Protocol—the Paris Agreement is a legally binding agreement (Rei et al., 2017).

Following its adherence to the Paris Agreement in September 2016, Brazil's national government duly presented its NDC in 2017, with an emissions reduction target of 37% below the national emissions average until 2025, and a complementary reduction target of 43% below 2005 levels until 2030 (Brasil. Sumário Executivo, 2017), as well as adaptation measures, all based in the 2016 National Adaptation Plan (PNA) (Ministério do Meio Ambiente (MMA), 2016). The 2016 PNA main objectives are to promote risk management and reduction at the national level; to guide the expansion and dissemination of knowledge; to foster cooperation and coordination between the public and private sectors, and to identify and propose adaptation measures. PNA covers a total of 11 strategic sectors: industry, vulnerable social groups, water, infrastructure, coastal zone, food security, agriculture, biodiversity, cities, health, and disaster prevention, and mitigation.

With regard specifically to the linkages between climate change and public health, the Ministry of Health has edited a Sectoral Health Plan for Mitigation and Adaptation to Climate Change in 2013 (Ministério da Saúde, 2013). This plan's main objective is to organise the public health services response to climate change in terms of both adaptation and mitigation, through four lines of action: health surveillance, health care, health promotion and education, and health research, each pointing out objectives, indicators, expected results, and the governmental agency responsible for each action.

The 2013 Sectoral Health Plan addresses the critical challenges of climate change to public health. Firstly, deaths and morbidities due to disasters and communicable diseases such as dengue fever, yellow fever, Zika virus, and leishmaniosis. Secondly, deaths and morbidities due to non-communicable diseases such as cardiovascular and respiratory diseases, linked to air pollution, and also skin cancer, linked to the destruction of the ozone layer. The plan highlights the vital role to be played by local authorities regarding resilience, adaptation, and mitigation. In terms of resilience, the plan focuses on natural disasters caused by extreme weather events. In terms of adaptation, the plan points out local actions in the fields of housing, sanitation, and green areas. Lastly, for mitigation, the plan highlights the importance of active urban mobility in order to target air pollution and promote physical activity.

The 1988 Brazilian Constitution expressly grants municipalities the powers to act on environmental preservation as well as on urban planning. Thus, climate actions are expected to permeate both environmental and urban policies at the local level, either in local climate plans or in local regulations in general, such as master plans, land use regulations, and sectoral policy plans.

Nonetheless, at the national level, the topics "climate change," "sustainable urban development," "public health," and "disasters management" are under the responsibility of separate federal bodies, respectively, Ministry of Environment, Ministry of Cities (the Ministry of Cities was terminated by President Jair Bolsonaro in January 2019—its former attributions have been partially absorbed by the Ministry of Regional Development, Ministry of Health, and Ministry of Science, Technology, Innovation and Communications). There is roughly no common agenda specifically focused on urban climate policy, which results in the lack of consistent monitoring of the local policies, plans, programmes, and climate actions currently conducted by the Brazilian municipalities (Maglio & Philippi, 2018; Yigitcanlar et al., 2018).

8.4.2 Local Emissions Inventories and Climate Plans

Firstly, four of the five surveyed cities have published GHG emissions inventories in the context of decentralised cooperation agreements: Curitiba, with the support of ICLEI; Goiânia, with support from the Inter-American Development Bank; Salvador, with support from ICLEI and WRI Brazil and São Paulo, with support from the World Bank. All four investigated cities, having compiled GHG inventories, have subsequently edited their respective Local Climate Plans. Thus, preparation of GHG inventories seems to be, in fact, a necessary preparatory step for climate plans. Moreover, all the investigated Local Climate Plans came to light as of 2009, the year of the edition of the National Policy on Climate Change (PNMC).

Manaus was the only city not to assemble an emissions inventory or a specific Climate Plan. Nevertheless, its 2011 Environmental Code contains references to Adaptation and Mitigation actions, allowing its use as a proxy for a climate plan in this investigation. Of the five cities investigated, only Curitiba edited documents specifically dedicated to Adaptation and Resilience, following the internationally recommended standards. Also, it has contemplated climate actions in its Master Plan, Land Use Regulation, and all Sectoral Plans. Hence, this city is at the forefront of climate action, ahead of the other surveyed cities. Decentralised cooperation and the edition of the PNMC can be pointed out as the two main engines for the structuring of local climate actions, considering, firstly, the participation of international actors in the local emissions inventories and climate plans; and, secondly, the adherence of the investigated cities to national guidelines—PNMC included—while structuring their local public policies.

8.4.3 Contribution to Greenhouse Gas Emission Reduction

The PNMC emphasises the city's role in climate Adaptation, first of all, and in urban Resilience, but not so much in climate Mitigation, which is consistent with the data contemplated in the National GHG inventory. National emissions estimates show that the contribution of the most relevant sectors to urban activities—e.g., waste and energy—is relatively low when compared to the agriculture and land use sectors. In line with the PNMC, adaptation was the climate action axis with the most significant number of references in the master plans, land use regulations, and sectoral plans. Mitigation, however, had the highest number of references in the local climate plans, urban mobility and waste management, which enhances the cities' concerns with fossil fuel consumption and sustainable management of waste residues and effluents.

Furthermore, in line with the PNMC, resilience was the second most referenced climate action axis in land use regulations, disaster management, drainage, health and urban mobility plans, demonstrating that the investigated cities have made efforts to incorporate the coping of shock and stress factors into their planning and management processes. Decentralised cooperation was overall the least referenced climate action. Nonetheless, it was possible to detect references to cooperation actions in the compiled documents from all investigated cities, confirming their engagement with the international community regarding climate change issues.

8.4.4 Climate Action in Public Health and Disaster Management

Qualitative analysis results reported in this chapter revealed that health promotion effectively integrates local climate plans, master plans, land use regulations and sectoral plans of the five investigated cities, with a more significant number of references to Health promotion actions in Sanitation and Waste Management Plans, which reinforces the critical connection between sanitation and public health policies in urban areas. Municipal Public Health Plans and the Disaster Management regulations of the five surveyed cities do incorporate climate action, mainly in the Adaptation axis, but not so much in Resilience. The analysed Local Public Health and Disaster Management Policies did present some connections, but not, as it would be desirable, effective coordination of projects, and actions in the investigated cities. Municipality of Manaus, at this point, provided innovative treatment to the issue, incorporating disaster management into the Municipal Public Health Plan.

8.4.5 Limitations of the Study

This study encountered a number of limitations. The following issues should be considered when interpreting the findings of this study: (a) The data collection process, even undertaken thoroughly, may have unintentionally omitted some policy documents—e.g., the ones that are approved after the data is collected; (b) The study does not factor in the maturity of the policy—e.g., whether just released or in place for quite some time; (c) The study does not factor in whether the policy is implemented; and if so how effective results it generated; (d) The content analysis technique, even conducted with the aid of a software, may contain unconscious bias—e.g., selection of keywords and nodes; (e) The results of the content analysis, even reviewed with utmost care, may have interpreted differently—e.g., due to unconscious bias of the researchers; (f) The findings could not be crosschecked with any other empirical research, since this is a pioneering study; (g) The study portrays the institutional framework before President Bolsonaro's inauguration, as comprised in the years of 2017 and 2018; meaning it discloses a much more positive standing of Brazil in the climate action than what it is actually today; and (h) The study did not investigate the preparedness of the planning and emergency frameworks in combatting major climate-related catastrophes, such as the recent Amazon rainforest fires (Figure 8.17). In our prospective studies, we will overcome these limitations.

FIGURE 8.17
A snapshot from August 2019 Amazon rainforest fires (Sopala, 2019).

8.5 Conclusion

The study at hand focused on tackling an important question of: "Are, and if yes how, cities considering the climate emergency in their local planning mechanisms?" The results of the study revealed that despite the existing federal presidential-level climate crisis denial and efforts towards deforestation of Amazonia, the investigated municipalities have attempted to address the issue at least through municipal legislations, regulations, and plans. These municipalities not only incorporate climatic issues in their urban planning and urban management regulations, but also contribute to the fulfilment of Brazil's climate commitments, more in terms of adaptation than in terms of mitigation, consistently with the findings of local and national emissions inventories.

Concerning possible policy improvements at the municipal level, the study recommends greater coordination between local health promotion and disaster management actions by the investigated cities. This is of utmost importance in order to foster urban resilience and sustainability. However, these municipalities' policies concentrate more on adaptation rather than mitigation (Galán-Martín et al., 2018). Meaning that there is much more room for a better policy to adequately tackle the biggest problem of our time—i.e., the climate crisis—further developing strategies in urban adaptation and resilience. According to the obtained results from our analysis, it is possible to draw the following insights and conclusions.

Firstly, decentralised cooperation and the edition of the National Policy on Climate Change (PNMC) were the main drivers for the structuring of local climate actions by the five investigated cities. However, with the new federal administration in office, the national climate change policy has the risk of staying current or not being endorsed and implemented (Rochedo et al., 2018).

Secondly, by including climate actions in their Master Plans, Land Use regulations, and Sectoral Plans, the investigated cities do contribute to the compliance of Brazil's NDC, more in terms of Adaptation than Mitigation, coherently with the data of national and local GHG emissions inventories and with the guidelines of the PNMC and the National Adaptation Plan (PNA). It was not possible, however, to determine precisely to what extent—i.e., emissions percentages—the investigated cities can contribute to the national target of 37% of GHG emissions. It can only be inferred that local mitigation actions concentrate on Mobility and Waste Management policies.

Thirdly, even though the investigated Brazilian cities have included adaptation and resilience measures in their master plans, land use regulations, and sectoral policies, there is significant room for improvement, especially considering that mitigation, not adaptation, was by far the most referenced axis in four of the five investigated Local Climate Plans.

Fourthly, even though cooperation was overall the least referenced climate action axis, all investigated cities have presented some references to cooperation actions apart from their respective GHG inventories, which confirms some degree of engagement with the international community regarding climate change issues.

Fifthly, health, as expected, effectively integrates the climate plans, master plans, land use regulations, and sectoral plans of the five investigated cities, especially in Waste Management and Sanitation. Besides the climate emergency, due to inadequate infrastructure and informal housing—so-called Favelas of Brazil—health has been a major policy focus and concern in the nation.

Sixthly, the study findings reveal that Brazilian cities are rich in terms of policy but poor in terms of their implementation. Although the five investigated cities showed some connection points between their climate, public health, and disaster management policies, they should further coordinate and integrate these policies in order to develop urban resilience and foster SDGs fulfilment at the local level. These local-level policy and implementation attempts are hoped to showcase promising results on the ground and generate motivation for other Brazilian municipalities, states, and the federal government to further invest in tackling the climate crisis, which is upon Brazil (Eduardo & Franchini, 2017; Giannini et al., 2017) and the entire globe. The results indicate that the investigated cities have the necessary tools to act on climate mitigation, adaptation, and resilience, but assessing the effectiveness of these actions goes beyond the objectives and scope of the proposed research and could be developed by further studies on the matter.

Lastly, the literature clearly indicates that climate change and its catastrophic consequences are the inevitable harsh reality of our time (Hartmann, 2010; Zheng et al., 2002). In such dire straits, we need to move urgently from climate policy/agenda to climate action (Spratt & Sutton, 2008; Tang et al., 2010). This is to say, the lip service (e.g., rhetoric and/or policy without effective implementation and impact) will not get the desired outcomes. Cities need to urgently lead the way in climate change action (Rosenzweig et al., 2010); however, they need to find ways to successfully implement policies as well. The road to successful implementation is highly bumpy; we need to find ways to transform barriers into enablers of action on climate change and sustainable urban development (Burch, 2010; Yigitcanlar, 2010a, 2010b). Citizens, scientists, institutions, policymakers, and multilevel governance structures, along with good/honest politics, are required for such immediate action and practice turnaround (i.e., changing the attitude, behaviour, and choice)—both industrial and household (Selin & VanDeveer, 2009; Shove, 2010).

Acknowledgements

This chapter, with permission from the copyright holder, is a reproduced version of the following journal article: Sotto, D., Philippi, A., Yigitcanlar, T., & Kamruzzaman, M., (2019). Aligning urban policy with climate action in the global south: are Brazilian cities considering climate emergency in local planning practice? Energies, 12(8), 3418.

References

Abreu, V. (2017). *Produção de CO2 em Goiânia é 28.5% acima da média nacional.* O Popular: Goiânia.

Aldieri, L., Carlucci, F., Vinci, C. P., & Yigitcanlar, T. (2019). Environmental innovation, knowledge spillovers and policy implications: A systematic review of the economic effects literature. *Journal of Cleaner Production, 239,* 118051.

Arbolino, R., Carlucci, F., Cira, A., Ioppolo, G., & Yigitcanlar, T. (2017). Efficiency of the EU regulation on greenhouse gas emissions in Italy: The hierarchical cluster analysis approach. *Ecological Indicators, 81,* 115–123.

Arbolino, R., De Simone, L., Carlucci, F., Yigitcanlar, T., & Ioppolo, G. (2018). Towards a sustainable industrial ecology: Implementation of a novel approach in the performance evaluation of Italian regions. *Journal of Cleaner Production, 178,* 220–236.

Azevedo, T. (2017). ; Observatório do Clima:; *Emissões de GEE do Brasil e suas Implicações para Políticas Públicas e a Contribuição Brasileira para o Acordo de Paris; Documento síntese; Período 1970–2015*. SEEG.

Brasil. Lei 6.938 de 31 de Agosto de 1981. Available online: http://www.planalto.gov.br/ccivil_03/Leis/L6938.htm (accessed on 24 August 2019).

Brasil. Constituição da Republica Federativa do Brasil de 1988. Available online: http://www.planalto.gov.br/ccivil_03/Constituicao/Constituicao.htm (accessed on 24 August 2019).

Brasil. Lei 10.257 de 10 de Julho de 2001—Estatuto da Cidade. Available online: http://www.planalto.gov.br/ccivil_03/Leis/LEIS_2001/L10257.htm (accessed on 24 August 2019).

Brasil. Lei 11.124 de 16 de Junho de 2005. Available online: http://www.planalto.gov.br/ccivil_03/_Ato2004-2006/2005/Lei/L11124.htm (accessed on 24 August 2019).

Brasil. Lei 11.445 de 5 de Janeiro de 2007. Available online: http://www.planalto.gov.br/ccivil_03/_Ato2007-2010/2007/Lei/L11445.htm (accessed on 24 August 2019).

Brasil. Lei 12.187 de 29 de Dezembro de 2009. Available online: http://www.planalto.gov.br/ccivil_03/_Ato2007-2010/2009/Lei/L12187.htm (accessed on 24 August 2019).

Brasil. Lei 12.305 de 2 de Agosto de 2010. Available online: http://www2.mma.gov.br/port/conama/legiabre.cfm?codlegi=636 (accessed on 24 August 2019).

Brasil. Lei 12.587 de 3 de Janeiro de 2012. Available online: http://www.planalto.gov.br/ccivil_03/_Ato2011-2014/2012/Lei/L12587.htm (accessed on 24 August 2019).

Brasil. Lei 12.608 de 10 de Abril de 2012. Available online: http://www.planalto.gov.br/ccivil_03/_Ato2011-2014/2012/Lei/L12608.htm (accessed on 24 August 2019).

Brasil. Lei 8.080 de 19 de Setembro de 1990. Available online: http://www.planalto.gov.br/ccivil_03/leis/L8080.htm (accessed on 24 August 2019).

Brasil. Sumário Executivo (2017). Documento-Base para Subsidiar os Diálogos Estruturados Sobre a Elaboração de uma Estratégia de Implementação e Financiamento da Contribuição Nacionalmente Determinada do Brasil ao Acordo de Paris. Ministério do Meio Ambiente.

Broto, V. C. (2017). Urban governance and the politics of climate change. *World Development, 93*, 1–15.

Burch, S. (2010). Transforming barriers into enablers of action on climate change: Insights from three municipal case studies in British Columbia, Canada. *Global Environmental Change, 20*, 287–297.

Burke, M., Dykema, J., Lobell, D. B., Miguel, E., & Satyanath, S. (2015). Incorporating climate uncertainty into estimates of climate change impacts. *The Review of Economics and Statistics, 97*, 461–471.

Caviglia-Harris, J. L. (2018). Agricultural innovation and climate change policy in the Brazilian Amazon: Intensification practices and the derived demand for pasture. *Journal of Environmental Economics and Management, 90*, 232–248.

De Azevedo, T. R., Junior, C. C., Junior, A. B., dos Santos Cremer, M., Piatto, M., Tsai, D. S., Barreto, P., Martins, H., Sales, M., Galuchi, T., et al. (2018). SEEG initiative estimates of Brazilian greenhouse gas emissions from 1970 to 2015. *Scientific Data, 5*, 180045.

Eduardo, V., & Franchini, M. (2017). *Brazil and climate change: Beyond the Amazon*. Routledge.

Engle, N. L., & Lemos, M. C. (2010). Unpacking governance: Building adaptive capacity to climate change of river basins in Brazil. *Global Environmental Change, 20*, 4–13.

Farias, A. R., Mingoti, R., Valle, L. D., Spadotto, C. A., & Lovisi Filho, E. (2017). *Comunicado Técnico. Identificação, Mapeamento e Quantificação das Áreas Urbanas do Brasil.* Embrapa.

Galán-Martín, A., Pozo, C., Azapagic, A., Grossmann, I. E., Mac Dowell, N., & Guillén-Gosálbez, G. (2018). Time for global action: An optimised cooperative approach towards effective climate change mitigation. *Energy & Environmental Science, 11,* 572–581.

Giannini, T. C., Costa, W. F., Cordeiro, G. D., Imperatriz-Fonseca, V. L., Saraiva, A. M., Biesmeijer, J., & Garibaldi, L. A. (2017). Projected climate change threatens pollinators and crop production in Brazil. *PLoS ONE, 12,* e0182274.

Governo do Estado da Bahia (2010). *Primeiro Inventário de Emissões Antrópicas de Gases de Efeito Estufa do Estado da Bahia.* SEMA. Available online: http://www.consultaesic.cgu.gov.br/busca/dados/Lists/Pedido/Attachments/456145/RESPOSTA_PEDIDO_1%2020100915%20inventario%20emissoes%20BA.pdf (accessed on 24 August 2019).

Governo do Estado do Paraná. (2015). Inventário de Emissões Antrópicas Diretas e Indiretas de Gases de Efeito Estufa 2005–2012. Resumo Executivo; SEMA: Curitiba, Brazil.

Governo do Estado de São Paulo. Comunicação Estadual (2011). *1o Inventário de Emissões Antrópicas de Gases de Efeito Estufa Diretos e Indiretos do Estado de São Paulo.* CETESB.

Hale, T. (2016). The Paris Agreement and nonstate climate action. *Global Environmental Politics, 16,* 12–22.22.

Hartmann, B. (2010). Rethinking climate refugees and climate conflict: Rhetoric, reality and the politics of policy discourse. *Journal of International Development: Journal of Development Studies Association, 22,* 233–246.

IBGE (2018). *População em Áreas de Risco no Brasil.* IBGE.

Ingrao, C., Messineo, A., Beltramo, R., Yigitcanlar, T., & Ioppolo, G. (2018). How can life cycle thinking support sustainability of buildings? Investigating life cycle assessment applications for energy efficiency and environmental performance. *Journal of Cleaner Production, 201,* 556–569.

Instituto Ekos Brasil & Geolock Consultoria e Engenharia Ambiental (2013). *Inventário de Emissões e Remoções Antrópicas de Gases de Efeito Estufa do Município de São Paulo de 2003 a 2009 com Atualização para 2010 e 2011 nos Setores Energia e Resíduos.* ANTP: São Paulo.

Intergovernmental Panel on Climate Change—IPCC (2018). Global warming of 1.5°C: An IPCC special report on the impacts of global warming of 1.5°C above pre-industrial levels and related global greenhouse gas emission pathways. In: *Context of strengthening the global response to the threat of climate change, sustainable development, and efforts to eradicate poverty.* Intergovernmental Panel on Climate Change.

Jacobi, P., & Sulaiman, S. (2016). Governança ambiental urbana em face das mudanças climáticas. *Revista USP, 109,* 133–142.

Jakob, M., & Steckel, J. C. (2016). Implications of climate change mitigation for sustainable development. *Environmental Research Letters, 11,* 104010.

Lahsen, M. (2009). A science–policy interface in the global south: The politics of carbon sinks and science in Brazil. *Climate Change, 97,* 339.

Liu, Y., Brito, J., Dorris, M. R., Rivera-Rios, J. C., Seco, R., Bates, K. H., & Goldstein, A. H. (2016). Isoprene photochemistry over the Amazon rainforest. *Proceedings of the National Academy of Sciences of the United States of America, 113,* 6125–6130.

Maglio, I., & Philippi, A. Jr. (2018). Sustentabilidade Ambiental e Mudanças Climáticas. In *Gestão Urbana e Sustentabilidade*; Philippi, A., Jr., and Bruna, G., Eds.; Manole,; pp. 428–453.

Malhi, Y., Roberts, J. T., Betts, R. A., Killeen, T. J., Li, W., & Nobre, C. A. (2008). Climate change, deforestation, and the fate of the Amazon. *Science, 319*, 169–172.

Manowska, A., & Nowrot, A. (2019). The importance of heat emission caused by global energy production in terms of climate impact. *Energies, 12*, 3069.

Ministério da Ciência (2018). *Tecnologia, Inovação e Comunicação—MCTIC. Terceira Comunicaçãeo Nacional do Brasil à Convenção-Quadro das Nações Unidas sobre Mudança do Clima*. MCTIC.

Ministério da Saúde (2013). *Plano Setorial da Saúde para Mitigação e Adaptação à Mudança do Clima*. Ministério da Saúde.

Ministério do Meio Ambiente (MMA) (2016). *Plano Nacional de Adaptação à Mudança do Clima—Sumário Executivo*. Ministério do Meio Ambiente.

Nobre, C. (2001). Mudanças climáticas globais: Possíveis impactos nos ecossistemas do país. *Parcer. Estratégicas, 6*, 239–258.

Nobre, C. A., Sampaio, G., Borma, L. S., Castilla-Rubio, J. C., Silva, J. S., & Cardoso, M. (2016). Land-use and climate change risks in the amazon and the need of a novel sustainable development paradigm. *Proceedings of the National Academy of Sciences of the United States of America, 113*, 10759–10768.

Octaviano, C., Paltsev, S., & Gurgel, A. C. (2016). Climate change policy in Brazil and Mexico: Results from the MIT EPPA model. *Energy Economics, 56*, 600–614.

Owen, G. T. (2014). Qualitative methods in higher education policy analysis: Using interviews and document analysis. *Qualitative Report, 19*, 1–19.

Pacheco, R., & Silva, C. (2019). Global warming potential of biomass-to-ethanol: Review and sensitivity analysis through a case study. *Energies, 12*, 2535.

Parmesan, C., & Yohe, G. (2003). A globally coherent fingerprint of climate change impacts across natural systems. *Nature, 421*, 37.

Pecl, G. T., Araújo, M. B., Bell, J. D., Blanchard, J., Bonebrake, T. C., Chen, I. C., & Falconi, L. (2017). Biodiversity redistribution under climate change: Impacts on ecosystems and human well-being. *Science, 355*, eaai9214.

Pessini, L., & Sganzerla, A. (2016). As mudanças climáticas e seus impactos no reino a vida: Perspectivas para um futuro não apocalíptico. *Revista Iberoamericana de Bioética, 2*, 1–13.

Prefeitura de Curitiba (2016). *2o e 3o Inventários de Gases de Efeito Estufa para a Cidade de Curitiba*. Prefeitura Municipal de Curitiba.

Prefeitura de Salvador (2016). *Inventário de Emissões dos Gases do Efeito Estufa de Salvador*. Secretaria Cidade Sustentável.

Rei, F., Gonçalves, A. F., & de Souza, L. P. (2017). Acordo de Paris: Reflexões e Desafios para o Regime Internacional de Mudanças Climáticas. *Veredas Do Direito, 14*, 81–99.

Rochedo, P. R., Soares-Filho, B., Schaeffer, R., Viola, E., Szklo, A., Lucena, A. F., & Rathmann, R. (2018). The threat of political bargaining to climate mitigation in Brazil. *Nature Climate Change, 8*, 695–698.

Rosenzweig, C., Solecki, W., Hammer, S. A., & Mehrotra, S. (2010). Cities lead the way in climate–change action. *Nature, 467*, 909–911.

Rottz, M., Sell, D., Pacheco, R., & Yigitcanlar, T. (2019). Digital commons and citizen coproduction in smart cities: Assessment of Brazilian municipal e-government platforms. *Energies, 12*, 2813.

Saldiva, P. (2018). *Vida Urbana e Saúde. Os Desafios dos Habitantes das Metrópoles*. Contexto.

Scranton, R. (2015). *Learning to die in the anthropocene: Reflections on the end of a civilization*. City Lights Publishers.

SEEG. Plataforma SEEG Brasil. (2019). Available online: http://plataforma.seeg.eco.br/ (accessed on 24 August 2019).

Selin, H., & VanDeveer, S. D. (2009). *Changing climates in North American politics: Institutions, policymaking, and multilevel governance*. MIT Press.

Shove, E. (2010). Beyond the ABC: Climate change policy and theories of social change. *Environment and Planning A*, *42*, 1273–1285.

Sopala, R., A copyright free image by of 2019 Amazon fires. Available online: https://pixabay.com/photos/fire-forest-fire-children-fear-4429478/ (accessed on 3 September 2019).

Souza, S. (2017). *Frota de Manaus em Circulação Supera Marca de 713 mil Veículos*. A Crítica. Available online: https://www.acritica.com/channels/manaus/news/frota-de-manaus-superamarca-de-713-mil-veiculos (accessed on 24 August 2019).

Spratt, D., & Sutton, P. (2008). *Climate code red: The case for emergency action*. Scribe Publications.

Suframa. O que é o Projeto ZFM? Available online: http://site.suframa.gov.br/assuntos/modelo-zona-franca-de-manaus/o-que-e-o-projeto-zfm (accessed on 24 August 2019).

Tang, Z., Brody, S. D., Quinn, C., Chang, L., & Wei, T. (2010). Moving from agenda to action: Evaluating local climate change action plans. *Journal of Environmental Planning and Management*, *53*, 41–62.

United Nations, General Assembly (2012 on 27 July 2012. 66/288). *Resolution Adopted by the General Assembly*. The Future We Want.

United Nations, General Assembly (2015 on 25 September 2015). *Resolution Adopted by the General Assembly*. United Nations.

United Nations, General Assembly (2016 on 23 December 2016; 71/256). *Resolution Adopted by the General Assembly*. New Urban Agenda.

Van Loon, A. F., Gleeson, T., Clark, J., Van Dijk, A. I., Stahl, K., Hannaford, J., & Hannah, D. M. (2016). Drought in the anthropocene. *Nature Geoscience*, *9*, 89–91.

World Health Organization—WHO. (2016). *Ambient air pollution: A global assessment of exposure and burden of disease*. World Health Organization.

Yigitcanlar, T. (2009). Planning for smart urban ecosystems: Information technology applications for capacity building in environmental decision making. *Theor. Empir. Res. Urban Manag*, *4*, 5–21.

Yigitcanlar, T. (2010a). *Rethinking sustainable development: Urban management, engineering, and design*. IGI Global.

Yigitcanlar, T. (2010b). *Sustainable urban and regional infrastructure development: Technologies, applications and management: Technologies, applications and management*. IGI Global.

Yigitcanlar, T. (2014). Position chapter: Benchmarking the performance of global and emerging knowledge cities. *Expert Systems with Applications*, *41*, 5549–5559.

Yigitcanlar, T., Foth, M., & Kamruzzaman, M. (2019a). Towards post-anthropocentric cities: Reconceptualising smart cities to evade urban ecocide. *Journal of Urban Technology*, *26*, 147–152.

Yigitcanlar, T., & Kamruzzaman, M. (2018. Does smart city policy lead to sustainability of cities?. *Land Use Policy*, *73*, 49–58.

Yigitcanlar, T., Kamruzzaman, M., Buys, L., Ioppolo, G., Sabatini-Marques, J., da Costa, E., & Yun, J. (2018a). Understanding 'smart cities': Intertwining development drivers with desired outcomes in a multidimensional framework. *Cities, 81*, 145–160.

Yigitcanlar, T., Kamruzzaman, M., Foth, M., Sabatini-Marques, J., Costa, E., & Ioppolo, G. (2019b). Can cities become smart without being sustainable? A systematic review of the literature. *Sustainable Cities and Society, 45*, 348–365.

Yigitcanlar, T., Metaxiotis, K., & Carrillo, F. J. (2012). *Building prosperous knowledge cities: Policies, plans and metrics.* Edward Elgar.

Yigitcanlar, T., Sabatini-Marques, J., da-Costa, E. M., Kamruzzaman, M., & Ioppolo, G. (2019c). Stimulating technological innovation through incentives: Perceptions of Australian and Brazilian firms. *Technological Forecasting and Social Change, 146*, 403–412.

Yigitcanlar, T., Sabatini-Marques, J., Lorenzi, C., Bernardinetti, N., Schreiner, T., Fachinelli, A., & Wittmann, T. (2018b). Towards smart florianópolis: What does it take to transform a tourist island into an innovation capital?. *Energies, 11*, 3265.

Yin, R. K. (2003). *Case study research; Sage publications.* Thousand Oaks.

Zapf, M., Pengg, H., & Weindl, C. (2019). How to comply with the Paris agreement temperature goal: Global carbon pricing according to carbon budgets. *Energies, 12*, 2983.

Zheng, B., Xu, Q., & Shen, Y. (2002). The relationship between climate change and quaternary glacial cycles on the Qinghai–Tibetan plateau: Review and speculation. *Quaternary International, 97*, 93–101.

9

Transit-Oriented Development and Urban Heat

9.1 Introduction and Background

This research aims to assess the urban heat island (UHI) effect of transit-oriented developments (TODs). TOD is a relatively new neighbourhood planning concept, developed as a part of the smart growth movement, to prevent sprawling development (Rohe, 2009). Urban sprawl is defined as low-density outward expansion of cities and is associated with a range of urban problems such as car-dependency, long commute trips, congestion, greenhouse gas emissions, high infrastructure costs, and the loss of community (Pacione, 2009). TODs, on the other hand, are conceptualised as moderate-high density, diverse land use patterns, well-connected street networks (as opposed to cul-de-sacs), and centred on high-frequency transit nodes (Bertolini et al., 2009; Cervero & Kockelman, 1997; Renne, 2009a; Rohe, 2009). High density generates enough people to support the viability of frequent public transport services in a TOD (Cervero & Kockelman, 1997; Oakes et al., 2007). Land use diversity promotes walking because of the availability of various destination types (Duncan et al., 2010; Kamruzzaman et al., 2014a). Street connectivity makes walking comfortable to destinations and transit nodes and, therefore, increases transit ridership (Lin & Gau, 2006; Oakes et al., 2007). High-frequency public transport services increase its utility to travel over competitive modes (cars) (Curtis et al., 2009).

Research studies have investigated various impacts of TODs such as satisfaction with living (Lund, 2006), health benefits (Brown & Werner, 2008), accessibility benefits in terms of finding a job (Cervero & Day, 2008), social sustainability (Kamruzzaman et al., 2014b), property value gain (Bowes & Ihlanfeldt., 2001; Kay et al., 2014), parking demand (Broaddus, 2010), changes in transit ridership, car use/ownership, and active transport uses (Broaddus, 2010; Crowley et al., 2009; Dill, 2008; Faghri & Venigalla, 2013; Kamruzzaman et al., 2015; Lee et al., 2010; Nasri & Zhang, 2014; Pan et al., 2011; Renne, 2009a). Generally, the findings from these studies are supportive of TOD policy. However, very little is known about the UHI effects of TOD neighbourhoods, whereas many studies have shown that sprawling development reduces vegetation cover and thus contributes to the UHI intensity (Al Kuwari et al., 2016; Lemonsu et al., 2015). While studies have shown that many of the individual

DOI: 10.1201/9781003737056-12

attributes of a TOD (e.g., high density and public transport services) can impact the UHI effect (Deilami et al., 2016b; Elsayed, 2012; Lemonsu et al., 2015), there has been little research about the aggregated impacts of these individual attributes which together form a TOD.

UHI effect is a phenomenon when urban areas experience a higher temperature compared to their surrounding non-urban areas (Ward et al., 2016). The adverse effect of UHI has been widely documented in the literature. For example, it increases the temperature of cities, contributes to global warming (EPA, 2016), initiates storms/precipitation events (Bornstein & Lin, 2000; Dixon & Mote, 2003), increases the energy demand of cities (Santamouris et al., 2015), and contributes to heat-related mortality (Hondula et al., 2014). As a result, an increasing emphasis is placed in the policy circle in terms of devising ways to mitigate the UHI effects in cities (Chow et al., 2012; Gago et al., 2013; Susca et al., 2011). In this regard, TOD policies could bring multiple co-benefits, including the mitigation of UHI effect. Nevertheless, for example, if TODs are likely to increase the UHI effect (similar to sprawling development), then the policy benefit of reducing greenhouse gas emissions through modal shift in TODs will be compromised by the increasing use of electricity. However, a lack of empirical evidence about the UHI effects of TODs hinders the justification of their operationalisation as a key policy tool worldwide (Kamruzzaman et al., 2016). As a result, the findings from this research possess a significant policy implication.

Against this background, this study seeks to address the following research question: to what extent do TODs reduce the UHI effect over sprawling development? The question is investigated using Brisbane, Queensland, Australia as a case study. Brisbane's high level of policy emphasis on TOD planning and design (Kamruzzaman et al., 2014d), along with its subtropical climate, and vulnerability to the impacts of global warming (Braganza et al., 2003), makes it an interesting case study.

9.2 Research Design

9.2.1 Data

Brisbane, a subtropical city with vulnerabilities to the impacts of global warming (Goonetilleke et al., 2014), was chosen as a case for the empirical investigation of this research because Brisbane City Council (2014) has implemented a TOD planning approach over the past decade around high-frequency public transport nodes. The city has adopted TOD strategies from a higher state-level strategy. The Queensland Government (2008) wishes to minimise congestion and carbon emissions in order to make the state as liveable place. In South East Queensland (SEQ), the government intends to double the share of active transport trips to 20%, double the share of public transport to 14%, and reduce the car use from 83% to 66% (Queensland Government, 2010). A more specific policy guide has been framed in the SEQ Regional Plan 2009–2031 aimed to achieve the targets (Queensland Government, 2009). One of the highlighted strategies is to restrict urban expansion and to facilitate development in well-defined transport nodes to promote self-contained activities. The nodes are being developed following the principles of TOD and to be served by a single or a combination of public transport modes operational in Brisbane, including train, bus rapid transit, and ferry (Sipe & Burke, 2011; Yang & Pojani, 2017). The key principles to be applied in these nodes are (Queensland Government, 2009, p. 102) as follows: diverse land uses, high residential density, high employment intensities, well-connected street networks, and fast and frequent public transport services.

Table 9.1 outlines the data used from various sources for the case study to answer the research question which can be grouped under four categories: (a) satellite imageries including Landsat downloaded from the USGS. website to monitor the UHI effects both within TOD neighbourhoods between 2004 and 2013, and between TOD and non-TOD neighbourhoods; (b) built environment (BE) data (e.g., land uses, road networks) downloaded from publicly available websites to derive different indicators used to identify TOD neighbourhoods; (c) population and employment data obtained from the Australian Bureau of Statistics (ABS) to complement the derivation of different indicators; and (d) travel behaviour data downloaded from the Queensland Government website to verify the validity of the identified TOD neighbourhoods.

9.2.1.1 Built Environment Data

This research used six BE indicators to identify TODs in Brisbane. The indicators were derived at the census collection district (CCD) level. Note that the CCDs were the lowest level of the administrative boundaries used to collect census data prior to 2011. However, this boundary structure has been revised, and within the current structure, the Statistical Area Level 1 (SA1) category roughly corresponds with the CCD boundary. This research aims to monitor the UHI effects of TODs over time. As a result, it was required to match the boundary spatially between censuses in order to utilise various census datasets (e.g., population density). A correspondence file is available at the ABS website to match the geographic boundaries between the censuses and was used in this research (http://www.abs.gov.au/websitedbs/D3310114.nsf/home/Correspondences). However, this operation correctly matched 1,661 out of 1,740 CCDs in Brisbane. As a result, the six BE indicators were derived for the matched 1,661 CCDs.

An 800-metre network buffer was derived from the centroid of each CCD, and the indicators were calculated within this spatial scale to represent the BE characteristics of each CCD. The 800-metre buffer was selected based on

TABLE 9.1

Sources and Functions of the Dataset Used in This Research

Dataset	Source	Function
Landsat 5/TM (2004/5/4)	USGS website	To extract UHI and land cover thematic maps
Landsat 8/OLI (2013/4/27)	USGS website	To extract UHI and land cover thematic maps
Google Earth images (2004 and 2013)	Google Earth	To assess the accuracy of the generated thematic maps
IKONOS (2001 and 2011)	DigitalGlobe Foundation	To complement the accuracy assessment of the thematic maps
Population and employment data	Australian Bureau of Statistics (ABS), Brisbane office	To derive population and employment density indicators; population density converted into dwelling density based on average household sizes
SA1 and CCD geographic boundaries	Australian Bureau of Statistics	To derive area-based indicators (e.g., population density)
SA1 to CCD correspondence	Australian Bureau of Statistics	To match SA1 and CCD boundaries
Parcel-level land use classification data	Brisbane City Council	To derive land use diversity index; extract residential-zoned land for net residential density; extract employment-generating land uses for net employment density
Physical road network data	Queensland Government (https://data.qld.gov.au/)	To derive intersection density and cul-de-sac density indicators
Public transport stops, routes, and timetable data	GTFS (https://developers.google.com/transit/gtfs/reference/)	To derive PTAL index
South East Queensland Household Travel Survey data	Queensland Government (https://data.qld.gov.au/)	To distinguish travel behaviour patterns between TOD and non-TOD areas

the Queensland Government (2009) recommendation of TOD precinct size (10 minutes' walking distance). The derived six indicators are as follows:

a. Net residential density (number of dwelling units per hectare of residential zone land) (Frank et al., 2005);
b. Net employment density (number of jobs per hectare of employment-generating land uses such as commerce and industry) (Center for Transit-Oriented Development, 2010);
c. Land use diversity (Simson's diversity index based on the area of five types of land uses, including residential, commercial, industrial,

recreational, and institutional) (Kamruzzaman & Hine, 2013; Simpson, 1949) ranges from 0 (single land use) to 1 (fully diversity—all five land uses are present in equal amount);

d. Intersection density (number of 3 or more ways of intersection per hectare of land) (Cervero & Gorham, 1995);
e. Cul-de-sac density (number of dead-ends per hectare of land) (Cerin et al., 2007; Stangl & Guinn, 2011); and
f. Public transport accessibility level (PTAL)—an index developed by Transport for London (2010) based on five attributes of public transport services: spatial reach (whether public transport services are located within the buffer), modes of public transport services available (bus and train in this case), frequency of the available services, ability to travel in different directions (routes), and reliability. This research derived PTAL for the morning peak period on weekdays using the General Transit Feed Specification (GTFS) data. The ferry was not considered due to a lack of availability of relevant data within the GTFS datasets.

The six indicators were selected based on a review of the 21 principles proposed for the construction of TODs in the research context (Queensland Government, 2009). The principles relate to the various components of a TOD, including BE (availability and intermodal connectivity of public transport services, land use mix, residential density, land use intensity for employment, and pedestrian connectivity), locational (new development, within existing urban areas), social (social diversity and inclusion), and process (coordination, timeframes, stakeholder engagement). This research extracted the BE–related principles and identified relevant indicators representing the principles based on the literature as discussed previously.

9.2.1.2 Travel Behaviour Data

Given that the main goal of TODs "is to create settings which prompt people to drive less and ride public transit more" (Cervero et al., 2004, p. 8), it was necessary to validate the identified TOD neighbourhoods by examining the travel behaviour of residents within these neighbourhood types (Kamruzzaman et al., 2014a). This research examines the differences in mode choice behaviour (the use of private, public, and active transport) between different TOD and non-TOD neighbourhoods.

The SEQ Household Travel Survey (SEQHTS) data was used to examine the travel behaviour differences. SEQHTS data is the main source of data used for transport decision-making in SEQ (including Brisbane) and has extensively been used by transport researchers in this context (Burke & Brown, 2007; Chia et al., 2016; Kamruzzaman et al., 2014c). The survey collects repeated cross-sectional data representative of the wider population based on two-stage stratified random sampling techniques. CCDs are randomly selected first, and then households are randomly selected from within each CCD. The survey collects one-day travel diary data from all members of a selected household. This research used the 2009–2012 version of the survey data downloaded from the Queensland Government website (https://data.qld.gov.au/dataset/2009-12-south-east-queensland-household-travel-survey).

The research analysed only weekday travel data of those respondents who lived in one of the CCDs used to identify TOD neighbourhoods. Respondents' home locations were found to match with 193 of the 1,661 CCDs. A total of 11,959 trips were made by 3,206 individuals from 1,377 households within the matched 193 CCDs. The SEQHTS collected travel-related variables (e.g., travel distance, journey time) of these trips based on 11 categories of travel modes. The modes were recoded into four categories for the analysis presented in this research: private vehicle (car driver, car passenger, taxi, motorcycle), public transport (bus, train, school bus, ferry), active transport (walk and cycle), and other.

9.2.1.3 Landsat Satellite Images

This research downloaded two Landsat images: firstly, to differentiate and monitor the UHI effect between different types of TODs and non-TOD areas; and secondly, to derive patterns of land cover changes within these neighbourhoods to explain the UHI effects. The two images, one Landsat TM acquired on 5 May 2004 and another Landsat OLI acquired on 4 June 2013, were selected based on three criteria: (a) minimum air temperature differences between the days of acquisition so that differences in UHI effect are less affected by air temperature differences; (b) maintaining sufficient interval between the periods to allow changes in land cover patterns; and (c) being relatively cloud free. The downloaded images have been corrected for radiometric, geometric, and topographic errors based on digital elevation model and ground control points (level 1T processing). This research also obtained IKONOS images of 2001 and 2011 and downloaded Google Earth images of 2004 and 2013 for the study area. The images were used as reference datasets for accuracy assessment of the thematic maps generated from the Landsat images. The IKONOS images were used in tandem with Google Earth to be certain about the land cover information in the reference maps. The images provided an opportunity to monitor the UHI effects of neighbourhoods between pre-TOD (2004) and TOD (2013) eras. Note that TOD as a neighbourhood planning concept was introduced in 2005 for the first time in this policy context through the approval of SEQ Regional Plan 2005–2026 (a regulatory strategic policy document).

9.2.2 Methods

9.2.2.1 Identification of TOD Neighbourhoods

This research applied the TwoStep cluster analysis technique to identify TOD neighbourhoods based on the literature (Kamruzzaman et al., 2014a;

Reusser et al., 2008; Zemp et al., 2011). The method identified natural groupings of the CCDs based on their BE attributes. All six indicators of the BE were used to operationalise the cluster analysis—net residential density, net employment density, land use diversity, intersection density, cul-de-sac density, and PTALs. The highest and lowest 5% scores of all these indicators were set to their 95th and 5th percentile values in order to lessen the effect of outliers in forming the clusters (De Vos et al., 2012; Schwanen & Mokhtarian, 2004).

9.2.2.2 Validation of the Identified TOD Neighbourhoods

As outlined in the literature, TOD neighbourhoods should reflect their locational efficiency in terms of mobility choices of people (Belzer & Autler, 2002). Therefore, it is logical to assume that an area with a good TOD would reduce car-dependency. This research identifies the efficiency of the selected TOD neighbourhoods by comparing the mode choice behaviour of people living in these neighbourhoods with non-TOD neighbourhoods. Based on the literature, individuals' average travel distances by mode were compared between TOD and non-TOD areas (Renne, 2009b). As indicated previously, conceptually TODs promote sustainable travel behaviour in two ways: (a) through facilitating participation in local activities using active transport, and therefore, a TOD reduces the need for motorised transport as a whole (Cervero & Kockelman, 1997; Lin & Gau, 2006); and (b) if people need to travel to access goods and services in other parts of a city, then they can choose fast, frequent, and well-connected PT services available at TOD and, thereby, reduce the need for car-based travel (Bertolini et al., 2009; Kamruzzaman et al., 2014a). Empirical evidence also suggests the same. For example, Nasri and Zhang (2014) showed that people living in TOD areas tend to drive less, reducing their vehicle miles travelled (VMT) by around 38% in Washington, DC and 21% in Baltimore, compared to residents living in non-TOD areas. For these reasons, this research tested the validity of the TODs based on both total and car-based travel distances.

9.2.2.3 Derivation of the UHI Effects

The digital number (DN) of the downloaded Landsat satellite images was converted into at-surface reflectance values based on Lu et al. (2014) and Jia et al. (2014) for the TM and OLI images, respectively. The derived reflectance values were then used to extract thematic land cover types using the support vector machine (SVM) classification method. The SVM method was selected over the maximum likelihood classifier, decision tree classifier, and neural network classifier by comparing the accuracy of the generated thematic maps. A similar finding has also been reported in other contexts (Huang et al., 2002; Otukei & Blaschke, 2010; Paneque-Gálvez et al., 2013; Poursanidis et al., 2015). A better performance of the SVM method is attributed to its low sensitivity to size, data distribution, and quality of training samples (Mountrakis et al., 2011). The method also performs better in case of mixed pixels (i.e., mistakenly considered as training samples) (Paneque-Gálvez et al., 2013).

This classification procedure generated four land cover classes in Brisbane: built-up, dense vegetation, upland natural, and water body areas which are consistent with previous research in this context (Shatu et al., 2014). The accuracy of the derived land cover classes was assessed against the reference maps (IKONOS and Google Earth images). One hundred pixels from each of the four land cover classes were selected randomly and compared against the reference maps in the respective year. The overall accuracy values were found to be 0.90 and 0.89 in 2004 and 2013, respectively (Table 9.2). These levels of accuracy possess an acceptable range in the literature (Thomlinson et al., 1999). The thematic maps were used to calculate the porosity level in each CCD. Unlike impervious surfaces such as roads, driveways, and parking lots, porous land is characterised as pervious surfaces through which water can penetrate and includes vegetation cover, bare soil, and garden (Lu & Weng, 2006). The porosity level of a CCD is calculated as the ratio of non–built up area to the total area of the CCD. Non–built up areas comprise three (dense vegetation, upland natural, and water body) of the four thematic classes generated in this research.

The derivation of the UHI intensities requires to derive land surface temperature (LST) initially. The thermal bands of the images (band 6 for the TM and band 10 for the OLI images) were used to derive LST for each year based on Li et al. (2016). Briefly, the at-satellite radiances of thermal bands were converted into at-sensor brightness temperatures using Equation (9.1).

$$T_B = \frac{K_2}{\ln\left(\frac{K_1}{L_\lambda} + 1\right)} \tag{9.1}$$

TABLE 9.2

Accuracy Assessment Results of the Generated Land Cover Classes (Confusion Matrix)

Year	Land Cover Class	Accuracy	Overall Accuracy	Kappa	Omission Error	Commission Error
2004	Built-up	0.90	0.87	0.91	0.86	–
2004	Dense vegetation	0.93	0.90	–	–	–
2004	Upland natural	0.79	0.93	–	–	–
2004	Water body	0.98	0.92	–	–	–
2013	Built-up	0.89	0.87	0.96	0.87	–
2013	Dense vegetation	0.82	0.92	–	–	–
2013	Upland natural	0.85	0.95	–	–	–
2013	Water body	0.98	0.85	–	–	–

where T_B is at-sensor brightness temperature in Kelvin; L_λ equals to at-satellite radiance derived from the preprocessing step; K_1 = 607.76 W/(m^2 sr μm) and K_2 = 1260.56 W/(m^2 sr μm) are the pre-launch calibration constants obtained from Landsat data users handbook (NASA, 2011). The generated brightness temperature values were then converted to emissivity-corrected LST (T_s) in Kelvin using Equation (9.2).

$$T_s = \frac{T_B}{1+\left[\left(\lambda \times \frac{T_B}{\alpha}\right)\ln\varepsilon\right]} \tag{9.2}$$

where λ is the wavelength of emitted radiance equal to 11.5 μm; α = hc/b (1.438×10^{-2} mk); b refers to the Boltzman constant as 1.38×10^{-23} J/K; h is the Planck's constant which is equal to 6.626×10^{-34} JS; c refers to the velocity of light 2.998×10^{8} m/s; and ε is surface emissivity.

Brisbane is a heterogeneous region; thus, it was necessary to consider the surface emissivity effect (ε) in estimating the LST. This research applied the normalised difference vegetation index (NDVI) threshold method to different land cover classes. Firstly, the emissivity of water body pixels was assigned as 0.990 (Deng & Wu, 2013). Then, an emissivity of 0.985 was assigned to each pixel with a NDVI value ≥ 0.5. These pixels were considered as fully vegetated. Finally, for the remaining pixels (i.e., covered with built-up and sparse vegetation), an emissivity was calculated based on Equations (9.3)–(9.5) (Deng & Wu, 2013).

$$\varepsilon_{mix} = \varepsilon_v p_v + \varepsilon_n (1 - P_v) + C_i \tag{9.3}$$

$$P_v = \left(\frac{NDVI - NDVI_{min}}{NDVI_{max} - NDVI_{min}}\right)^2 \tag{9.4}$$

$$C_i = (1 - \varepsilon_n)(1 - P_v) F \varepsilon_v \tag{9.5}$$

where ε_v= 0.985 (vegetation) and ε_n= 0.92 (urban surface) (Li et al., 2011); $NDVI_{max}$ and $NDVI_{min}$ are the values of full vegetation and non-vegetation, respectively; P_v is the scaled NDVI/fractional vegetation cover; C_i is the effect of geometric distribution of natural surfaces and internal reflections (Sobrino et al., 2004); and F = 0.55 which refers to the shape factor for geometric distribution. The calculated LST in Kelvin was then converted into Celsius for ease of interpretation.

The generated LST values were then used to derive the UHI intensity by subtracting the average LST value of all non-urban pixels from the LST value of each pixel (Weng & Lu, 2008). As outlined previously, the non-urban areas consist of water body, dense vegetation, and upland natural areas.

9.2.2.4 Factors Affecting the UHI Effect

This research estimated ordinary least square (OLS) regression model to identify significant factors contributing to the UHI effect in Brisbane. Firstly, an OLS model was estimated for the whole of Brisbane, then separate OLS models were estimated for the TOD and non-TOD areas to examine whether the model coefficients differ between different types of areas. Potential factors affecting the UHI intensity were identified from the literature and included both functional (residential density, employment density, land use mix) and compositional (proportion of porous land) aspects of the neighbourhood (Chun & Guldmann, 2014; Deilami et al., 2016a; Giridharan et al., 2004). The correlations of these factors were checked in order to avoid multicollinearity problem. The final model included only statistically significant coefficients after refining an initial model that included all possible factors.

9.3 Analysis and Results

9.3.1 TOD Neighbourhoods in Brisbane

Table 9.3 shows the BE characteristics of Brisbane. On average, there are 25 residential dwellings and almost an equal number of jobs located within a hectare of respective land use types. However, a high standard deviation value for these indicators, and particularly for employment density indicator, signifies that there is a large variation in the distribution of residences and jobs. This is also true for PTAL index. These findings clearly suggest that some CCDs in Brisbane are likely to qualify for a TOD and others do

TABLE 9.3

BE Characteristics of Brisbane

BE Indicator	Mean	Standard Deviation	Min	Max
Net residential density (dwellings/ha of residential zoned land)	24.78	50.21	0	899.37
Net employment density (jobs/ha of employment-zoned land)	25.55	149.53	0	2873.97
Land use diversity (Simpson's Diversity Index; range 0–1)	0.45	0.17	0	0.79
Cul-de-sac density (dead-ends/ha)	0.14	0.11	0	1.20
Intersection density (3+ way intersections/ha)	0.64	0.24	0	1.85
Public transport accessibility level (PTAL) index	5.30	9.32	0	130.93

not. It is also logical to hypothesise that CCDs that possess the quality of a TOD should have a higher score in the BE indicators (except cul-de-sac density) compared to the mean values as outlined in Table 9.3. The cluster analysis identified the natural groupings of these CCDs, as outlined in Figure 9.1.

The cluster analysis resulted in three unique clusters with an overall quality of the model in the "Fair" category (Figure 9.1). Net residential density contributed more to form the clusters, followed by net employment density and PTALs. The average scores of the indicators as shown in Figure 9.1 are useful in explaining the type of each cluster as outlined below:

1. **Cluster 1—Potential TODs**: About 52% of the CCDS (859 out of 1,661 CCDs analysed in this research) in Brisbane fall into this cluster. Figure 9.2 shows the geographical location of the cluster. They are labelled "Potential TODs" because although some BE indicators have a higher score (land use diversity, intersection density) compared to the average score of Brisbane, other indicators have lower than average scores (residential density, employment density, PTALs). This cluster has the lowest level of cul-de-sacs. These CCDs require a substantial policy effort to convert them into a complete TOD according to the definition of TOD as outlined in the Queensland Government policy document (Queensland Government, 2009).
2. **Cluster 2—Non-TODs**: About 29% of the CCDs are classified as having the characteristics of a non-TOD neighbourhood (Figure 9.2). Unlike cul-de-sac, the values of the remaining five BE characteristics of these CCDs are far less than the average of Brisbane. These CCDs also have a higher level of cul-de-sacs.
3. **Cluster 3—TODs**: A total of 327 CCDs in Brisbane (about 20%) were found to have the quality of a TOD (Figure 9.2). These CCDs not only have a greater score in the five BE indicators (except cul-de-sac), their average scores are well above the TOD characteristics as outlined in the Queensland Government policy document. For example, Queensland Government (2009) specified 30 dwellings per hectare as a minimum standard for a TOD. The average dwelling density was found to be 34/ha for this cluster. Also, these CCDs have a perfect job–housing balance which has been identified to serve as a critical role for the proper functioning of a TOD (Center for Transit-Oriented Development, 2010). Therefore, these CCDs can be considered as having the quality of a TOD in Brisbane.

However, despite having the quality of a TOD, Cluster 3 (also Cluster 1) cannot be labelled a TOD (potential), if they do not function well when measured by performance (Belzer & Autler, 2002). The following section presents the validation results of the identified clusters.

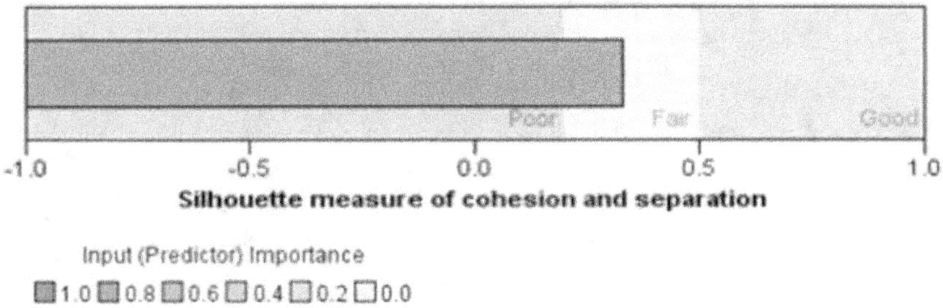

Cluster	1	2	3
Label	Potential TODs	Non-TODs	TODs
Description	Low residential density and PTALs, moderately diverse land use and street connectivity level, very low employment density	Very low residential and employment density, PTALs, diversity, and connectivity levels	High residential and employment density, good PTALs, highly diverse land use and street connectivity level
Size	51.7% (859)	28.6% (475)	19.7% (327)
Inputs	Net residential density 18.60	Net residential density 12.50	Net residential density 34.18
	Net employment density 3.87	Net employment density 3.00	Net employment density 33.13
	PTALs 4.20	PTALs 1.71	PTALs 8.26
	Cul-de-sac density 0.09	Cul-de-sac density 0.21	Cul-de-sac density 0.14
	Land use diversity 0.47	Land use diversity 0.33	Land use diversity 0.57
	Intersection density 0.66	Intersection density 0.51	Intersection density 0.75

FIGURE 9.1
Cluster analysis to differentiate between TOD and non-TOD areas in Brisbane.

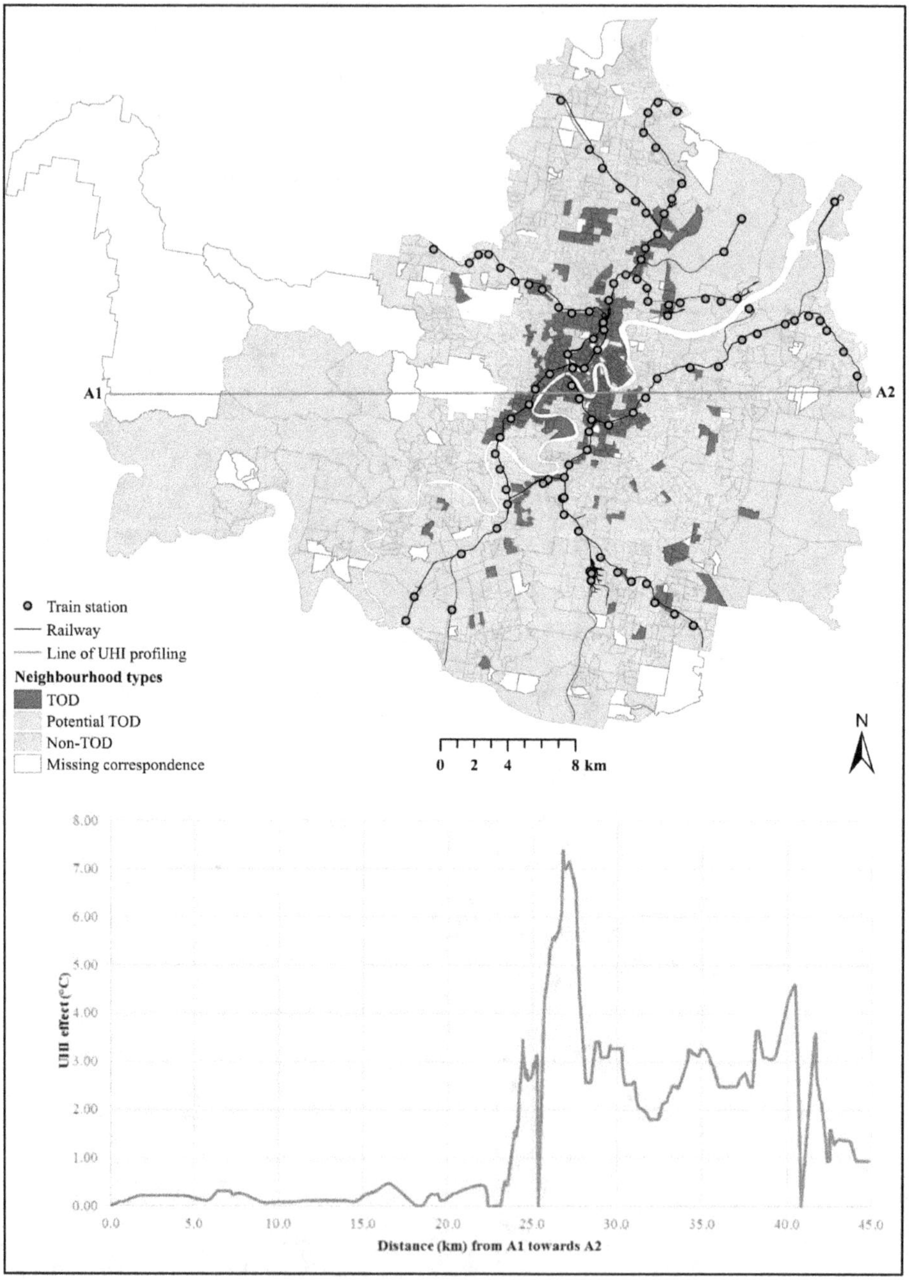

FIGURE 9.2
Spatial distribution of TOD and non-TOD neighbourhoods in Brisbane and their UHI profile.

TABLE 9.4
Distance Travelled by Mode per Person per Day

Classification of CCDs	Number of CCDs	% of CCDs	Total	Private Vehicle	Public Transport	Active Transport
Potential TOD	104	53.9	42.29	23.97	16.26	2.07
Non-TOD	46	23.8	58.90	34.93	21.04	2.94
TOD	43	22.3	38.88	25.43	10.45	2.99
Total	**193**	**100.0**	**45.49**	**26.91**	**16.10**	**2.48**

9.3.2 Mode Choice Behaviour in TOD and Non-TOD Areas

Table 9.4 outlines the results from the travel behaviour analysis of people living within the clusters. Note that this analysis was conducted for 193 CCDs given the sampling strategy as used in the SEQHTS. Despite this limitation, the distribution of the analysis CCDs closely matched the distribution of TOD clusters as identified in Section 3.1. For example, about 52% CCDs are identified as potential TODs (Figure 9.1). Table 9.4 shows that 54% of the analytical CCDs are located within this TOD cluster. Therefore, the findings from this analysis are expected to represent the travel behaviour of people living in non-sampled CCD clusters.

An ANOVA test results show that there are significant differences in the use of private vehicle ($F = 5.061$, $p < 0.01$) and public transport services ($F = 6.531$, $p < 0.01$) between the neighbourhood types. However, no significant differences were identified in the use of active transport between the clusters.

Table 9.4 shows that on average people living in non-TOD areas travelled 35km per day using private vehicles which is significantly higher compared to people living in TOD (25 km) and potential TOD (24 km) areas. No significant difference was found to exist in the use of private vehicles for people living between TOD and potential TOD areas.

People living in TODs also travelled significantly shorter distances per day using public transport compared to non-TOD and potential TOD areas. However, there is no significant difference in the use of public transport between potential TOD and non-TOD areas.

Overall, the ANOVA test results show that there is a significant difference in the total distance travelled by individuals living in the three types of neighbourhoods ($F = 9.312$, $p < 0.01$). People living in non-TOD areas travelled significantly longer distances per day than people living in the other two areas. However, the differences between TOD and potential TOD areas were found to be minimal (3 km), and the difference is statistically insignificant.

The results show that TODs reduce the travel need of people either by private vehicle or by public transport. Although people in potential TODs travel more using public transport, their overall travel was found to be similar to that TOD neighbourhood and significantly lower than those living in non-TOD areas. These findings signify the validity of the generated cluster in

TABLE 9.5

Changes in UHI Effect over the Periods (Paired Sampled t-Test)

Classification of CCDs	UHI Effect 2004 (°C)	UHI Effect 2013 (°C)	Change in UHI (°C)	Paired Samples t-test (t)	Sig
Potential TOD	2.24	3.01	0.77	49.67	0.001
Non-TOD	2.23	2.82	0.58	22.39	0.001
TOD	3.09	3.81	0.72	21.27	0.001
Total	**2.40**	**3.11**	**0.71**	**54.51**	**0.001**

this research. It also justifies the goal of TODs as a BE intervention: to reduce car-based travel (Cervero et al., 2004). The question remains, however, about their contribution to the UHI effect, and ultimately to the environmental sustainability of TODs as discussed in the following section.

9.3.3 The UHI Effects of TODs

Table 9.5 illustrates the average UHI effects (°C) of the three types of neighbourhoods—i.e., the differences in temperature between each of these neighbourhoods and non-urban areas (e.g., vegetation) in Brisbane. UHI effects are calculated for both 2004 and 2013 with the aim to monitor changes in UHI effects following the introduction of TOD policies. Paired samples *t* tests were conducted to investigate whether the UHI effects in 2013 are significantly different from the base period (2004). As shown in Table 9.5, on average Brisbane experienced a 0.71°C increase in UHI effect in 2013 compared to 2004. This is a significant increase from the base period (2004). This overall trend was observed in all three types of areas: the UHI effect increased by 0.72°C, 0.77°C, and 0.58°C for TOD, potential TOD, and non-TOD areas (Table 9.5).

An ANOVA test was also conducted to examine whether the changes in UHI effect in the three types of areas are significantly different from each other. Table 9.6 shows that the increase of UHI effect was similar for TOD and potential TOD areas (no significant difference). However, the increases

TABLE 9.6

ANOVA Test Results Showing Differences in UHI Effect Between TOD and Non-TOD Areas and Their Changes

Comparison (Classification of CCDs)	Mean Difference in UHI Change (2004–2013)	Mean Difference in UHI Effect (2013)
TOD vs. non-TOD	0.18[a]	−0.99[a]
Potential TOD vs. non-TOD	0.14[a]	0.19[a]
Potential TOD vs. TOD	0.04	−0.80[a]
F-value	**F = 19.20**[a]	**F = 79.37**[a]

[a] Coefficients are significant at the 0.01 level.

were significantly higher for TOD and potential TOD areas compared to non-TOD areas.

Table 9.5 indicates that TOD areas experienced the highest temperature differences in Brisbane (on average 3.81°C UHI effect) when compared to non-urban areas. Table 9.6 confirms that this UHI intensity was significantly higher compared to the intensity observed in potential TOD areas (by 0.80°C) and non-TOD areas (by 0.99°C). Figure 9.2 graphically shows the UHI profile of different neighbourhood types in relation to their geographical location in Brisbane.

In summary, TODs are the highest temperature zone in Brisbane, and the temperature in these neighbourhoods is increasing faster than the other neighbourhood types. However, given that all neighbourhood types experienced a significant increase in UHI effect, this suggests that the factors that contribute to the UHI effects were in operation in all neighbourhood types, albeit differently. The following section investigates this issue and identifies significant factors of UHI effects in each neighbourhood type.

9.3.4 Factors Affecting the UHI Intensities

Table 9.7 shows the OLS regression analyses results and identifies factors that contributed significantly to the UHI effect in Brisbane, and more specifically within each type of neighbourhood. The intent here is to identify which of the BE attributes of TODs have an association with the UHI effect. Therefore, all six indicators, as used to identify TOD neighbourhoods, were included

TABLE 9.7

Factors Affecting the UHI Effect in Brisbane

Explanatory Factors	All of Brisbane	Non-TOD	Potential TOD	TOD
Constant	5.91[a]	6.11[a]	5.76[a]	5.72[a]
% of porous land	−0.06[a]	−0.06[a]	−0.05[a]	−0.07[a]
Net residential density	NI	NI	NI	NI
Net employment density	NI	NI	NI	NI
Land use diversity	0.75[a]	1.00[a]	0.78[a]	1.93[a]
Intersection density	−0.32[a]	NS	−0.42[a]	NS
Cul-de-sac density	2.09[a]	1.52[a]	NS	NS
PTALs	NS	NS	NS	NS
F-value	341.53[a]	160.80[a]	69.16[a]	276.44[a]
R^2	0.45	0.51	0.20	0.63
N	1661	475	859	327

[a] Coefficients are significant at the 0.001 level.
NS = Not significant; NI = Not included.

in the model. In addition, previous research has identified that the amount of natural areas within a neighbourhood also affect the UHI intensities (Deilami et al., 2016b). As a result, the percentage of porous land within each CCD was also included in the model. A correlation analysis of the included factors, however, showed that a significant correlation exists between porosity and net residential density indicators (r = −0.42), between porosity and net employment density indicators (r = −0.32), and between net residential density and net employment density indicators (r = 0.56). A preliminary analysis showed that these correlations caused multicollinearity problem in the estimated models. As a result, both net residential density and net employment density indicators were excluded, and only the porosity factor was included given that this factor increased the explanatory power of the estimated models over the other two factors. The F coefficients of the overall model and the three cluster-specific models were found to be statistically significant, suggesting the usefulness of estimating the OLS model. The explanatory power of all the models was found to be defensible in this literature.

The estimated coefficients, as shown in Table 9.7, indicate that porosity is a significant factor in all neighbourhood types with a negative sign. This means, for example, that 1% increase in porous land in TODs is likely to reduce the UHI effect by 0.07°C. Note that porosity is negatively associated with residential density and employment density. This means that an increase of these two indicators is likely to increase UHI effect. Similarly, an increase of land use diversity level is likely to increase the UHI effect in general. However, the effect is more pronounced in TOD areas. Table 9.7 shows that one unit increase in land use diversity level (i.e., from zero diversity to fully diverse) is likely to increase the UHI effect by 2°C. Intersection density reduces the UHI effect, whereas the cul-de-sac density increases the UHI effect. However, these effects were not found to be significant in TOD areas.

9.4 Findings and Discussion

TODs have been proven to be a sustainable neighbourhood planning model in many respects such as the reduction of greenhouse gas emissions through lowering car-based travel, increase of social connectedness and well-being, and enhancing the economic productivity of places. This research examines an overlooked dimension of TODs—how much does it contribute to the UHI effect? Clearly, this research shows that TODs are the high-temperature zones in a city and increasingly being experienced high temperature over time. More importantly, the rate of increase of UHI effect in TODs is significantly higher than that of non-TOD areas. Nevertheless, all three types of areas (TODs, non-TODs, and potential TODs) in Brisbane experienced a higher level of UHI effect in 2013 compared to 2004.

The findings in this research point to two different types of factors that are in play for the increasing trend of UHI effect. Firstly, known factors such as those included in the model (e.g., changes in land use patterns) explained a majority proportion of the UHI effects in Brisbane. Secondly, unknown factors which could be external to the research context such as climate change vis-à-vis global warming (not included (NI) in the model) are potentially candidate factors to explain the unexplained variance in UHI data. However, such a climatic interaction did not affect the findings in relation to the research objective (To what extent do TODs reduce the UHI effect over sprawling development?) as presented in this research for two reasons: firstly, if climate change has an impact, it would have affected all types of areas equally. The results show that the rate of increase of UHI was higher in TODs compared to non-TOD areas—suggesting that some other mechanism is in play (e.g., the variables included); otherwise, the rate of increase should be equal in type of areas; and secondly, TODs experienced a higher UHI effect in both periods compared to non-TOD areas, suggesting that TODs are historically high-temperature zones in a city.

The findings from this research bear an important policy question: how should we design our TODs for environmental sustainability? The findings clearly show that TODs contribute to environmental sustainability by enabling travel self-containment—i.e., people travel significantly less mileage if they live in TODs and consequently reduce the greenhouse gas emissions, congestions, and so on. However, living in TODs means that people are likely to experience higher UHI intensities. This has several negative consequences. The more direct effect in the context of this research is to use more electricity for cooling. The generation of electricity again contribute to greenhouse gas emissions. Therefore, the policy benefits of reduced mileage are likely to diminish through increasing use of electricity. The findings also show that it is the BE factors that make a neighbourhood a TOD and thus reduce car-dependency, but these BE factors again contribute to the UHI intensity. Therefore, the question is an optimisation problem—how much of each factor should be adopted for a TOD that will reduce both car-dependency and the UHI effect. In this regard, clearly a compromise can be made between natural and built-up areas in the design of a TOD.

On that very point, strategies could also be adopted to make the built-up areas act more like a non–built up zone by effective use of sustainable building materials such as green roofs and living walls. This will ultimately enable planners to designate a higher proportion of land for built purposes. These technologies are emerging as important additions to the palette of construction techniques for creating healthy, ecologically responsible buildings. Today, in the major metropolitan cities of the world green roofs and living walls are becoming part of the urban texture. They act as insulation and to a degree as cooling devices, reducing the heat flux through the roof and walls and thus helping in energy saving and decreasing the UHI effect

(Castleton et al., 2010; Jim, 2015). The next generation of TOD design, therefore, should consider incorporating green roofs and living walls in their design.

A known limitation of this research is that transit nodes may not necessarily be located at the centre of each TOD because of the methodological approach applied to identify TODs. However, such a copybook example of TODs is hard to find in reality (Renne & Ewing, 2013). Nevertheless, a deviation from exactly in accordance with established criteria may have compromised the results presented in this research. Future research is planned to address this methodological issue and to identify the impact of green roofs and living walls in TODs on the UHI effect in Brisbane.

9.5 Conclusion

This chapter provides an important contribution to the growing discourse on the complex interactions between urban form, mobility-oriented planning, and microclimatic resilience. By examining the relationship between TOD and urban heat in Brisbane, the chapter challenges the widely held assumption that compact, transit-focused neighbourhoods inherently promote environmental sustainability across all dimensions. Through a rigorous analytical approach—combining cluster-based delineation of TOD areas, validation of transit orientation through residents' travel behaviour, and assessment of surface temperatures using satellite-derived indicators—the study reveals a counterintuitive yet consequential finding: TOD neighbourhoods in Brisbane exhibit higher levels of urban heat, as well as a faster rate of temperature increase, compared to non-TOD areas.

These results underscore that while TOD offers clear benefits in reducing car-dependency, supporting public transit use, and containing urban sprawl, its physical form and land use patterns can inadvertently intensify heat exposure. High land use diversity, extensive impervious surfaces, and limited porous land cover emerge as key contributors to elevated local temperatures within TOD precincts. This underscores a critical tension at the heart of contemporary sustainable planning practice: the environmental gains achieved through density and transit efficiency may be undermined if microclimatic considerations are not simultaneously integrated into development strategies.

The findings therefore highlight the necessity for a more holistic approach to TOD design—one that recognises the importance of urban greening, surface permeability, and balanced land cover in moderating local heat conditions. Integrating green infrastructure, enhancing tree canopy, and prioritising permeable materials can help ensure that TOD precincts deliver not only mobility and emissions benefits but also climatic comfort and resilience. Ultimately, this study demonstrates that achieving genuinely sustainable and climate-responsive urban environments requires aligning transit-oriented principles with deliberate, place-based strategies to mitigate urban heat.

Acknowledgements

This chapter, with permission from the copyright holder, is a reproduced version of the following journal article: Kamruzzaman, M., Deilami, K., & Yigitcanlar, T. (2018). Investigating the urban heat island effect of transit-oriented development in Brisbane. *Journal of Transport Geography, 66*(1), 116–124.

References

Al Kuwari, N., Ahmed, S., & Kaiser, M. (2016). Optimal satellite sensor selection utilized to monitor the impact of urban sprawl on the thermal environment in Doha City, Qatar. *Journal of Earth Science & Climatic Change, 2016.*

Belzer, D., & Autler, G. (2002). *Transit oriented development: Moving from rhetoric to reality.*

Bertolini, L., Curtis, C., & Renne, J. L. (2009). Introduction. In C., Curtis, J. L. Renne, & L. Bertolini (Eds.), *Transit oriented development: Making it happen*. Ashgate Publishing Limited.

Bornstein, R., & Lin, Q. (2000). Urban heat islands and summertime convective thunderstorms in Atlanta: Three case studies. *Atmospheric Environment, 34,* 507–516.

Bowes, D. R., & Ihlanfeldt, K. R. (2001). Identifying the impacts of rail transit stations on residential property values. *Journal of Urban Economics, 50,* 1–25.

Braganza, K., Reynolds, A., Karoly, D., & Risbey, J. (2003). Global warming contributes to Australia's worst drought. *Australasian Science, 24*(3), 14–17.

Brisbane City Council (2014). *Brisbane City plan 2014.*

Broaddus, A. (2010). Tale of two ecosuburbs in Freiburg, Germany. *Transportation Research Record: Journal of the Transportation Research Board, 2187,* 114–122.

Brown, B. B., & Werner, C. M. (2008). Before and after a new light rail stop: Resident attitudes, travel behavior, and obesity. *Journal of the American Planning Association, 75,* 5–12.

Burke, M., & Brown, A. L. (2007). Distances people walk for transport. *Road & Transport Research, 16,* 16–29.

Castleton, H. F, Stovin, V, Beck, S. B. M., & Davison, J. B. (2010). Green roofs; building energy savings and the potential for retrofit. *Energy and Buildings, 42,* 1582–1591.

Center for Transit-Oriented Development (2010). *Performance-based transit-oriented development typology guidebook.*

Cerin, E., Leslie, E., Owen, N., & Bauman, A. (2007). Applying GIS in physical activity research: Community 'Walkability' and walking behaviors. In P. Lai & A. H. Mak (Eds.), *GIS for health and the environment* (pp. 72–89). Springer Berlin Heidelberg.

Cervero, R., & Day, J. (2008). Suburbanization and transit-oriented development in China. *Transport Policy, 15*, 315–323.

Cervero, R., & Gorham, R. (1995). Commuting in transit versus automobile neighborhoods. *Journal of the American Planning Association, 61*, 210.

Cervero, R., & Kockelman, K. (1997). Travel demand and the 3Ds: Density, diversity, and design. *Transportation Research Part D: Transport and Environment, 2*, 199–219.

Cervero, R., Murphy, S., Ferrell, C., Goguts, N., Tsai, Y. H., Arrington, G. B., Boroski, J., Smith-Heimer, J., Golem, R., Peninger, P., Nakajima, E., Chui, E., Dunphy, R., Myers, M., Mckay, S., & Witenstein, N. (2004). *Transit-oriented development in the United States: Experiences, challenges, and prospects.*

Chia, J., Lee, J., & Kamruzzaman, M. (2016). Walking to public transit: Exploring variations by socioeconomic status. *International Journal of Sustainable Transportation, 10*, 805–814.

Chow, W. T., Brennan, D., & Brazel, A. J. (2012). Urban heat island research in Phoenix, Arizona: Theoretical contributions and policy applications. *Bulletin of the American Meteorological Society, 93*, 517.

Chun, B., & Guldmann, J. M. (2014). Spatial statistical analysis and simulation of the urban heat island in high-density central cities. *Landscape and Urban Planning, 125*, 76–88.

Crowley, D., Shalaby, A., & Zarei, H. (2009). Access walking distance, transit use, and transit-oriented development in North York City Center, Toronto, Canada. *Transportation Research Record: Journal of the Transportation Research Board, 2110*, 96–105.

Curtis, C., Renne, J. L., & Bertolini, L. (2009). *Transit oriented development: Making it happen*. Ashgate Publishing Limited, Surrey.

Deilami, K., Kamruzzaman, M., & Hayes, J. (2016a). Correlation or causality between land cover patterns and the urban heat island effect? Evidence from Brisbane, Australia. *Remote Sensing, 8*, 716.

Deilami, K., Kamruzzaman, M., & Hayes, J. F. (2016b). Correlation or causality between land cover patterns and the urban heat island effect? Evidence from Brisbane, Australia. *Remote Sensing, 8*, 716.

Deng, C., & Wu, C. (2013). Examining the impacts of urban biophysical compositions on surface urban heat island: A spectral unmixing and thermal mixing approach. *Remote Sensing of Environment, 131*, 262–274.

De Vos, J., Derudder, B., Van Acker, V., & Witlox, F. (2012). Reducing car use: Changing attitudes or relocating? The influence of residential dissonance on travel behavior. *Journal of Transport Geography, 22*, 1–9.

Dill, J. (2008). Transit use at transit-oriented developments in Portland, Oregon, Area. *Transportation Research Record: Journal of the Transportation Research Board, 2063*, 159–167.

Dixon, P. G., & Mote, T. L. (2003). Patterns and causes of Atlanta's urban heat island-initiated precipitation. *Journal of Applied Meteorology, 42*, 1273–1284.

Duncan, M. J., Winkler, E., Sugiyama, T., Cerin, E., duToit, L., Leslie, E., & Owen, N. (2010). Relationships of land use mix with walking for transport: Do land uses and geographical scale matter? *Journal of Urban Health, 87*, 782–795.

Elsayed, I. S. (2012). *Effects of population density and land management on the intensity of urban heat islands: A case study on the city of Kuala Lumpur.* INTECH Open Access Publisher.

EPA (2016). *Heat island impacts.* United States Environmental Protection Agency. http://www.epa.gov/hiri/impacts/index.htm.

Faghri, A., & Venigalla, M. (2013). Measuring travel behavior and transit trip generation characteristics of transit-oriented developments. *Transportation Research Record: Journal of the Transportation Research Board, 2397*, 72–79.

Frank, L. D., Schmid, T. L., Sallis, J. F., Chapman, J., & Saelens, B. E. (2005). Linking objectively measured physical activity with objectively measured urban form: Findings from SMARTRAQ. *American Journal of Preventive Medicine, 28*, 117–125.

Gago, E. J., Roldan, J., Pacheco-Torres, R., & Ordóñez, J. (2013). The city and urban heat islands: A review of strategies to mitigate adverse effects. *Renewable and Sustainable Energy Reviews, 25*, 749–758.

Giridharan, R., Ganesan, S., & Lau, S. (2004). Daytime urban heat island effect in high-rise and high-density residential developments in Hong Kong. *Energy and Buildings, 36*, 525–534.

Goonetilleke, A., Yigitcanlar, T., Ayoko, G. A, & Egodawatta, P. (2014). *Sustainable urban water environment: Climate, pollution and adaptation*. Edward Elgar Publishing.

Hondula, D. M., Georgescu, M., & Balling, R. C. (2014). Challenges associated with projecting urbanization-induced heat-related mortality. *Science of the Total Environment, 490*, 538–544.

Huang, C., Davis, L., & Townshend, J. (2002). An assessment of support vector machines for land cover classification. *International Journal of Remote Sensing, 23*, 725–749.

Jia, K., Wei, X., Gu, X., Yao, Y., Xie, X., & Li, B. (2014). Land cover classification using landsat 8 operational land imager data in Beijing, China. *Geocarto International, 29*, 941–951.

Jim, C. Y. (2015). Assessing climate-adaptation effect of extensive tropical green roofs in cities. *Landscape and Urban Planning, 138*, 54–70.

Kamruzzaman, M., Baker, D., Washington, S., & Turrell, G. (2014a). Advance transit oriented development typology: Case study in Brisbane, Australia. *Journal of Transport Geography, 34*, 54–70.

Kamruzzaman, M., Baker, D., Washington, S., & Turrell, G. (2014d). Advance transit oriented development typology: Case study in Brisbane, Australia. *Journal of Transport Geography, 34*, 54–70.

Kamruzzaman, M., Baker, D., Washington, S., & Turrell, G. (2016). Determinants of residential dissonance: Implications for transit-oriented development in Brisbane. *International Journal of Sustainable Transportation, 10*, 960–974.

Kamruzzaman, M., & Hine, J. (2013). Self-proxy agreement and weekly school travel behaviour in a sectarian divided society. *Journal of Transport Geography, 29*, 74–85.

Kamruzzaman, M., Shatu, F. M., Hine, J., & Turrell, G. (2015). Commuting mode choice in transit oriented development: Disentangling the effects of competitive neighbourhoods, travel attitudes, and self-selection. *Transport Policy, 42*, 187–196.

Kamruzzaman, M., Wood, L., Hine, J., Currie, G., Giles-Corti, B., & Turrell, G. (2014b). Patterns of social capital associated with transit oriented development. *Journal of Transport Geography, 35*, 144–155.

Kamruzzaman, M., Yigitcanlar, T., Washington, S., Currie, G., & Turrell, G. (2014c). Australian baby boomers switched to more environment friendly modes of transport during the global financial crisis. *International Journal of Environmental Science and Technology, 11*, 2133–2144.

Kay, A. I., Noland, R. B., & DiPetrillo, S. (2014). Residential property valuations near transit stations with transit-oriented development. *Journal of Transport Geography, 39*, 131–140.

Lee, S., Lee, S., Park, D., & Lee, C. (2010). Analysis of travel patterns between road and transit-oriented development areas. *Journal of Advanced Transportation, 44*, 3–10.

Lemonsu, A., Viguié, V., Daniel, M., & Masson, V. (2015). Vulnerability to heat waves: Impact of urban expansion scenarios on urban heat island and heat stress in Paris (France). *Urban Climate, 14*, 586–605.

Li, J., Song, C., Cao, L., Zhu, F., Meng, X., & Wu, J. (2011). Impacts of landscape structure on surface urban heat islands: A case study of Shanghai, China. *Remote Sensing of Environment, 115*, 3249–3263.

Li, X., Li, W., Middel, A., Harlan, S. L., Brazel, A. J., & Turner Ii, B. L. (2016). Remote sensing of the surface urban heat island and land architecture in Phoenix, Arizona: Combined effects of land composition and configuration and cadastral–demographic–economic factors. *Remote Sensing of Environment, 174*, 233–243.

Lin, J. J., & Gau, C. C. (2006). A TOD planning model to review the regulation of allowable development densities around subway stations. *Land Use Policy, 23*, 353–360.

Lu, D., Song, K., Zang, S., Jia, M., Du, J., & Ren, C. (2014). The effect of urban expansion on urban surface temperature in Shenyang, China: An analysis with Landsat Imagery. *Environmental Modeling & Assessment, 20*, 197–210.

Lu, D., & Weng, Q. (2006). Use of impervious surface in urban land-use classification. *Remote Sensing of Environment, 102*, 146–160.

Lund, H. (2006). Reasons for living in a transit-oriented development and associated transit use. *Journal of the American Planning Association, 72*, 357–366.

Mountrakis, G., Im, J., & Ogole, C. (2011). Support vector machines in remote sensing: A review. *ISPRS Journal of Photogrammetry and Remote Sensing, 66*, 247–259.

NASA (2011). *Landsat 7 science data users handbook*. NASA.

Nasri, A., & Zhang, L. (2014). The analysis of transit-oriented development (TOD) in Washington, DC and Baltimore metropolitan areas. *Transport Policy, 32*, 172–179.

Oakes, J. M., Forsyth, A., & Schmitz, K. H. (2007). The effects of neighborhood density and street connectivity on walking behavior: The Twin Cities walking study. *Epidemiologic Perspectives & Innovations, 4*.

Otukei, J., & Blaschke, T. (2010). Land cover change assessment using decision trees, support vector machines and maximum likelihood classification algorithms. *International Journal of Applied Earth Observation and Geoinformation, 12*, S27–S31.

Pacione, M. (2009). *Urban geography: A global perspective*. Routledge.

Pan, H., Shen, Q., & Liu, C. (2011). Transit-oriented development at the urban periphery. *Transportation Research Record: Journal of the Transportation Research Board, 2245*, 95–102.

Paneque-Gálvez, J., Mas, J. F., Moré, G., Cristóbal, J., Orta-Martínez, M., Luz, A. C., Guèze, M., Macía, M. J., & Reyes-García, V. (2013). Enhanced land use/cover classification of heterogeneous tropical landscapes using support vector machines and textural homogeneity. *International Journal of Applied Earth Observation and Geoinformation, 23*, 372–383.

Poursanidis, D., Chrysoulakis, N., & Mitraka, Z. (2015). Landsat 8 vs. Landsat 5: A comparison based on urban and peri-urban land cover mapping. *International Journal of Applied Earth Observation and Geoinformation, 35*, 259–269.

Queensland Government (2008). *Toward Q2: Tomorrow's Queensland*.

Queensland Government (2009). *South East Queensland Regional Plan 2009–2031*.

Queensland Government (2010). *Connecting SEQ 2031 – An integrated regional transport plan for South East Queensland*.

Renne, J. L. (2009a). From transit-adjacent to transit-oriented development. *Local Environment, 14,* 1–15.

Renne, J. L. (2009b). Measuring the success of transit oriented development. In C. Curtis, J. L. Renne, & L. Bertolini (Eds.), *Transit oriented development: Making it happen* (pp. 241–255). Ashgate Publishing Limited.

Renne, J. L., & Ewing, R. (2013). *Transit-oriented development: An examination of america's transit precincts in 2000 & 2010.*

Reusser, D. E., Loukopoulos, P., Stauffacher, M., & Scholz, R. W. (2008). Classifying railway stations for sustainable transitions – Balancing node and place functions. *Journal of Transport Geography, 16,* 191–202.

Rohe, W. M. (2009). From local to global: One hundred years of neighborhood planning. *Journal of the American Planning Association, 75,* 209–230.

Santamouris, M., Cartalis, C., Synnefa, A., & Kolokotsa, D. (2015). On the impact of urban heat island and global warming on the power demand and electricity consumption of buildings—A review. *Energy and Buildings, 98,* 119–124.

Schwanen, T., & Mokhtarian, P. (2004). The extent and determinants of dissonance between actual and preferred residential neighborhood type. *Environment and Planning B: Planning and Design, 31,* 759–784.

Shatu, F. M., Kamruzzaman, M., & Deilami, K. (2014). Did Brisbane grow smartly? Drivers of city growth 1991–2001 and lessons for current policies. *SAGE Open, 4.*

Simpson, E. H. (1949). Measurement of diversity. *Nature, 163,* 688.

Sipe, N., & Burke, M. (2011). Can river ferries deliver smart growth? *Transportation Research Record: Journal of the Transportation Research Board, 2217,* 79–86.

Sobrino, J. A., Jiménez-Muñoz, J. C., & Paolini, L. (2004). Land surface temperature retrieval from LANDSAT TM 5. *Remote Sensing of Environment, 90,* 434–440.

Stangl, P., & Guinn, J. M. (2011). Neighborhood design, connectivity assessment and obstruction. *Urban Design International, 16,* 285–296.

Susca, T., Gaffin, S. R., & Dell'Osso, G. R. (2011). Positive effects of vegetation: Urban heat island and green roofs. *Environmental Pollution, 159,* 2119–2126.

Thomlinson, J. R., Bolstad, P. V., & Cohen, W. B. (1999). Coordinating methodologies for scaling landcover classifications from site-specific to global: Steps toward validating global map products. *Remote Sensing of Environment, 70,* 16–28.

Transport for London (2010). *Measuring public transport accessibility levels (PTALs): Summary.*

Ward, K., Lauf, S., Kleinschmit, B., & Endlicher, W. (2016). Heat waves and urban heat islands in Europe: A review of relevant drivers. *Science of the Total Environment, 569–570,* 527–539.

Weng, Q., & Lu, D. (2008). A sub-pixel analysis of urbanization effect on land surface temperature and its interplay with impervious surface and vegetation coverage in Indianapolis, United States. *International Journal of Applied Earth Observation and Geoinformation, 10,* 68–83.

Yang, K., & Pojani, D. (2017). A decade of transit oriented development policies in Brisbane, Australia: Development and land-use impacts. *Urban Policy and Research, 35,* 347–362.

Zemp, S., Stauffacher, M., Lang, D. J., & Scholz, R. W. (2011). Classifying railway stations for strategic transport and land use planning: Context matters! *Journal of Transport Geography, 19,* 670–679.

10

Policy and Technology Responses to Urban Heat

10.1 Introduction and Background

Amidst the climate crisis, cities worldwide are increasingly affected by the urban heat island (UHI) effect, which describes the phenomenon of higher temperatures in urban areas compared to their surrounding rural areas (Gunawardena et al., 2017; Levermore et al., 2018; Li et al., 2019; Sanchez-Guevara et al., 2019; Stone et al., 2012; Yu et al., 2020). With regard to UHI mitigation strategies, various measures were identified and conceptualised including urban design, cool building envelopes, high albedo materials, green roofs, green facades, shade trees, ground vegetation, and waste heat (Akbari et al., 2012; Aleksandrowicz et al., 2017; Mirzaei & Haghighat, 2010; Ulpiani, 2021). Further, the political implementation of UHI mitigation outcomes is still under-researched due to a lack of understanding how research corresponds to the needs, priorities, and limitations of the political domain (Aleksandrowicz et al., 2017). There are three stakeholder groups most relevant for mitigating UHI effects: (a) the government, including politicians acting at a local or national level; (b) the city administration, including city planners and public health officials; and (c) the inhabitants, as individuals or as societal groups living in the urban area (Hintz et al., 2018). The private sector is also important since most buildings are privately held, and many of the UHI mitigation solutions are produced by industry. Various methods are used to enhance UHI analytical techniques. These include using types of satellite images, which can be used to analyse temperature patterns (Deilami et al., 2018), or to capture parameters that explain characteristics of UHI patterns related to the heat magnitude, the surface UHI magnitude, and the land surface temperature (Ward et al., 2016). The UHI research has focused on different spatial scales, such as South Asia (Kotharkar et al., 2018), India (Veena et al., 2020), or Greater Kuala Lumpur (a.k.a. Klang Valley) (Ramakreshnan et al., 2018). Other studies focus on policy responses to UHI mitigation (Parsaee et al., 2019) as well as technological developments including active technologies, such as heating, ventilation and air conditioning (HVAC) systems related to building energy consumption (Li et al., 2019), and passive technologies, like thermal properties of asphalt concrete for cool pavements and roofs (Mohajerani et al., 2017; Santamouris, 2013, 2014), which can substantially contribute to reducing temperatures in urban environments.

 DOI: 10.1201/9781003737056-13

We develop a novel framework drawing on existing literature reviews such as Santamouris (2014) and other academic works that discuss UHI policy and technology responses. Our framework explains how joint policy and technology interventions reduce anthropogenic heat and alleviate the incremental environmental burden. We include two different aspects to anthropogenic heat: (a) heat pollution from human activities, and (b) exacerbated heat "when vegetation and water bodies are replaced by materials such as concrete and asphalt, which have higher heat capacities and thermal conductivity" (Estrada et al., 2017, p. 403). Building on the proposed framework, we present 13 research questions to help guide future research on the synergistic effects of combined policy and technology interventions. This is much needed in UHI research because policy and technology are often developed in isolation of each other and not in conjunction; therefore, they have limited reaffirming effects when aimed at solving global issues. This is particularly true for UHIs, where both policy and technology responses are vital for mitigating their effect. To analyse this aspect, we conducted a systematic literature review with a focus on capturing the current understanding of policy and technology responses in mitigating the UHI effect. Previous literature reviews predominantly have focused on policy and technology aspects separately, such as mitigation strategies for improved policymaking (Aleksandrowicz et al., 2017; Parsaee et al., 2019), or urban materials production and technical properties of cool surfaces (Qin, 2015; Santamouris, 2014). The current state of UHI research vastly looks at either policy or technology aspects, which we scrutinise in our literature review. We aim at combining both aspects for joint optimisation in a complex city context, where the isolation of policy and technology tends to increase injurious effects to the city's performance (Patorniti et al., 2017).

This review conceptualises trends in policy and technology responses for UHI mitigation. It contributes to UHI research by combining both streams of mitigation strategies and proposing directions for future research to enable effective interactions between policy and technology. It affords the potential to reinforce the benefits of conjoining policy and technology responses and generate coordination effects (Arthur, 1988; Jänicke, 2015). To this end, policy refers to any course or principle of action to guide decisions to mitigate UHIs, while technology relates to any kind of technology, that is, any kind of object crafted by human intervention. We include the policy arenas of urban planning, public health, as well as transport and land use policies. Our framework informs policymakers on the interaction of policy and technology, for which we propose action research as an adequate approach to develop policy processes to implement the different policies in city-level decision-making. We present study suggestions and key approaches at the end of our study.

The chapter is structured as follows. After this introduction, we provide an overview of our approach to conducting the review, present the results of our literature search, and synthesise the literature by developing a concept matrix. Then, we discuss identified concepts including evidence base for policymaking identified in the literature review as well as policy and technology responses. Finally, we build on the insights from our literature review and present a conceptual framework of policy and technology responses in mitigating the UHI effect. Based on the proposed conceptual framework, we present directions for future research and close with conclusions.

10.2 Research Design

We followed a four-stage guide (Okoli, 2015) for conducting a systematic literature review and begin with Stage 1, "planning," where planning required the formulation of the purpose and the search strategy. In Stage 2, "selection," we set up the selection criteria and justified the records for consideration and elimination. We compiled a database of records for consideration in the systematic review. In Stage 3, "extraction," the review team labelled the records and extracted essential information for the systematic review. Finally, in Stage 4, "execution," the systematic review was produced after processing the labelled records (see Figure 10.1).

In Stage 1, the purpose of the review was to identify the trends in policy and technology responses to mitigating the effects of UHIs. To ensure a clear scope within the boundaries of the chosen literature review methodology, we focused on academic literature. Nevertheless, we acknowledge existing practical literature produced for policymakers. For example, the *Cooling Singapore* project is developing solutions to address UHI challenges in Singapore and

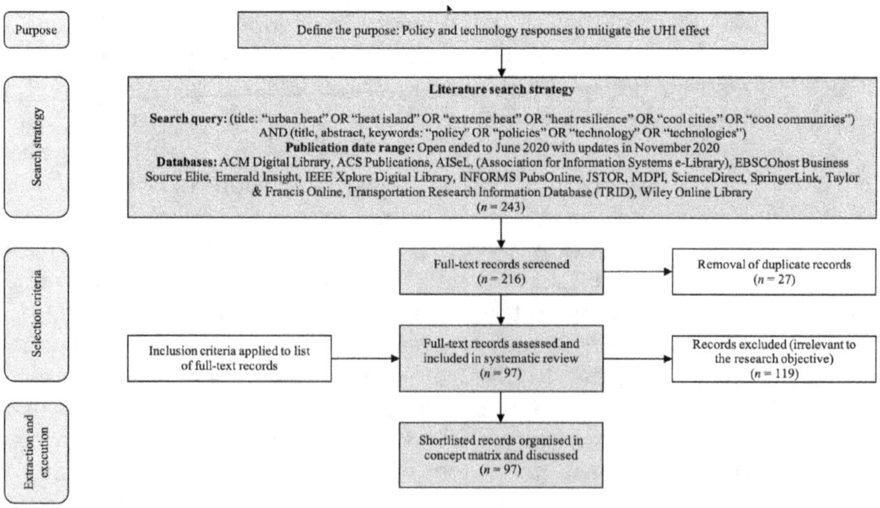

FIGURE 10.1
The literature review search strategy.

has identified 86 strategies to measure and mitigate the UHI effect, which are grouped into seven clusters in a published report: vegetation, urban geometry, water features and bodies, materials and surfaces, shading, transport and energy (Ruefenacht & Acero, 2017). Another report of the *Cool Neighborhoods NYC* project outlines efforts of the New York City (NYC) Government to tackle extreme heat, where mitigation, adaptation, and monitoring strategies are discussed, including tree planting, cool roof and pavement implementation, climate risk training for home health aides, and data collection and analysis strategies for health-focused climate policymaking (NYC Government, 2017). Since government reports are beyond the scope of our review of academic literature, we suggest future research to review and synthesise such reports and city policies, which we expect will provide practical recommendations for city policymakers to mitigate the UHI effect. As such, policy documents can list measures that have already been taken, measures that will be taken in the future, or measures that should be taken according to administration and policymakers.

In Stage 2, the literature search was conducted in June 2020 with updates in November 2020. The following search query string was used in database searches: (title: "urban heat" OR "heat island" OR "extreme heat" OR "heat resilience" OR "cool cities" OR "cool communities") AND (title, abstract, keywords: "policy" OR "policies" OR "technology" OR "technologies"). The following databases were selected and searched: ACM Digital Library; ACS Publications, AISeL (Association for Information Systems eLibrary), EBSCOhost Business Source Elite, Emerald Insight, IEEE Xplore digital library, INFORMS PubsOnline, JSTOR, MDPI, ScienceDirect, SpringerLink, Taylor & Francis Online, Transportation Research Information Database (TRID), and Wiley Online Library. We excluded journals and conferences that were not indexed in these research databases to establish a valid search parameter. We acknowledge that as a result some outlets were not covered by our review such as the International Conference on Countermeasures to Urban Heat Islands (IC2UHI). However, including those would exceed the scope of our review; therefore, we decided it would be reasonable not to consider them. Our search returned 243 articles in total, and 27 duplicates were removed.

We defined selection criteria based on the review's stated purpose, and records (n = 119) were excluded based on the following reasons:

- The chapters articulated the findings of the measurement, modelling, simulation, and tracking of a UHI using a variety of research methods. We excluded this body of research since it has been discussed in detail in other systematic reviews (Huang & Lu, 2018). These records also did not make distinctive policy or technology inferences to mitigation tactics and were excluded (n = 83).
- The chapters discussed other issues related to UHIs that did not directly relate to policies or technologies deployed to mitigate the UHI effect (n = 36).

Records were captured in a database during Stage 3 and were processed and labelled. We followed guidelines by vom Brocke et al. (2015) to screen the literature for applicability based on the keywords that we used to search for literature as a coding scheme. We documented the screening process in Microsoft Excel and as we screened each article, we compiled a concept matrix identifying the following categories: (a) evidence base for policymaking including timescale analysis, effective policymaking instruments as well as decision support and scenario planning; (b) policy responses including landscape and urban form, green and blue area ratio, albedo enhancement policies, transport modal split as well as public health and participation; (c) passive technologies including green building envelopes and development of cool surfaces; and (d) active technologies including sustainable transport as well as energy consumption, HVAC, and waste heat. We also included research aim, region, and Köppen climate classification (where applicable). We utilised Microsoft Excel's Pivot Table functionality to view the different relationships among the records under review.

10.3 Analysis and Results

Based on our systematic literature review, the shortlisted records (n = 97) were considered in this study. Most records were retrieved from the following sources: ScienceDirect (n = 38), SpringerLink (n = 35), and Taylor & Francis Online (n = 11) (see Figure 10.2). Most of the shortlisted articles

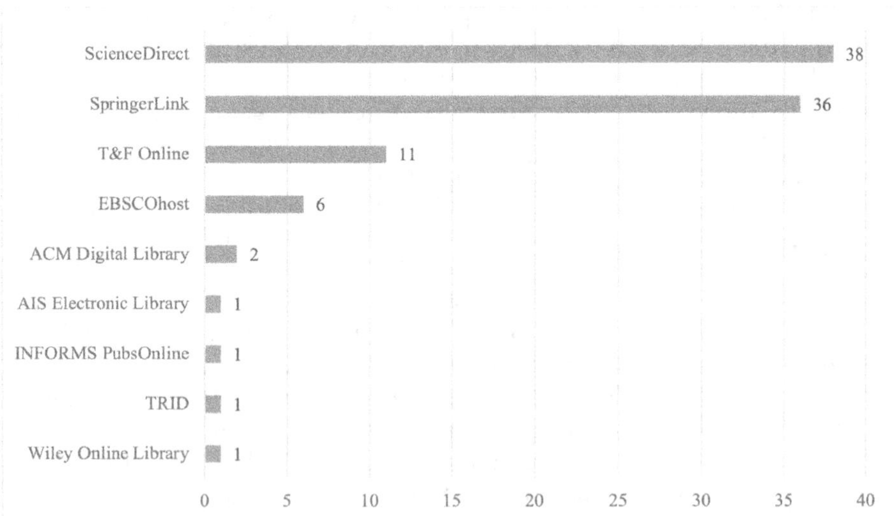

FIGURE 10.2
Records retrieved from selected databases.

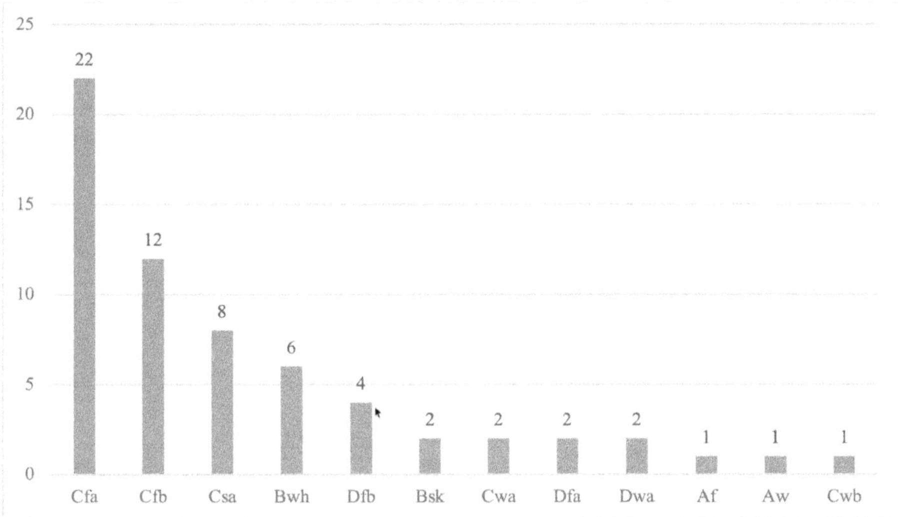

FIGURE 10.3
Köppen climate classification of selected records.

(n = 72, or 74%) were published from 2015 on and can, therefore, be considered very recent.

Most records (n = 63%, or 65%) focused on a specific region and these were further classified under the Köppen climate classification. The studies were mostly conducted in Cfa/*Humid subtropical climate* (n = 22), Cfb/*Temperate oceanic climate* (n = 12), Csa/*Hot-summer Mediterranean climate* (n = 8), BWh/*Hot deserts climate* (n = 6), and Dfb/*Warm-summer humid continental climate* (n = 4) (see Figure 10.3).

We further classified the identified regions under the World Bank Country Income Classification. The studies were carried out in high-income (n = 54), upper-middle-income (n = 9), lower-middle-income (n = 2), and low-income (n = 1) countries. We further classified the geographic distribution of authors (see Figure 10.4). Based on the 97 articles, we identified 362 authors from 30 countries of whom five had multiple affiliations.

In Stage 4, we synthesised the selected literature by discussing each identified concept, which allowed us to compile a concept matrix (Webster & Watson, 2002). We followed a logical approach to grouping the key concepts and uncovered further dimensions to detect units of analysis. To that end, we included concepts with a major focus that we found in five or more of the reviewed articles. Additionally, we included transportation because currently transportation plays an underrated role in the UHI context, which is anticipated to increase in the future (Yigitcanlar et al., 2007) (Louiza et al., 2015; Wagner & Viswanathan, 2016). The concept matrix in Table 10.A1 provides an overview of the literature and the concepts.

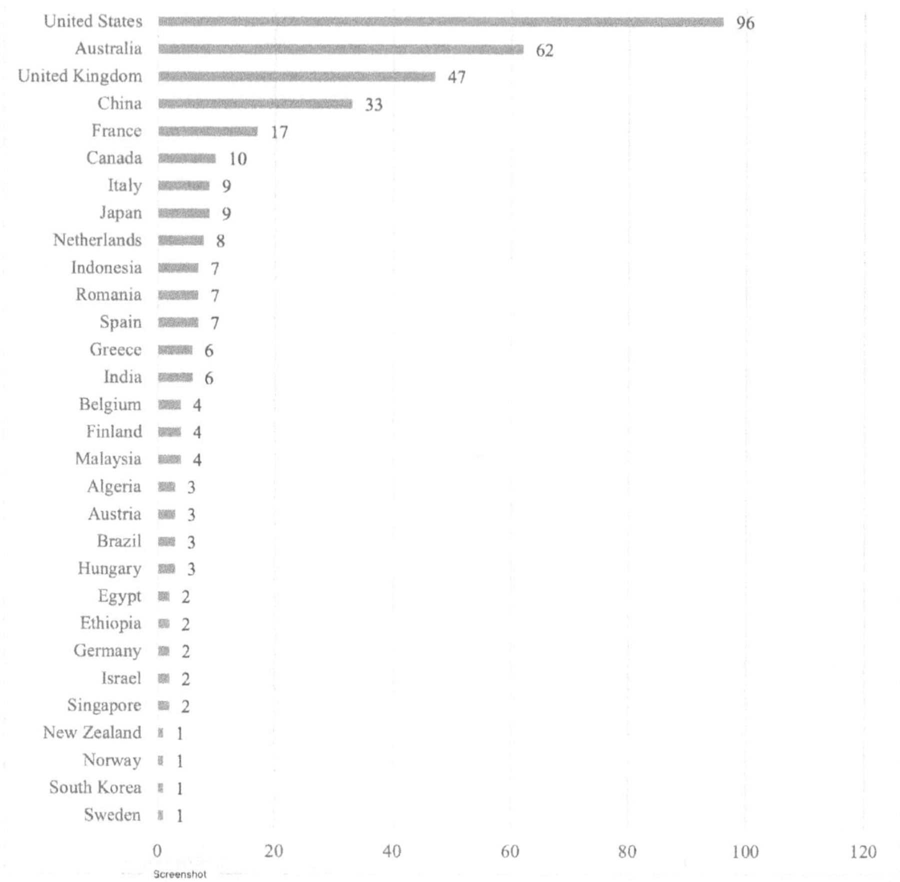

FIGURE 10.4
Geographic distribution of authors.

10.3.1 Analysis of Identified Literature

10.3.1.1 Evidence Base for Policymaking

The majority of UHI research over the past three decades has been in the domain of modelling, simulating and tracking the UHI phenomena (Deilami & Kamruzzaman, 2017; Huang & Lu, 2018). Due to our focus on policymaking, we excluded these technical aspects of UHI research and only focused on the technology aspects that are relevant for policymaking. In this regard, we have identified three areas of evidence base for policymaking: timescale analysis, effective policymaking instruments, as well as decision support and scenario planning.

10.3.1.2 Timescale Analysis

Policymakers have a range of methodologies to track the spatiotemporal changes in UHIs. It is essential to note the variety of timescale analyses deployed. Timescale analysis allows policymakers to understand spatiotemporal variable conditions of the UHI over the long term and track the UHI intensity and heat stress over shorter periods.

For example, a short-term numerical simulation in a Chinese city reveals a significant peak in heat stress during rush hour in the diurnal cycle of UHI intensity (Gao et al., 2017). Policymakers could use these data to improve early warning systems in such scenarios. Longitudinal studies are helpful to measure and track UHIs over space and time. In Delhi, researchers utilised the UrbClim (urban climate) model to project short-term climate change projections to forecast near-future (2026–2045) and far-future (2081–2100) changes with a focus on ten extreme heat indices (Sharma et al., 2019). These data were useful in the drafting of shorter-term climate resilience and adaptation strategies.

Timescale analysis also supports policymakers to identify the land uses that have the most influence on UHIs. For example, research showed that an industrial park was the most vulnerable urban precinct when changes in the UHI were measured over the period 1980–2011 by spectral mixture analysis based on Landsat images (Henits et al., 2017). A study on the UHI of two American cities over the period 1984–2016 argued for closer attention to urban growth patterns, also called urban morphology (Mbuh et al., 2019). Policymakers are therefore required to understand the thermal qualities of different land uses and the variance that atmospheric humidity plays (see, Roy & Singh, 2015).

10.3.1.3 Effective Policymaking Instruments

Concerted efforts to combine climate resilience interventions with UHI mitigation strategies result in multi-sector and multidimensional policy interventions (Santamouris et al., 2019). In China, the so-called sponge cities policy adopted in 2014 seeks to implement a range of surface water retention and related runoff management approaches. He et al. (2019) investigated the co-benefits of this policy as it pertains to UHI mitigation and discussed their findings in five dimensions: social, economic, political/institutional, climate-related, and building. A framework based on the policy analysis informs stakeholders to address the urban flooding problem and manage increasing urban temperatures. Pathways to co-benefits from the implementation of sponge city and UHI mitigation strategies are proposed because policymakers should develop a synergistic technical system, define integrative plans and policies, communicate among different departments and various stakeholders, and implement pilot projects for quantifying and optimising a co-benefits approach. He et al. (2019) conclude that the sponge city approach is an opportunity to technically mitigate UHI effects with higher compatibilities in cool pavement and green infrastructures.

Other studies focus on specific policies, such as an analysis of a tree protection policy in Woodlands Township in Texas, which consists of tree removal permits and minimum tree and shrub cover regulations (Sung, 2013). Here, the effect of the policy is analysed on UHI mitigation by comparing the mean land surface temperatures derived from thermal infrared bands of Landsat TM images between the Woodlands' neighbourhoods and nearby control neighbourhoods without a tree protection policy. Results show that the Woodlands' neighbourhoods were 1.5°C–3.9°C lower on average than those of the control neighbourhoods, providing evidence of the mitigative impact of such policies. The study concludes that a local tree protection policy is effective in UHI mitigation at the neighbourhood scale and further suggests that the cooling effect of the Woodlands' tree protection policy was more prominent in summer when UHI mitigation was needed most. There are other projects assessing the role of trees in addressing air pollution and UHIs; for example, The Nature Conservancy published a report where tree planting policies are discussed as a cost-effective solution to improve health and tackle UHIs (McDonald et al., 2016).

These exemplary studies show that policy analysis has been conducted mostly on a regional scale. Thus, we propose that our framework of policy and technology responses will help policymakers to mitigate the UHI effect on a broader scale and guide researchers for future policy analysis.

10.3.1.4 Decision Support and Scenario Planning

Policymakers require active community engagement, decision support, and scenario planning systems premised on scientific data used in the simulation and assessment of UHIs. With such scientific data at hand, it is argued that a more dynamic relationship is required between urban climate knowledge and the formulation of urban development plans and actions (Parsaee et al., 2019). Community participation systems informing local mitigation strategies remain weak in Canadian cities (Parsaee et al., 2019). The portrayal of the UHI in popular media was studied in Australia and overall there was positive reportage on actions to mitigate the UHI effect (Iping et al., 2019). The tone and language of communicating scientific knowledge are marked by words such as "the apparent" or "so-called" UHI, which "highlights that climate change is still a politicised term within media at various scales" (Iping et al., 2019, p. 427). In New York, localising global climate change science requires co-production between stakeholders since interventions are not divorced from their social settings (Corburn, 2009). Including essential stakeholders such as utility companies in decision-making can account for shared benefits of UHI mitigation strategies, in this case, "cool city" programs, and can be quantified in cost–benefit tests (Shickman & Rogers, 2019).

It is therefore essential that urban climate knowledge is incorporated within the planning and action strategies of local councils, which has an implication on budgeting and performance management. A response to this challenge shows that the design of an optimisation model for decision support can maximise revenue for selling or leasing land to developers while limiting the UHI intensity to find an optimal balance between buildings and vegetation in urban planning (Tuczek et al., 2019). It is argued that such a tool must "show city planners, based on quantitative values, that it is necessary to integrate appropriate countermeasures of the UHI effect, such as the integration of vegetation zones in the relevant districts" (Tuczek et al., 2019, p. 11). Social computing could support the response to the wicked problem of UHIs in terms of data sharing and interactive forums (Kuznetsov & Tomitsch, 2018).

A similar model was developed to assess the impact of different urbanisation patterns on Amsterdam's UHI (Koomen & Diogo, 2017). Such decision support models are usually underpinned by defining the appropriate indicators "rolled up" into a performance framework, which is utilised to compare results and create a community of practice to refine these management strategies (Echevarría Icaza et al., 2016). Decision-makers should also be aware that a mix of science and user-generated data might enhance the responsiveness of urban interventions. One such example is Smart Community Centric Urban Thermal Sensing, first tested in Georgia, United States, where crowdsensing data and a community engagement strategy were incorporated into the data mix for improved decision-making (Tonekaboni et al., 2018).

10.3.2 Policy Responses

In this section of the literature review, we discuss the policies deployed in mitigating UHIs. The five overarching topics in this discussion are as follows: (a) landscape and urban form, (b) green and blue area ratio, (c) albedo enhancement policies, (d) transport modal split, and (e) public health and participation.

10.3.2.1 Landscape and Urban Form

The natural landscape and form of cities have a causal relationship with land surface temperatures observed over space and time. Urban densities, the direction and strength of seasonal variations in wind, the sky-view factor (a measurement of urban microclimate at scales below 100 m) and the proportion of open spaces among buildings have an influence on the intensity of the UHI (Chun & Guldmann, 2018; Wang & Akbari, 2016). Denser city centres tend to have higher levels of UHI intensity due to the compact nature of build-to-open-space ratio (Connors et al., 2013). Urban planners and policymakers utilise urban design, land use controls, and spatial planning strategies to ensure the balance between the built environment, street widths, open spaces, permeable surfaces, and transportation options, resulting in a naturally ventilated and cool city centre (Koomen & Diogo, 2017). The architectural form in cities with an already existing UHI is also vital to ensure thermal comfort, as researchers found in their assessment of indoor climate variations in London (Oikonomou et al., 2012) and Paris (Dhalluin & Bozonnet, 2015).

Cities characterised by low-density residential sprawl might measure a lower UHI effect than more compact and dense city centres, explained by the higher proportion of open spaces and permeable surfaces, but this does not necessarily result in more sustainable cities and energy use (Kohler et al., 2017). In Brisbane, researchers tested the land use scenarios of transit-oriented development—or intensifying land use in compact development in close proximity to public transport interchanges—and found that infill development or brownfields regeneration were the most productive forms of land use to maximise compact development within the bounds of the UHI (Deilami & Kamruzzaman, 2017; Kamruzzaman et al., 2018). When policy-makers consider UHI mitigation measures early in the design process, as observed in the case of the London Olympic Village, urban regeneration can contribute to the cooling of cities (Hamilton et al., 2014). Such positive outcomes benefit greatly from multi-disciplinary teams, including built environment professionals, economists, and climate scientists, who improve the UHI policy response in terms of materials, technologies, and practices (Golden, 2004). For example, from an economic point of view, considering the benefit–cost ratio for urban planning policies helps to calculate differences of net benefits and aggregated costs of implementing various policy options to assess economic impacts caused by UHIs (Estrada et al., 2017).

10.3.2.2 Green and Blue Area Ratio

The so-called cool islands, including green and blue spaces but also urban design and other strategies, are responsible for offsetting UHIs. Consider the thermal qualities of Central Park in New York and the River Thames in London. With regard to green spaces, Gaffin et al. (2008) looked at the effect of Central Park and studied the UHI of New York during the twentieth century (i.e., 1900–2000). They found that the average annual temperature in the vicinity of Central Park increased from 11.6°C in 1900 to 13.1°C in 2000. The historical data revealed that the UHI was responsible for 33% of the overall warming of the city. An essential contribution of this study was to contest the findings of Peterson (2003), who disputed the formation of UHIs in other American cities such as Boston, Dallas, and Seattle. In those cities, the influence of other microclimate "inhomogeneities" such as rural elevation was suggested would dispel the UHI effect.

Regardless of such variations in microclimatic factors, it has been proven that provisions for green (e.g., parks, forests, riverbanks, and biodiverse regions) and blue (e.g., water bodies such as rivers, lakes, ponds, and wetlands) infrastructure remain the most effective measures in cooling UHIs (Gunawardena et al., 2017; Teferi & Abraha, 2017). During periods of heatwaves and the intensification of UHIs through anthropogenic heat generation, healthy ecosystems provide essential services to moderate the microclimates found in cities (Jenerette et al., 2011; Marando et al., 2019). Not only do these existing natural environments contribute to UHI cooling, but large-scale landscaping projects act as natural heat sinks (Li et al., 2014; Vasilakopoulou et al., 2014). Cities have recognised these benefits; for example, Paris has added nearly 70 ha of green infrastructure and rooftop gardens to reduce the UHI effect (C40, 2015). However, the effect of parks and water on a UHI depends on the size of these spaces, where depth plays a major role (Targino et al., 2019; Xiao et al., 2018; Zhu & Jia, 2011).

In environments of constrained resources and the limits of ecological services, policymakers strive to achieve co-benefit agreements when considering the impact of urban development. In China, a focus on sponge city policies has been an effective integrated planning approach to achieve such co-benefits (He et al., 2019). In other cases, adopting policies aimed at water-sensitive urban design and urban-greening policies can result in innovative projects such as Park Cool Islands and the reforestation of cities, and these "green lungs" effectively become networks of green spaces that are equally enjoyed by city dwellers as recreational areas (Hiemstra et al., 2017; Jamei & Tapper, 2019). Hence, policy instruments such as tree protection policies (Sung, 2013) and wetland protection policies (Cai et al., 2016) are essential in maintaining quality green and blue spaces. Policymakers should also consider the water use requirements of maintaining ecological services since scenarios of urban densification include evaporative cooling, shading, and reflectance, which are all factors that influence the general albedo of cities (Gober et al., 2009; Lemonsu et al., 2015).

10.3.2.3 Albedo Enhancement Policies

Albedo refers to the energy balance of the earth's surface, which is defined by the equation $(1 - \alpha_{abs})$, where α_{abs} is the absorptivity of the surface (Abd El-Hakim & El-Badawy, 2019). Therefore, albedo is measured on a scale between 0 and 1, where a surface with 0 albedo is perfectly absorptive and a surface with an albedo of 1 is perfectly reflective. The high proportion of dark paved surfaces (often asphalt), such as pavements and roads, which can account for up to 20%–40% of a typical city's area (Qin, 2015), results in heat storage and aggravates the UHI (Vasilakopoulou et al., 2014). Policymakers have a range of options to increase the albedo factor by using available and novel approaches. Urban greening policies have an albedo enhancing effect. In Montreal, researchers found that three main UHI mitigation measures of urban canopy, urban vegetation, and light paved areas increased the city's albedo factor when assessed against the sky-view factor control and wind speed (Wang & Akbari, 2016).

Microclimate simulations can help to generate a multi-criteria "cool city model" when the variables of cool pavements, cool roofs, and vegetation variables are considered (Wang et al., 2016). Such simulation models can support urban planners and policymakers to justify the investment in albedo enhancing projects.

10.3.2.4 Transport Modal Split

Cities prioritising investments in public transport aim to achieve a modal split where more commuters opt for public transport options. An overreliance on private car policies, which reinforce road construction projects and low residential density sprawling neighbourhood extensions, results in increased fuel consumption, greenhouse gas emissions, and pollution in cities (Kamruzzaman et al., 2015). It has been established that heat transfer and radiation of these emitting vehicles are negative side effects of congestion and intensify the UHI (Louiza et al., 2015). A simulation of transportation options in Beijing found that the greatest UHI intensity mitigation effect for transport was to replace conventional vehicles with energy-efficient public transport (Kolbe, 2019).

Transport policy also has an influence on land use planning. In Brisbane, Australia, five alternative neighbourhood planning scenarios aligned with smart growth policies were simulated against UHI mitigation: (a) business as usual, (b) transit-oriented development, (c) infill development, (d) motorway corridor-oriented development, and (e) sprawl development (Deilami & Kamruzzaman, 2017). Results from a geographically weighted regression analysis showed that the infill development scenario as a smart growth policy has a marginally better potential to mitigate the UHI effect in Brisbane in 2023 compared to the sprawl development scenario.

Transport policies such as transit-oriented development are generally preferred over low-density urban sprawl but do not necessarily reduce the UHI. A balancing act is therefore required between natural and built-up areas (Kamruzzaman et al., 2018). Numerical and multi-criteria models are also useful to test various scenarios informing policy choices related to power consumption, transportation options, and their relationship with land use (Silva & Fillpot, 2018). However, the issue is much more complex due to the transportation system's high dependence on the city's layout, which in turn varies across the globe (Xu et al., 2020; Yang & Gakenheimer, 2007).

10.3.2.5 Public Health and Participation

During climatic periods where the UHI intensifies through heatwaves, vulnerable populations are at risk, and this poses a public health policy challenge. This is particularly relevant in pandemic outbreaks such as COVID-19, which require an urgent need for enhanced urban monitoring and coping strategies to improve urban resilience and environmental quality (Ulpiani,

2021). In this regard, extreme heat exposure has important implications for health costs and the quality of healthcare, imposing a substantial economic burden on healthcare systems (Si et al., 2019; Wondmagegn et al., 2019). A review of the literature suggests that UHI mitigation strategies have a positive impact on public health (Heaviside et al., 2017). In Finland, a risk mapping exercise was conducted, considering five different zoning options; 11 different weighting options were applied to assess vulnerability, for example, to elderly people prone to dehydration and heat stress, which leads to serious health implications (Räsänen et al., 2019). Such data supports policymakers to target relief efforts. Vulnerability is also in the focus of another study that finds tensions between addressing heat as an emergency and heat as a source of chronic stress, which affects people's health as well as physical and mental well-being (Bolitho & Miller, 2017). In Madrid, it was found that low-income households that cannot afford to regulate their indoor temperatures using, for example, air conditioning, face overheating problems in periods of UHI intensity, which has an adverse effect on public health (Sanchez-Guevara et al., 2019).

Heatwaves caused by anthropogenic emissions of greenhouse gases and the expansion of UHIs through rapid urbanisation (Sun et al., 2014) have a substantial effect on a variety of health outcomes from mortality to mental stress. For example, the 2011 heatwave in Houston, Texas, led to a 3.6% excess risk in emergency department visits and a 0.6% increase in mortality risk (Zhang et al., 2015). Elderly patients are particularly at risk of exposure to adverse health conditions during hot periods, such as dehydration, hyperthermia, malaise, hyponatremia, renal colic, and renal failure (Josseran et al., 2009). Heatwaves have also been shown to lead to mental health illness (Thompson et al., 2018), particularly during the COVID-19 pandemic, where heat stress affects cognitive and physical performance (Davey et al., 2020). An increase in research dealing with heatwave resilience and health implications can be expected due to the urgency of pandemic outbreaks.

10.3.3 Technology Responses

In this section of the discussion, we consider technology responses to the UHI effect. We classify these technologies as passive and active and indicate the interrelationships between policy formation and technology development in attempts to mitigate UHI effects. In this regard, passive technology usually has a single function, such as pavements and roofs in the urban context, while active technology is multifunctional, such as HVAC systems, allowing users to interact with multifaceted functions. In our policy-focused review, we identified green building envelopes and cool surfaces as the main passive technologies discussed in the identified literature, and sustainable transport and HVAC systems as the main active technologies.

10.3.3.1 Passive Technologies

10.3.3.1.1 Green Building Envelopes

The thermal comfort and qualities of buildings can either neutralise or aggravate the UHI since HVAC systems are energy-intensive and create anthropogenic heat (Oikonomou et al., 2012). In the UK, researchers tested passive design solutions for natural ventilation by using thermal modelling to measure the performance of non-domestic buildings (Short et al., 2004). In California, a review of "cool community" interventions resulted in community co-benefits including reduced utility bills, improved air quality, and enhanced urban liveability (Gilbert et al., 2016). Greening building envelopes are therefore not only important to increase urban sustainability and metabolic flows but innovations such as green roofs in combination with albedo modification (for example, lighter walls) deliver a cooling effect (Lehmann, 2014; Price et al., 2015). Four building materials (brick, aerated concrete, wood with glass-wool insulation, and glass fibre-reinforced concrete with glass-wool insulation) were tested to assess thermal qualities of buildings in existing UHIs in a tropical climate (Wonorahardjo et al., 2020).

10.3.3.1.2 Development of Cool Surfaces

The albedo of buildings can also be increased through reflective building materials, and the effectiveness of these materials coincides with other factors such as building characteristics, urban environment, meteorological, and geographical conditions (Dizdaroglu et al., 2012; Yang et al., 2015). A low albedo and high ratio of impervious surfaces perpetuate the UHI, and technologies have been developed to cool down urban surfaces. To develop effective technology responses to heating surfaces, it is important to address the thermal properties of urban materials. Besides absorptivity and emissivity, which are independent of each other and describe the absorption of radiation energy from the sun and the ability to emit infrared energy, albedo plays an important role in the UHI context. It is important to take albedo into consideration in passive technology development because the albedo of pavement surfaces differs substantially given the materials used in construction. In Atlanta, future urban development could increase the proportion of dark-coloured pavements by 45% over ten years, reaching two-thirds of impervious surfaces. To counteract this low albedo, researchers recommend that light surfaces be selected to increase albedo (Lee & French, 2009). Cool pavements with technologies such as sustainable urban drainage (consisting of bioswales, permeable paving, wet basins, and flower beds) are effective in mitigating the UHI effect (Dwivedi et al., 2019). Studies in Paris have demonstrated that regular pavement watering can also increase albedo and mitigate heat stress (Hendel et al., 2016).

Furthermore, anthropogenic heat generation can be mitigated by increasing the earth's albedo factor through renewable energy infrastructure such as solar panels (Rossi et al., 2013). Cool roofs are perhaps the most globally applicable cooling intervention to mitigate UHI and increase the reflection of incoming solar radiation in urban areas by increasing the albedo of roof surfaces (see, Li et al., 2014). Technology development can look into the design and materials of cool roofs, which are single ply or liquid applied, including white paints, elastomeric, polyurethane, or acrylic coatings (Santamouris, 2014). In Beijing, researchers tested a willingness-to-pay valuation approach for cool roofs with building tenants, and it was found that government credibility and education are important factors in promoting public participation (Zhang et al., 2019).

10.3.3.2 Active Technologies

10.3.3.2.1 Sustainable Transport

Since urban transport contributes to the formation of UHIs, making transportation more sustainable is a technological challenge to support UHI mitigation (Louiza et al., 2015). A comparison of different transport technologies including internal combustion engine vehicles, electric vehicles, hydrogen vehicles, and public transportation shows the impact of different mobility concepts on UHI mitigation and carbon dioxide emissions (Kolbe, 2019). It is argued that the release of heat from combustion engines intensifies the UHI effect, and a scenario analysis in Kolbe's study finds that the largest UHI mitigation is in a scenario where conventional vehicles are replaced with energy-efficient metro travel; the saving potential for carbon dioxide emissions is highest for electric vehicles with renewable electricity generated from wind turbines, closely followed by hydrogen vehicles that are fuelled with hydrogen generated from the electrolysis of water using wind electricity (Kolbe, 2019). While Kolbe's study only looked at wind and hydro energy, we acknowledge that other renewable sources such as solar energy might lead to similar results. In our review, we found very little research on the impact of transportation on UHIs. Therefore, we suggest that future research delves deeper into the role of transport in mitigating the UHI effect because two areas that are considered to be the largest contributors to an increase in UHI effects are the built environment and transportation (Wagner & Viswanathan, 2016).

10.3.3.2.2 Energy Consumption, HVAC and Waste Heat

HVAC systems play an important role in urban energy consumption because through the densification of cities, the overall energy consumption increases, and HVAC systems significantly contribute to the increase of energy consumption as their load increases with warming (Louiza et al., 2015). Hence, there is a requirement to advance the technological capabilities of HVAC standards to enhance energy efficiency. Based on a study in Mumbai, a combination of HVAC usage and structure cooling (which prevents heat from entering living spaces through a system of pipes that absorb heat from the building structure) can significantly reduce the load on HVAC systems, resulting in a substantial reduction of energy consumption (Dwivedi et al., 2019). Other studies look into how urban-sprawl countermeasures influence

UHI intensities and energy demands for space heating buildings (Kohler et al., 2017), or how UHI impacts state residential energy costs and carbon dioxide emissions (Roxon et al., 2020).

From a technology development point of view, it is promising for UHI mitigation in hot regions to design HVAC systems that are more efficient under hot conditions to mitigate peak load during extreme heatwaves (Burillo et al., 2019). The current standards EER (energy efficiency ratio) and SEER (seasonal energy efficiency ratio) are primarily for air temperatures at or below 35°C. Another important aspect is waste heat from HVAC systems, which requires technological upgrades for the waste heat collection through improved insulation and tightness to limit heat loss (He, 2019). Waste heat intensifies the local air temperature contributing to UHI effects and increasing the probability of extreme heat (Olivo et al., 2017). Our review reveals that energy consumption, HVAC, and waste heat appear to be marginal topics in the UHI literature. Due to the critical importance of these topics for UHI intensification, we suggest that future research should place greater emphasis on this area.

10.3.3.3 Literature Including Both Aspects of Policy and Technology Responses

Our literature review reveals that policy and technology responses are loosely interlinked in the current UHI literature. Out of the 97 identified literature, we found only five articles that have a major focus on both policy and technology aspects. Four of these articles focus on landscape and urban form, green and blue area ratio, and albedo enhancement policies from a policy response perspective; and from a technology response view, the articles focus on green building envelope and development of cool surfaces. One article has a focus on transport modal split as a policy response, and sustainable transport from a technology aspect. Below, we describe how policy and technology are integrated in these articles. A count of policy and technology aspects for all articles is included in the concept matrix in Table 10.A1.

- Dhalluin and Bozonnet (2015): Analysis of the awareness and consideration level of UHI constraints by organisations that influence the design of building hosting sensitive people to high temperature (nursery and retirement homes).
- *Policy focus:* Landscape and urban form. Regulatory issues of national and international policies are analysed. The focus is on different schemes and plans including SRCAE (Regional Climate, Air and Energy Scheme), PCET (Territorial Climate Energy Plan), SCOT (Territorial Coherence Scheme), and PLU (Local Urbanism Plan). The study is conducted for the French Ministry of Ecology, Sustainable

Development and Energy (MEDDE), and recommendations are given to improve policymaking.

- *Technology focus:* Green building envelope, development of cool surfaces. Passive technologies in ten French cities are analysed including vegetation (green space, green screens and roofs) and specific coatings on facades and roofs (variation of radiative properties by using different colours and materials).
- Gober et al. (2009): Investigation of trade-offs between water use and nighttime cooling inherent in urban form and land use choices.
- *Policy focus:* Landscape and urban form, green and blue area ratio, albedo enhancement policies. The study builds on identified shortcomings of the downtown redevelopment policy of the city of Phoenix, AZ, and provides implications for the City of Phoenix Water Services Department.
- *Technology focus:* Development of cool surfaces. Passive technologies are investigated regarding attributes of the surface in Phoenix with a focus on planting more trees and other irrigated vegetation to prevent daytime heat storage and facilitate nighttime cooling, which requires water resources that are limited in a desert city like Phoenix.
- He et al. (2019): Examination of co-benefits for policymaking of UHI and urban flooding management.
- *Policy focus:* Landscape and urban form, green and blue area ratio. The study builds on the policy of the Ministry of Housing and Urban-Rural Development of China (MOHURD) "Construction guideline of sponge city in China–low impact development of the stormwater system (trail)."
- *Technology focus:* Development of cool surfaces. Technical codes for roof engineering and technical specifications for asphalt and concrete pavements as well as green roofs are examined. To facilitate co-benefits, a synergistic technical system is suggested due to commonalities but also differences between the techniques and performances of UHI and sponge city management.
- Hendel et al. (2016): Conduct of an experiment in Paris, France, on differences between watered and dry portions of a street to examine the effects of watering to reduce maximum daily heat stress.
- *Policy focus:* Landscape and urban form. The study is providing recommendations for improved policymaking for the Water and Sanitation Department of the City of Paris.
- *Technology focus:* Development of cool surfaces. Pavement-watering materials are analysed including road materials and impervious asphalt concrete.

- Kolbe (2019): Examination to what extent a range of different mobility concepts can lead to a reduction in UHI intensity and carbon dioxide emissions in the city of Beijing, China.
- *Policy focus:* Transport modal split. Recommendations for policymakers are given in three different scenarios to discuss the impact of a transition from conventional vehicles to electric vehicles.
- *Technology focus:* Technical characteristics between conventional vehicles and electric vehicles are compared including fuel efficiency and heat emission to assess a reduction of UHI intensity.

Further, our literature review shows that in the identified literature the evidence base of timescale analysis, effective policymaking instruments as well as decision support and scenario planning are mainly used. Nevertheless, using this evidence base comes with limited possibilities to analyse policy and technology responses in conjunction. We propose that further research methods and approaches will be necessary to help establish intervention cycles to evaluate coordination effects of policymaking in conjunction with technology development in order to address responses to UHI mitigation more effectively. We explain our directions for future research along with our conceptual framework in the next section regarding the use of different types of research methods to analyse policy and technology responses, as well as regarding how to overcome the lack of interaction between policy and technology responses to UHI mitigation.

10.4 Findings and Discussion

10.4.1 Framework and Purpose of Research Questions

Based on the systematic literature review, we conceptualise the impacts of policy and technology interventions on anthropogenic heat and incremental environmental burden (Figure 10.5).

We propose that a lack of interaction between policy and technology responses will lead to limited effects to mitigate heat and environmental burden, while effective interactions will reinforce the benefits of conjunction and generate coordination effects (Arthur, 1988; Jänicke, 2015). The framework informs future research on the direct effects of policy and technology on anthropogenic heat (Mortoja & Yigitcanlar, 2020), and the moderating effects on the impact of anthropogenic heat on the incremental environmental burden. Our main assertion is that policy and technology are often developed in isolation, which is evident in the current state of UHI research (see, Aleksandrowicz et al., 2017; Hintz et al., 2018; Parsaee et al., 2019).

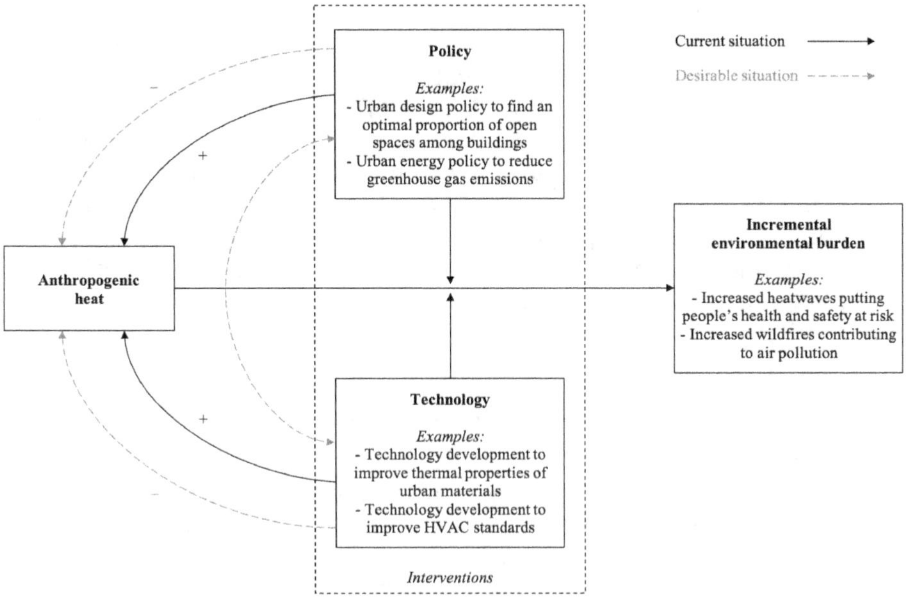

FIGURE 10.5
Conceptual framework of the interplay between policy and technology interventions to reduce anthropogenic heat and alleviate incremental environmental burden.

Our proposed questions will inform future UHI research in four main areas: (a) coordination effects of policy and technology responses, (b) synergistic effects of combined policy and technology interventions, (c) aggravating effects of poor policy and technology responses, and (d) moderating effects of policy and technology responses on the impact of anthropogenic heat on the incremental environmental burden. Our study revealed that all of these four areas are currently under-researched and require focal attention due to two reasons. First, an extension of knowledge in these areas will help to tackle UHI-related issues because policy and technology synergies can increase the effectiveness of UHI mitigation strategies, aggravating effects of poor policy and technology responses are real-world phenomena that need our attention, and our understanding of the incremental environmental burden due to UHI intensification is currently limited. Second, in UHI research, we have currently reached a certain level of saturation in other important areas including the conceptualisation of UHI mitigation strategies, the classification of UHI analytical techniques, and the analysis and comparison of global UHI characteristics. However, there is still a wide gap in our regional/local specific understanding of what mix of strategies should be implemented in the most effective and viable way. While we believe that these important areas need continuous attention, we suggest that our proposed areas can be combined because coordination, synergistic, aggravating, and moderating effects can be looked at from different angles, including conceptualising UHI mitigation strategies, classifying UHI analytical techniques, and analysing global UHI characteristics. For example, one combination opportunity is to conceptualise policy and technology synergies for UHI mitigation strategies; another opportunity involves investigating analytical techniques to counteract the aggravating effects of poor policy and technology responses. Regarding global UHI characteristics, we can expect that there will be regional differences regarding coordination, synergistic, aggravating, and moderating effects, which provide further opportunities for future UHI research.

10.4.2 Effects of Policy and Technology Responses

The first question refers to the use of different types of research methods and approaches to analyse policy and technology responses to UHI mitigation. Our study revealed that current research methods in the UHI literature provide limited possibilities to analyse coordination effects of policy and technology responses. We suggest that further types of research methods are required, such as action research, which is considered to be "a research process that is relevant for both the practitioner who is struggling with a system of problems, as well as for the scholar whose purpose is to advance the current state of knowledge" (Baburoglu & Ravn, 1992, p. 19). We expect that such robust practice approaches can include the researcher as an active participant rather than a passive observer to meet the double hurdle of rigour and relevance (Eden & Ackermann, 2018; Järvinen, 2007). This is particularly important in our case of policy and technology responses to UHI mitigation because local actors constantly shape the research process and design by the values, knowledge, and perspectives of regional differences in UHI dynamics (Douglas et al., 2018). The action research process requires the researcher to complete four stages, where (a) the practice-inspired and theory-ingrained problem is formulated, (b) the problem framing and theoretical premises are used to build and evaluate the interventions, (c) the intervention results are reflected to capture lessons learned, and (d) the lessons learned are formalised to generalise outcomes (Sein et al., 2011). With regard to the two types of interventions in our context of UHI mitigation (i.e., policy and technology), which is central in our conceptual framework, we expect that action research will help to establish intervention cycles through action-planning with a policy focus and action-taking with a technology focus (Lindgren et al., 2004). Other research approaches, such as the socio-technical systems approach (Leavitt, 1965), might also help to explain the interrelatedness of social and technical aspects of systems, which will clarify the structure and roles of policy formation, and the task and physical conditions of technology responses (Bostrom & Heinen, 1977). This will allow researchers to evaluate the coordination effects of policymaking in conjunction with technology development in order to address responses to UHI mitigation more effectively and provide contributions to knowledge and utility (Sein et al., 2011). There are examples where UHI interventions in the form of policy formation and technology development have been built and evaluated such as the *Cooling Singapore* project or the *Cool Neighborhoods NYC* project, where UHI mitigation strategies have been developed to address UHI challenges in Singapore and NYC (NYC Government, 2017; Ruefenacht & Acero, 2017).

The next five questions refer to the combined intervention of policy and technology responses, as depicted in the framework. Most studies address either policy or technology responses from a separate viewpoint; however, a combined approach has the potential to increase the effectiveness of the intervention through coordinated policy and technology responses (Arthur, 1988). Such coordination effects can contribute to a reinforced diffusion of both UHI-mitigating technology and the supporting policy. While the diffusion of technologies and supporting policies is typically interlinked (Jänicke, 2015), our knowledge on interactions between policy and technology is currently limited. Thus, we propose a set of research questions to help guide future research on synergistic effects of a combined approach of policy and technology interventions.

The next set of research questions relates to the direct impact of policy and technology responses on anthropogenic heat. We propose that policy and technology are both direct contributors to anthropogenic heat; for example, a lack of policy on HVAC efficiency will contribute to heat, whereas an effective HVAC efficiency standard will lower the effect of HVAC heat exchange. The absence of or poor policy directly contributes to heat generation. While the current state of UHI research has the main focus of UHI mitigation through policy responses or technology intervention, little is known about aggravating effects through a lack of or poor policy and technology response. A recent study suggests that UHI warming will be equivalent to about half the warming caused by climate change by the year 2050 (Harvey, 2019), which makes research on aggravating effects even more important. An example of research tackling aggravating effects is a group of scientists at the University of Georgia in the USA, who developed a network of sensors deployed on buses and citizens' bodies for measuring local temperatures (Lapidus, 2019). Such a network of local sensors provides an early warning system for city planners, for example, to decide where to plant trees. Communities can also use such systems, for instance, to ensure that people at risk are less exposed to UHI effects and have access to water and air-conditioned spaces. Besides increased environmental degradation, UHI effects can also worsen economic loss; this could amount to a loss of over 10% of gross domestic product in cities by 2100 (Estrada et al., 2017). For example, the economic loss reached 2.5 million EUR for each heatwave day in Cluj-Napoca city, Romania (Herbel et al., 2018). In particular, this shows the importance of transitioning from current situations where a lack of or poor policy and technology responses are aggravating anthropogenic heat to desirable situations where improved policies and technologies are needed to help mitigate the UHI effect. While we acknowledge that a continuation of research on mitigation effects is necessary, we propose that future research explores aggravating effects on heat as an additional stream of research to develop a better understanding of poor policy and technology responses and how governments can improve those.

Our final set of research questions pays attention to the widely neglected moderating effects of policy and technology responses on the impact of anthropogenic heat on the incremental environmental burden. While heat mitigation is mainly covered in the current state of UHI research, the anthropogenic heat–environmental burden relationship is under-researched, particularly regarding the moderating effects of UHI policy and technology. Thus, we call for research that goes beyond short-term outcomes regarding heat generation and explores longer-term outcomes related to environmental degradation leading to natural hazards, such as heatwaves, wildfires, storms, floods, and droughts. From a health perspective, elderly people are considered to be most vulnerable to heatwaves but are also most vulnerable to infectious diseases such as COVID-19 (Batcheller, 2020; Yigitcanlar et al., 2020). A recent study found that environmental hazards and the lack of access to green spaces affect low-income neighbourhoods of colour, leading them to be hotter than their wealthier, whiter counterparts (Karlson, 2020).

There are various reasons to analyse policy and technology interventions in different contexts including health outcomes, pandemic shocks, wealth distribution, and racism, which can be a result of the incremental environmental burden caused by anthropogenic heat, and link government policies to unequal exposure to extreme heat. For example, the incremental environmental burden of increased heatwaves put people's health and safety at risk, even more so during the COVID-19 pandemic (Davey et al., 2020). We propose that research on the moderation of UHI policy and technology interventions will extend our knowledge on how anthropogenic heat leads to incremental environmental burden.

10.4.3 Future Research and Overview of Research Questions

We suggest research that informs on the "policy → technology" and "technology → policy" relationship nexus will require interdisciplinary teams. We understand that such a nexus implies complex interdependencies, which will be better addressed with expertise from diverse research disciplines, such as management to assist the decision-making process of policy formulation and implementation, or engineering in driving technology development. Other disciplines such as urban planning will help to form the nexus from a holistic approach. To support the information flow between policy and technology instantiations, the information systems discipline will enable an integrated network of people, processes, and computer systems to be established to support organisational and societal goals of UHI mitigation. Table 10.1

TABLE 10.1

Emergent Research Questions of Policy and Technology Responses to UHI Mitigation

Research Area	Research Question	Study Suggestion	Key Approach
Coordination effects of policy and technology responses	RQ1: What are the appropriate research methods/approaches such as action research or the socio-technical systems approach to analyse the coordination effects of policy and technology responses to UHI mitigation?	Review of various research methods/approaches to identify the level of appropriateness for the analysis of coordination effects	Identify strengths and weaknesses of appropriate research methods/approaches
Synergistic effects of a combined approach of policy and technology interventions	RQ2: What are appropriate policy responses to enable effective UHI mitigation through technology development and diffusion?	Review of governmental policies with a focus on technology development for UHI mitigation	Identify best practices of policy responses to UHI mitigation
	RQ3: To what extent is a coordination between policies leading to a set of harmonised, reinforcing policies that will maximise the impact of UHI mitigation through technology development and diffusion?	Conduct a comparative analysis of governmental policies with a focus on technology development for UHI mitigation	Identify reinforcement effects between policy responses to UHI mitigation
	RQ4: How can governments effectively prioritise policy options to maximise the impact of UHI mitigation through technology development and diffusion?	Expert interviews and focus group analysis with governmental organisations on policy prioritisation	Create a list of priorities using a heat map analysis to determine the level of attention for UHI mitigation
	RQ5: How can governments improve policy design and implementation to support the development and diffusion of technologies for UHI mitigation?	Conduct an action research study to evaluate policy and technology intervention cycles	Identify shortcomings in the design and implementation of governmental policies to elaborate improvements
	RQ6: Which mechanisms should deliberately be considered to learn from policy feedback on UHI mitigation to reinforce technology development and diffusion?	Conduct a case study on policy feedback mechanisms to explain why policies change over time	Identify factors that explain how existing policies affect the probability and design of future policies

(Continued)

TABLE 10.1 *(Continued)*

Emergent Research Questions of Policy and Technology Responses to UHI Mitigation

Research Area	Research Question	Study Suggestion	Key Approach
Aggravating effects of poor policy and technology responses	RQ7: How does a lack of or poor policy and technology response contribute to an aggravation of anthropogenic heat?	Comparative study of differences in anthropogenic heat in various regions	Compare existing policies as the independent variable to determine the effect on anthropogenic heat as the dependent variable
	RQ8: How can governments effectively detect and counteract a lack of or poor policy and technology response to alleviate UHI aggravation?	Risk analysis to identify and assess factors that lead to UHI aggravation	Develop and implement early warning systems
	RQ9: Which decisions/processes lead to poor policy and technology responses that aggravate anthropogenic heat?	Analysis of decision-making to analyse aggravation effects of policymaking in conjunction with technology development	Develop and implement decision support systems
	RQ10: To what extent are multiple, conflicting policies hampering UHI mitigation through deficient technology development and diffusion?	Conduct a comparative analysis of governmental policies with a focus on technology development for UHI mitigation	Identify diminishing effects between policy responses to UHI mitigation
Moderating effects of policy and technology responses on the impact of anthropogenic heat on the incremental environmental burden	RQ11: What are the specific variations of the impact of anthropogenic heat on the incremental environmental burden in the presence and absence of UHI policy and technology interventions?	Experimental study to manipulate the presence and absence of UHI policy and technology interventions	Analyse different scenarios to extend the knowledge about the impact of anthropogenic heat on the incremental environmental burden
	RQ12: How can governments design and implement UHI policy and technology interventions that fundamentally address both anthropogenic heat and the incremental environmental burden?	Conduct an action research study to evaluate policy and technology intervention cycles	Advance knowledge on the design and implementation of governmental policies
	RQ13: What is an appropriate approach for governments to make decisions about whether the development of UHI policy and technology interventions will help alleviate the impact of anthropogenic heat on the incremental environmental burden?	Analysis of decision-making to determine the moderation effects of policymaking in conjunction with technology development	Develop and implement decision support systems

summarises the 13 research questions that emerged from our literature review along with suggestions on how to study the questions and key approaches to consider, which we propose will help guide future research on UHI mitigation.

10.5 Conclusion

UHIs are a growing concern for policymakers and city planners due to natural degradation, health risks, and economic loss (Estrada et al., 2017). While UHI policy and technology responses are considered separately in current reviews, a synthesis of combined interventions is missing. We take an initial step towards a unified policy and technology approach and present a systematic literature review including 97 UHI research articles that we conceptualise in various dimensions. A closer look at the current state of UHI research reveals that the following four areas are underrepresented:

- Coordination effects of policy and technology responses.
- Synergistic effects of a combined approach of policy and technology interventions.
- Aggravating effects of poor policy and technology responses.
- Moderating effects of policy and technology responses on the impact of anthropogenic heat on the incremental environmental burden.

We derived 13 research questions to guide future research from the framework of UHI policy and technology responses. These questions will inform policymakers and city planners in their tasks to mitigate the UHI effect on a broader scale and guide scholarship to enable effective interactions between policy and technology responses.

Acknowledgement

This chapter, with permission from the copyright holder, is a reproduced version of the following journal article: Degirmenci, K., Desouza, K., Fieuw, W., Watson, R., & Yigitcanlar, T. (2021). Understanding policy and technology responses in mitigating urban heat islands: A literature review and directions for future research. *Sustainable Cities and Society, 70*(1) 102873.

Appendix

TABLE 10.A1

Concept Matrix of the Literature Review

Article	Evidence Base			Policy Responses					Technology Responses: Passive		Technology Responses: Active		Policy	Technology	Research Aim	Region	Climate
	Timescale Analysis	Effective Policymaking Instruments	Decision Support and Scenario Planning	Landscape and Urban Form	Green and Blue Area Ratio	Albedo Enhancement Policies	Transport Modal Split	Public Health and Participation	Green Building Envelopes	Development of Cool Surfaces	Sustainable Transport	Energy Consumption, HVAC and Waste Heat					
Abd El-Hakim and El-Badawy (2019)		○							●	●	●	●		●	This strategic review of technologies utilised to mitigate the UHI is presented as a framework with recommendations.	N/A	N/A
Aflaki et al. (2017)		●		●	●	●							●		A review of the influence of urban vegetation on reducing UHI intensity; proposes practical guidelines for East Asian cities.	Kuala Lumpur, Malaysia	Af
Akbari and Kolokotsa (2016)				○					●	●			○	●	A review over the past three decades of the development and evaluation of mitigation measures, including cool roofs, cool pavements and urban vegetation.	N/A	N/A

Aleksandrowicz et al. (2017)		●		○	○	●		○	○		●	○	This systematic review of UHI mitigation strategies for the period 2009–2013 found that research inclines to urban cooling interventions and most studies are conducted in subtropical climates.	N/A	N/A
Bolitho and Miller (2017)			●				●				●		Reducing heat vulnerability involves many government departments and external relations. Improving these governance arrangements could help address extreme heat as a chronic stressor in the longer term rather than an emergency in the shorter term.	Melbourne, Australia	CfB
Burillo et al. (2019)				●						●		●	An estimation of vulnerabilities of neighbourhood-scale electricity infrastructure due to increasing air temperatures found that trade-offs are required to prevent substation overloading.	Los Angeles, USA	Csa
Cai et al. (2016)	●			●	●						●		Establishes a causal relationship between land surface temperature (LST) and land use change caused by urbanisation and proposes the protection of wetlands.	Fuzhou City, China	Cfa
Chen and You (2019)				●	●	●		○	○		●	○	By utilising a combination of a computational fluid dynamics model with remote sensing, it was established that urban greening is the most natural and effective mitigation strategy.	Tianjin, China	Dfa
Chen et al. (2013)	●			●	○	○					●		Using a forecasting model, the LST was projected to 2050 and 2090 and the cooling benefit of urban greening was assessed for present-day and future climate.	Melbourne, Australia	Cfb
Chun and Guldmann (2018)	●			●	●						●		The impacts of urban greening are influenced by seasonal variations of the UHI.	Columbus, Ohio, USA	Cfa
Connors et al. (2013)		○	○	●	●						●		Land use policy and its subsequent implications for urban form and design have an influence on LST and UHIs.	Phoenix, Arizona, USA	BWh

(*Continued*)

TABLE 10.A1 *(Continued)*

Concept Matrix of the Literature Review

Article	Evidence Base			Policy Responses					Technology Responses: Passive		Technology Responses: Active		Policy	Technology	Research Aim	Region	Climate
	Timescale Analysis	Effective Policymaking Instruments	Decision Support and Scenario Planning	Landscape and Urban Form	Green and Blue Area Ratio	Albedo Enhancement Policies	Transport Modal Split	Public Health and Participation	Green Building Envelopes	Development of Cool Surfaces	Sustainable Transport	Energy Consumption, HVAC and Waste Heat					
Corburn (2009)		●	●	●		●			○	○			●	○	Localising global climate change science for effective policymaking is not divorced from its social setting and requires a multi-stakeholder co-production of solutions.	New York City, USA	Cfa
Deilami and Kamruzzaman (2017)		●	●	●	○		○		○	○	○		●	○	Addresses the question of whether smart growth policies reduce the UHI and proposes five growth scenarios. In this case, infill development could be marginally effective to control the UHI.	Brisbane, Australia	Cfa
Deilami et al. (2018)		●													A systematic review of the spatiotemporal factors contributing to the UHI; develops an understanding of causalities and improves urban policymaking.	N/A	N/A

Dhalluin and Bozonnet (2015)		●	○	●	○	○		●	●		●	●	Situational analysis of ten French cities' response to the UHI impact revealed that mitigation strategies are not yet mainstreamed in decision-making.	Ten French cities	N/A
Dwivedi et al. (2019)		○						●	●	●		●	An investigation into passive architectural design to lessen the burden on HVAC systems.	Mumbai, India	Aw
Echevarría Icaza et al. (2016)		●				●					●		The development of indices of indicators to effectively measure and track the UHI in six Dutch cities.	Six Dutch cities	N/A
Fu and Weng (2018)		○	○	○			●				○	●	Alternative growth scenarios are required for Atlanta after the UHI simulation revealed that the status quo is harmful to public health.	Atlanta, Georgia, USA	Cfa
Gaffin et al. (2008)	●			○	○						○		A study of historical climate data revealed that the UHI effect is responsible for one-third of climate change observed in New York City over the past century.	New York City, USA	Cfa
Gao et al. (2017)	●			○	○						○		Researchers conducted numerical simulation, which revealed significant peak during rush hour in the diurnal cycle of UHI intensity.	Xi'an, China	BSk
Gilbert et al. (2016)			●										A review of recent "cool community" interventions across California indicates that strong leadership, broad support and policy complementarity are critical success factors.	California, USA	Csa
Gober et al. (2009)				●	●	●		○	●		●	●	In this study, trade-offs between water use and nighttime cooling were modelled by considering urban form and land use choices, finding that water bodies contribute to UHI mitigation.	Phoenix, Arizona.	BWh
Golden (2004)			●	○	○	○					○		This contribution makes a link between built environment professions, urban economics and climate science to improve the UHI policy response in terms of materials, technologies and practices.	Phoenix, Arizona.	BWh
Guindon and Nirupama (2015)		●	○			○					○		Local land use plans are essential to modify the urban form and reduce heat storage.	Montréal, Toronto, Canada	Dfb

(Continued)

TABLE 10.A1 *(Continued)*

Concept Matrix of the Literature Review

Article	Evidence Base: Timescale Analysis	Evidence Base: Effective Policymaking Instruments	Evidence Base: Decision Support and Scenario Planning	Policy Responses: Landscape and Urban Form	Policy Responses: Green and Blue Area Ratio	Policy Responses: Albedo Enhancement Policies	Policy Responses: Transport Modal Split	Policy Responses: Public Health and Participation	Technology Responses (Passive): Green Building Envelopes	Technology Responses (Passive): Development of Cool Surfaces	Technology Responses (Active): Sustainable Transport	Technology Responses (Active): Energy Consumption, HVAC and Waste Heat	Policy	Technology	Research Aim	Region	Climate
Gunawardena et al. (2017)		●		○	○	●			○	○			●	○	Green and blue spaces mitigate UHI intensity effectively and have positive outcomes on the microclimates of urban centres.	N/A	N/A
Hamilton et al. (2014)		●		○	○	●			○	○			●	○	Substantial urban regeneration such as the London Olympic Village has the potential to modify urban microclimates when UHI mitigation is considered in the design process.	London	Cfb
He et al. (2019)		●		●	●				○	●			●	●	Co-benefits can be achieved when urban policy, in this case that of Sponge City, and UHI-mitigation strategies are jointly considered.	China (numerous)	N/A
Heaviside et al. (2017)		●		○	○	●	●	●	○	○		○	●	○	A review of the literature suggests that UHI mitigation strategies have a positive impact on public health.	N/A	N/A

Hendel et al. (2016)				●	○					●	●	●	Data collected from two cases of pavement watering in Paris as a means of increasing shortwave absorptivity indicates this technique to be effective in reducing heat stress.	Paris, France	Cfb
Henits et al. (2017)	●			●							●		Land use changes increase the impervious surface ratio and result in UHI intensification.	Szeged, Hungary	Dfb
Herbel et al. (2018)	●			●				○			●	○	Land use changes can intensify existing UHIs and impact local economies during periods of heatwaves.	Cluj-Napoca city, Romania	Dfb
Hiemstra et al. (2017)		●		○	●	●			○	○	●	○	Park Cool Islands and the reforestation of cities can be planned as a network of green spaces.	Europe (Numerous)	N/A
Huang and Lu (2018)	●												Based on a comprehensive literature survey, UHI determination (intensity, heat source, modelling and remote sensing) remains the dominant topic of UHI studies.	N/A	N/A
Iping et al. (2019)		●	●					●			●		Analysis of media reporting on the UHI effect found that, overall, there was positive reportage on actions to mitigate the UHI effect; however, climate denialism appears to still be socially construed.	Australian cities	N/A
Jamei and Tapper (2019)			●		○			●		○	●	○	The link between climate science and planning for green and blue space is essential in UHI mitigation strategies.	N/A	N/A
Jenerette et al. (2011)	●			○	○					○	○	○	A systems evaluation is required to measure the water loss requirements associated with land surface cooling.	Phoenix, Arizona, USA	BWh
Jones (2018)			●					●			●		Early warning and detection systems alongside longer-term planning and preparedness can help reduce extreme heat risks.	N/A	N/A
Kamruzzaman et al. (2018)	●			○	○		●			○	●	○	Transit-oriented development strategies are preferred over low-density urban sprawl but do not necessarily reduce the UHI. A balancing act is therefore required between natural and built-up areas.	Brisbane	Cfa

(*Continued*)

TABLE 10.A1 *(Continued)*

Concept Matrix of the Literature Review

Article	Evidence Base			Policy Responses					Technology Responses				Policy	Technology	Research Aim	Region	Climate
									Passive		Active						
	Timescale Analysis	Effective Policymaking Instruments	Decision Support and Scenario Planning	Landscape and Urban Form	Green and Blue Area Ratio	Albedo Enhancement Policies	Transport Modal Split	Public Health and Participation	Green Building Envelopes	Development of Cool Surfaces	Sustainable Transport	Energy Consumption, HVAC and Waste Heat					
Kohler et al. (2017)				○					●			●	○	●	An urban growth simulation was developed to ascertain floor-space heating energy demands, and it was found that a scenario of a compact city is similar to that of a sprawling city. Therefore, urban form does not influence the energy demands required to heat buildings.	Strasbourg–Kehl urban region (France–Germany)	Cfb
Kolbe (2019)							●				●		●	●	A simulation found that the greatest UHI intensity mitigation effect for transport was to replace conventional vehicles with energy-efficient public transport.	Beijing, China	Dwa
Koomen and Diogo (2017)				●			●						●		Land use controls and spatial planning instruments should contain urban growth as sprawling land use cover perpetuates UHI.	Amsterdam, Holland	Cfb

Kuznetsov and Tomitsch (2018)			●					●			●		The UHI effect presents as a wicked problem, and no single intervention can effectively mitigate this effect. It is proposed that social computing can assist through data sharing and interactive forums.	Sydney, Australia	Cfa
Lee and French (2009)			○			●			○		●	○	In Atlanta, it was estimated that two-thirds of the impervious surface would be dark-coloured pavement, an increase of 45% over ten years. It is recommended that light surfaces be selected to increase albedo.	Atlanta, Georgia, USA	Cfa
Lehmann (2014)			○			●			○		●	○	In addition to green roofs, urban farming, roof-top gardening and green walls mitigate the UHI effect through the process of evapotranspiration.	Sydney, Australia	Cfa
Lemonsu et al. (2015)			○		●				○		●	○	Scenarios of urban densification require an engagement with the water requirements to maintain urban greenery in UHI mitigation.	Paris, France	Cfb
Li et al. (2014)	●		○		●				○		●	○	Large-scale landscaping projects including green belts, forests and parklands have had a positive mitigating influence over the UHI of Shanghai.	Shanghai, China	Cfa
Louiza et al. (2015)							○	○		○	○	○	Heat transfer and radiation are negative side effects of congestion and intensify the UHI.	Paris, France	Cfb
Marando et al. (2019)			○	○	●				○		●	○	Green infrastructure is an effective ecosystem-based climate adaptation strategy.	Rome, Italy	Csa
Mavrogianni et al. (2011)			●	○	●			○			●	○	A decision support tool was developed to guide UHI mitigation measures since urban development processes are interlinked and complex.	London, UK	Cfb
Mbuh et al. (2019)			○	○									This simulation proves that the denser urban centre of Chicago has a higher UHI than St. Paul.	Chicago, USA	Dfa
Mees et al. (2015)		●											The "mainstreaming" climate mitigation measures into existing policies, programs and projects could result in efficiency gains and increased collaboration in public and private interests in "cool" governance.	The Netherlands	CfB

(*Continued*)

TABLE 10.A1 *(Continued)*

Concept Matrix of the Literature Review

Article	Evidence Base			Policy Responses					Technology Responses: Passive		Technology Responses: Active		Policy	Technology	Research Aim	Region	Climate
	Timescale Analysis	Effective Policymaking Instruments	Decision Support and Scenario Planning	Landscape and Urban Form	Green and Blue Area Ratio	Albedo Enhancement Policies	Transport Modal Split	Public Health and Participation	Green Building Envelopes	Development of Cool Surfaces	Sustainable Transport	Energy Consumption, HVAC and Waste Heat					
Mirzaei (2015)			○												This literature review categorises and explains different types of models suitable for various objectives and scales of UHI studies.	N/A	N/A
Oikonomou et al. (2012)		●		●		○			○			○	●	○	Variation of higher indoor temperatures can be explained by architectural form and the existence of a UHI.	London, UK	Cfb
Parsaee et al. (2019)		●	●					●					●		Information management systems are required to improve decision-making processes considering urban action plans and spatial maps. Such collaborative platforms can increase participation.	N/A	N/A
Pomerantz (2018)						●							●		Increasing the albedo factor by 0.2 due to cool surfaces results in a reduction in electrical energy by <1 kWh per modified m^2 per year. The size of the benefit is small and hence decision-makers need to select mitigation measures carefully.	California, USA	Csa

Price et al. (2015)							○	○			○	Vertical green walls complement green roofs in mitigating UHI in green building envelopes.	N/A	N/A
Qin (2015)					●			○		●	○	It has been established that cool pavements are an effective UHI-mitigation factor. Less is known of heat-harvesting pavements and should be reviewed against energy output and durability over time.	N/A	N/A
Räsänen et al. (2019)		●	○			●				●		In this risk mapping, 5 different zoning options and 11 different weighting options were applied to assess vulnerability.	Helsinki, Finland	Dfb
Richards and Edwards (2018)				●				○		●	○	Co-benefits can be achieved by using water infrastructure for flood management and mitigating the UHI.	N/A	N/A
Romero Rodríguez et al. (2020)		●	○			○				○	○	A generic methodology is proposed for the estimation of the UHI and allows improved mitigation strategies and economic investments into vulnerable areas.	Seville, Spain	Cfa
Rosenfeld et al. (1998)	●											Investments in "cool communities" can result in $1.2b savings for the Los Angeles basin.	Los Angles, USA	Csa
Rossi et al. (2013)					●		○		○	●	○	Anthropogenic heat generation can be mitigated by increasing the earth's albedo factor through renewable energy infrastructure.	N/A	N/A
Roxon et al. (2020)							●		●		●	UHI studies have tended to focus on peak summer temperatures, and this study suggests that UHI could save energy costs for regions in cold climates.	USA (Various)	N/A
Roy and Singh (2015)			○	●						●		Water bodies have an influence on the humidity levels observed within Delhi's UHI.	Delhi, India	Cwa
Sánchez-Guevara Sánchez et al. (2017)			○			●			○	●	○	Low-income households can face overheating problems in periods of UHI intensity, which has an adverse effect on the public health of Madrid's vulnerable populations.	Madrid, Spain	Csa

(*Continued*)

TABLE 10.A1 *(Continued)*

Concept Matrix of the Literature Review

Article	Evidence Base			Policy Responses					Technology Responses				Policy	Technology	Research Aim	Region	Climate
									Passive		Active						
	Timescale Analysis	Effective Policymaking Instruments	Decision Support and Scenario Planning	Landscape and Urban Form	Green and Blue Area Ratio	Albedo Enhancement Policies	Transport Modal Split	Public Health and Participation	Green Building Envelopes	Development of Cool Surfaces	Sustainable Transport	Energy Consumption, HVAC and Waste Heat					
Santamouris (2013)										●				●	Reflective and permeable/water-retentive "cool pavements" hold the greater promise for cool pavements to make a substantial contribution to UHI mitigation.	N/A	N/A
Santamouris et al. (2018)										●				●	Building cooling demand increases three times at the peak of UHI intensity in Sydney, which is up to a 6°C increase in local temperature. Cool pavements have reduced peak ambient temperature by 3°C and reduced building cooling demand by 20%.	Sydney, Australia	Cfa
Santamouris et al. (2019)				●									●		Three case studies from Australian cities indicate the positive synergies between investments in green infrastructure and reduction of the UHI effect.	Sydney, Australia	Cfa
Sharma et al. (2019)				○				●					●		This study reports on the UrbClim (urban climate) model used to forecast near-future (2026–2045) and far-future (2081–2100) changes, with a focus on ten extreme heat indices.	Delhi, India	Cwa

Shickman and Rogers (2019)		○			●	○	○	○	●	○	Utility providers can account for benefits of UHI mitigation/cool city programs provided in cost–benefit tests.	USA (Various)	N/A
Short et al. (2004)						○		○			Passive design for natural ventilation is explored by using thermal modelling to measure the performance of non-domestic buildings in the UK.	UK (Various)	N/A
Si et al. (2019)					●				●		Analysis of the quality of health care provided during heatwaves indicates that a higher quality of care is provided during extreme heat but can be explained by fewer patients.	Xi'an, China	BSk
Silva and Fillpot (2018)	●					○		○		○	A zero-dimensional energy balance model was utilised to measure energy cost savings by taking account of four modelled UHI-mitigation strategies; conductivity was found to be the most effective mitigation strategy.	Phoenix, Arizona, USA	BWh
Silva et al. (2010)	●			●		○	○	○	●	○	A zero-dimensional energy balance model was utilised to measure the impact of mitigation strategies, and it was found that four strategies would lead to a 48% reduction in heat-related emergency service calls, where increasing albedo is the most effective mitigation strategy.	Phoenix, Arizona	BWh
Stone et al. (2010)		●							●		A study of metropolitan areas in the USA over five decades revealed that sprawling suburbs are two times more susceptible to extreme heat events than compact city centres.	USA	N/A
Stone and Rodgers (2001)	●	○	○			○		○	○	○	Low-density residential developments contribute to radiant UHI formation, and compact medium-to-high density development and green area ratio are recommended.	Atlanta, Georgia, USA	Cfa
Su et al. (2020)					●				●		Occupational injuries and insurance payouts decreased by 13% and 24% respectively over a two-year period after the labour protection policy, Administrative Measures on Heatstroke Prevention (AMHP2012), was adopted by the Chinese Government in 2012.	China	N/A

(Continued)

TABLE 10.A1 *(Continued)*

Concept Matrix of the Literature Review

Article	Evidence Base			Policy Responses					Technology Responses: Passive		Technology Responses: Active		Policy	Technology	Research Aim	Region	Climate
	Timescale Analysis	Effective Policymaking Instruments	Decision Support and Scenario Planning	Landscape and Urban Form	Green and Blue Area Ratio	Albedo Enhancement Policies	Transport Modal Split	Public Health and Participation	Green Building Envelopes	Development of Cool Surfaces	Sustainable Transport	Energy Consumption, HVAC and Waste Heat					
Sung (2013)		●	●		○								○		By comparing the LST of neighbourhoods, it was found that the cooling effect in Woodland can be attributed to its tree protection policy.	Woodland, Texas, USA	Cfa
Susca and Pomponi (2020)			●												Life cycle assessments are expanded by coupling the land use and local climate as a means of measuring the UHI.	N/A	N/A
Takebayashi and Moriyama (2012)									●					●	A study of the heat transfers in an urban street canopy revealed that mitigation measures are most effective when prioritised for large roofs, followed by smaller roofs and then wall surfaces.	Japan	Cfa
Takebayashi et al. (2014)						●							●		Tree canopies, green walls and reflective pavements should consider the orientation of streets and solar penetration at the urban block scale.	Japan	Cfa

Targino et al. (2019)		●		○			○	○	○	A simulation study found that a park would be sevenfold cheaper than building a city pond and is more effective in mitigating the UHI.	Londrina, Brazil	Cfa
Teferi and Abraha (2017)		●		○			○	○	○	This longitudinal study confirms that green spaces are one of the most promising local climate change adaptation strategies.	Addis Ababa, Ethiopia	Cwb
Tonekaboni et al. (2018)		●		○		○	○	○	○	A propositional framework combines a variety of human, vehicle, and drone-borne sensors which is joined with data from satellite and weather stations in a decision-making support system.	N/A	N/A
Tuczek et al. (2019)		●		○		○	○		○	A propositional framework for decision support that maximises revenue while maintaining an optimal building/vegetation balance within the UHI limits.	N/A	N/A
Vargo et al. (2016)		●		○		○	○		○	Comparison of heat-related avoided mortality rate between albedo, green and combined scenarios	N/A	N/A
Vasilakopoulou et al. (2014)		●		○	○		○		○	This systematic review of albedo enhancements concludes that such interventions successfully mitigate the UHI.	N/A	N/A
Wang and Akbari (2016)			○	○	○					The three main UHI interventions used in Montréal (urban canopy, urban vegetation, and albedo of paved-over areas) is reviewed against the sky-view factor control and wind speed.	Montréal, Canada	Cfb
Weber et al. (2015)	●	●								This case study provides insights into policy formation based on vulnerability indices informed by the modelling of the UHI.	Philadelphia	Cfa
Wilhelmi and Hayden (2016)						●		●		The System for Integrated Modelling of Metropolitan Extreme Heat Risk project funded by NASA is a good example of an iterative, evidence-based and responsive stakeholder engagement platform at the science-policy interface of heat-health research.	Houston, Texas, USA	Cfa

(*Continued*)

TABLE 10.A1 *(Continued)*

Concept Matrix of the Literature Review

Article	Evidence Base			Policy Responses					Technology Responses: Passive		Technology Responses: Active		Policy	Technology	Research Aim	Region	Climate
	Timescale Analysis	Effective Policymaking Instruments	Decision Support and Scenario Planning	Landscape and Urban Form	Green and Blue Area Ratio	Albedo Enhancement Policies	Transport Modal Split	Public Health and Participation	Green Building Envelopes	Development of Cool Surfaces	Sustainable Transport	Energy Consumption, HVAC and Waste Heat					
Wondmagegn et al. (2019)								●					●		Extreme heat events place a demand on the healthcare system when reviewed against indicators such as ER visits, hospitalisation and ambulance call-outs.	N/A	N/A
Wonorahardjo et al. (2020)									●					●	Four building materials (brick, aerated concrete, wood with glass-wool insulation, and glass fibre-reinforced concrete with glass-wool insulation) were reviewed to assess thermal qualities in an observed UHI.	N/A (Tropical Area)	N/A
Xiang et al. (2016)								●					●		Survey results indicate that workplace heat exposure during extreme heat events requires risk management and prevention strategies.	Adelaide, Australia	Csa
Yamaguchi and Ihara (2020)															Energy efficiency strategies have proven useful to combat UHI intensity in Japan, when considered alongside other countermeasures, such as greening.	Tokyo, Japan	Cfa

Yang et al. (2015)				○		○			○				○	○	The potential of increasing albedo by means of reflective materials and mitigating the UHI effect depends on a set of factors, including building characteristics, urban environment, meteorological and geographical conditions.	N/A	N/A
Zhang et al. (2019)			●			○			○	○			○	○	A willingness-to-pay valuation approach was tested for cool roofs, and it was found that government credibility and education are important factors in promoting public participation.	Beijing, China	Dwa
Zinzi et al. (2014)			●			○			○	○			○	○	The study presents an energy-rating scheme for cool roofs and, based on simulations, the performance of the cool roof's radiative properties was assessed.	Italy	Csa
Total ●	12	19	27	18	14	18	4	17	8	9	2	5	54	16			
Total ○	0	4	11	26	23	10	2	7	21	29	2	9	18	39			

Notes: ● major focus, ○ minor focus; Af = Tropical rainforest climate; Aw = Tropical savannah, wet; BSk = Cold semi-arid (steppe) climate; BWh = Hot deserts climate; Cfa = Humid subtropical climate; Cfb = Temperate oceanic climate; Csa = Hot-summer Mediterranean climate; Cwa = Monsoon-influenced humid subtropical climate; Cwb = Subtropical highland climate or temperate oceanic climate with dry winters; Dfa = Hot-summer humid continental climate; Dfb = Warm-summer humid continental climate; Dwa = Monsoon-influenced hot-summer humid continental climate.

References

Abd El-Hakim, R., & El-Badawy, S. (2019). Quantifying effects of urban heat islands: State of the art. In S. Badawy & D. H. Chen (Eds.), *Recent developments in pavement engineering* (pp. 42–69). Springer. https://doi.org/10.1007/978-3-030-34196-1_4

Aflaki, A., Mirnezhad, M., Ghaffarianhoseini, A., Ghaffarianhoseini, A., Omrany, H., Wang, Z.-H., & Akbari, H. (2017). Urban heat island mitigation strategies: A state-of-the-art review on Kuala Lumpur, Singapore and Hong Kong. *Cities, 62*, 131–145. https://doi.org/10.1016/j.cities.2016.09.003

Akbari, H., & Kolokotsa, D. (2016). Three decades of urban heat islands and mitigation technologies research. *Energy and Buildings, 133*, 834–842. https://doi.org/10.1016/j.enbuild.2016.09.067

Akbari, H., Matthews, H. D., & Seto, D. (2012). The long-term effect of increasing the albedo of urban areas. *Environmental Research Letters, 7*(2), 1–10. https://doi.org/10.1088/1748-9326/7/2/024004

Aleksandrowicz, O., Vuckovic, M., Kiesel, K., & Mahdavi, A. (2017). Current trends in urban heat island mitigation research: Observations based on a comprehensive research repository. *Urban Climate, 21*, 1–26. https://doi.org/10.1016/j.uclim.2017.04.002

Arthur, W. B. (1988). Self-reinforcing mechanisms in economics. In P. W. Anderson, K. J. Arrow, & D. Pines (Eds.), *The economy as an evolving complex system* (pp. 9–31). CRC Press. https://doi.org/10.1201/9780429492846

Baburoglu, O. N., & Ravn, I. (1992). Normative action research. *Organization Studies, 13*(1), 19–34. https://doi.org/10.1177/017084069201300104

Batcheller, P. (2020). *Citizen scientists to map urban heat islands in Detroit this summer*. WDET. https://wdet.org/posts/2020/07/13/89810-citizen-scientists-to-map-urban-heat-islands-in-detroit-this-summer

Bolitho, A., & Miller, F. (2017). Heat as emergency, heat as chronic stress: Policy and institutional responses to vulnerability to extreme heat. *Local Environment, 22*(6), 682–698. https://doi.org/10.1080/13549839.2016.1254169

Bostrom, R. P., & Heinen, J. S. (1977). MIS problems and failures: A socio-technical perspective. Part I: The causes. *MIS Quarterly, 1*(3), 17–32. https://doi.org/10.2307/248710

Burillo, D., Chester, M. V., Pincetl, S., & Fournier, E. (2019). Electricity infrastructure vulnerabilities due to long-term growth and extreme heat from climate change in Angeles County. *Energy Policy, 128*, 943–953. https://doi.org/10.1016/j.enpol.2018.12.053

C40. (2015). *Cities100: Paris – Green spaces keep the city cool*. https://www.c40.org/case_studies/cities100-paris-green-spaces-keep-the-city-cool

Cai, Y., Zhang, H., Zheng, P., & Pan, W. (2016). Quantifying the impact of land use/land cover changes on the urban heat island: A case study of the natural wetlands distribution area of Fuzhou city, China. *Wetlands, 36*(2), 285–298. https://doi.org/10.1007/s13157-016-0738-7

Chen, D., Wang, X., Khoo, Y. B., Thatcher, M., Lin, B. B., Ren, Z., Wang, C.-H., & Barnett, G. (2013). Assessment of urban heat island and mitigation by urban green coverage. In A. Khare & T. Beckman (Eds.), *Mitigating climate change: The emerging face of modern cities* (pp. 247–257). Springer. https://doi.org/10.1007/978-3-642-37030-4_13

Chen, R., & You, X.-y. (2019). Reduction of urban heat island and associated greenhouse gas emissions. *Mitigation and Adaptation Strategies for Global Change.* https://doi.org/10.1007/s11027-019-09886-1

Chun, B., & Guldmann, J.-M. (2018). Impact of greening on the urban heat island: Seasonal variations and mitigation strategies. *Computers, Environment and Urban Systems, 71,* 165–176. https://doi.org/10.1016/j.compenvurbsys.2018.05.006

Connors, J. P., Galletti, C. S., & Chow, W. T. L. (2013). Landscape configuration and urban heat island effects: Assessing the relationship between landscape characteristics and land surface temperature in Phoenix, Arizona. *Landscape Ecology, 28*(2), 271–283. https://doi.org/10.1007/s10980-012-9833-1

Corburn, J. (2009). Cities, climate change and urban heat island mitigation: Localising global environmental science. *Urban Studies, 46*(2), 413–427. https://doi.org/10.1177/0042098008099361

Davey, S. L., Lee, B. J., Robbins, T., Randeva, H., & Thake, C. D. (2020). Heat stress and PPE during COVID-19: Impact on health care workers' performance, safety and well-being in NHS settings. *Journal of Hospital Infection,* 1–10. https://doi.org/10.1016/j.jhin.2020.11.027

Deilami, K., & Kamruzzaman, M. (2017). Modelling the urban heat island effect of smart growth policy scenarios in Brisbane. *Land Use Policy, 64,* 38–55. https://doi.org/10.1016/j.landusepol.2017.02.027

Deilami, K., Kamruzzaman, M., & Liu, Y. (2018). Urban heat island effect: A systematic review of spatio-temporal factors, data, methods, and mitigation measures. *International Journal of Applied Earth Observation and Geoinformation, 67,* 30–42. https://doi.org/10.1016/j.jag.2017.12.009

Dhalluin, A., & Bozonnet, E. (2015). Urban heat islands and sensitive building design – A study in some French cities' context. *Sustainable Cities and Society, 19,* 292–299. https://doi.org/10.1016/j.scs.2015.06.009

Dizdaroglu, D., Yigitcanlar, T., & Dawes, L. (2012). A micro-level indexing model for assessing urban ecosystem sustainability. *Smart and Sustainable Built Environment, 1*(3), 291–315. https://doi.org/10.1108/20466091211287155

Douglas, E. M., Reardon, K. M., & Täger, M. C. (2018). Participatory action research as a means of achieving ecological wisdom within climate change resiliency planning. *Journal of Urban Management, 7*(3), 152–160. https://doi.org/10.1016/j.jum.2018.05.003

Dwivedi, A., Khire, M. V., Mohan, B. K., & Shah, S. (2019). The role of structure cooling to reduce the effect of urban heat island in Mumbai. *Advances in Building Energy Research, 13*(2), 174–192. https://doi.org/10.1080/17512549.2018.1488611

Echevarría Icaza, L., van der Hoeven, F. D., & van den Dobbelsteen, A. (2016). The urban heat island effect in Dutch city centres: Identifying relevant indicators and first explorations. In W. Leal Filho, K. Adamson, R. M. Dunk, U. M. Azeiteiro, S. Illingworth, & F. Alves (Eds.), *Implementing climate change adaptation in cities and communities: Integrating strategies and educational approaches* (pp. 123–160). Springer. https://doi.org/10.1007/978-3-319-28591-7_7

Eden, C., & Ackermann, F. (2018). Theory into practice, practice to theory: Action research in method development. *European Journal of Operational Research, 271*(3), 1145–1155. https://doi.org/10.1016/j.ejor.2018.05.061

Estrada, F., Botzen, W. J. W., & Tol, R. S. J. (2017). A global economic assessment of city policies to reduce climate change impacts. *Nature Climate Change, 7,* 403–406. https://doi.org/10.1038/nclimate3301

Fu, P., & Weng, Q. (2018). Responses of urban heat island in Atlanta to different land-use scenarios. *Theoretical and Applied Climatology, 133*(1), 123–135. https://doi.org/10.1007/s00704-017-2160-3

Gaffin, S. R., Rosenzweig, C., Khanbilvardi, R., Parshall, L., Mahani, S., Glickman, H., Goldberg, R., Blake, R., Slosberg, R. B., & Hillel, D. (2008). Variations in New York City's urban heat island strength over time and space. *Theoretical and Applied Climatology, 94*(1), 1–11. https://doi.org/10.1007/s00704-007-0368-3

Gao, M., Shen, H., Han, X., Li, H., & Zhang, L. (2017). Multiple timescale analysis of the urban heat island effect based on the community land model: A case study of the city of Xi'an, China. *Environmental Monitoring and Assessment, 190*(1), 8. https://doi.org/10.1007/s10661-017-6320-9

Gilbert, H., Mandel, B. H., & Levinson, R. (2016). Keeping California cool: Recent cool community developments. *Energy and Buildings, 114*, 20–26. https://doi.org/10.1016/j.enbuild.2015.06.023

Gober, P., Brazel, A., Quay, R., Myint, S., Grossman-Clarke, S., Miller, A., & Rossi, S. (2009). Using watered landscapes to manipulate urban heat island effects: How much water will it take to cool Phoenix? *Journal of the American Planning Association, 76*(1), 109–121. https://doi.org/10.1080/01944360903433113

Golden, J. S. (2004). The built environment induced urban heat island effect in rapidly urbanizing arid regions – A sustainable urban engineering complexity. *Environmental Sciences, 1*(4), 321–349. https://doi.org/10.1080/15693430412331291698

Guindon, S.-M., & Nirupama, N. (2015). Reducting risk from urban heat island effects in cities. *Natural Hazards, 77*(2), 823–831. https://doi.org/10.1007/s11069-015-1627-8

Gunawardena, K. R., Wells, M. J., & Kershaw, T. (2017). Utilising green and bluespace to mitigate urban heat island intensity. *Science of the Total Environment, 584–585*, 1040–1055. https://doi.org/10.1016/j.scitotenv.2017.01.158

Hamilton, I., Stocker, J., Evans, S., Davies, M., & Carruthers, D. (2014). The impact of the London Olympic Parkland on the urban heat island. *Journal of Building Performance Simulation, 7*(2), 119–132. https://doi.org/10.1080/19401493.2013.791343

Harvey, C. (2019). *Urban heat islands mean warming will be worse in cities*. Scientific American. https://www.scientificamerican.com/article/urban-heat-islands-mean-warming-will-be-worse-in-cities

He, B.-J. (2019). Towards the next generation of green building for urban heat island mitigation: Zero UHI impact building. *Sustainable Cities and Society, 50*, 1–14. https://doi.org/10.1016/j.scs.2019.101647

He, B.-J., Zhu, J., Zhao, D.-X., Gou, Z.-H., Qi, J.-D., & Wang, J. (2019). Co-benefits approach: Opportunities for implementing sponge city and urban heat island mitigation. *Land Use Policy, 86*, 147–157. https://doi.org/10.1016/j.landusepol.2019.05.003

Heaviside, C., Macintyre, H., & Vardoulakis, S. (2017). The urban heat island: Implications for health in a changing environment. *Current Environmental Health Reports, 4*(3), 296–305. https://doi.org/10.1007/s40572-017-0150-3

Hendel, M., Gutierrez, P., Colombert, M., Diab, Y., & Royon, L. (2016). Measuring the effects of urban heat island mitigation techniques in the field: Application to the case of pavement-watering in Paris. *Urban Climate, 16*, 43–58. https://doi.org/10.1016/j.uclim.2016.02.003

Henits, L., Mucsi, L., & Liska, C. M. (2017). Monitoring the changes in impervious surface ratio and urban heat island intensity between 1987 and 2011 in Szeged, Hungary. *Environmental Monitoring and Assessment, 189*(2), 86. https://doi.org/10.1007/s10661-017-5779-8

Herbel, I., Croitoru, A.-E., Rus, A. V., Roşca, C. F., Harpa, G. V., Ciupertea, A.-F., & Rus, I. (2018). The impact of heat waves on surface urban heat island and local economy in Cluj-Napoca city, Romania. *Theoretical and Applied Climatology, 133*(3), 681–695. https://doi.org/10.1007/s00704-017-2196-4

Hiemstra, J. A., Saaroni, H., & Amorim, J. H. (2017). The urban heat island: Thermal comfort and the role of urban greening. In D. Pearlmutter, C. Calfapietra, R. Samson, L. O'Brien, S. Krajter Ostoić, G. Sanesi, & R. Alonso del Amo (Eds.), *The urban forest: Cultivating green infrastructure for people and the environment* (pp. 7–19). Springer. https://doi.org/10.1007/978-3-319-50280-9_2

Hintz, M. J., Luederitz, C., Lang, D. J., & von Wherden, H. (2018). Facing the heat: A systematic literature review exploring the transferability of solutions to cope with urban heat waves. *Urban Climate, 24,* 714–727. https://doi.org/10.1016/j.uclim.2017.08.011

Huang, Q., & Lu, Y. (2018). Urban heat island research from 1991 to 2015: A bibliometric analysis. *Theoretical and Applied Climatology, 131*(3–4), 1055–1067. https://doi.org/10.1007/s00704-016-2025-1

Iping, A., Kidston-Lattari, J., Simpson-Young, A., Duncan, E., & McManus, P. (2019). (Re)presenting urban heat islands in Australian cities: A study of media reporting and implications for urban heat and climate change debates. *Urban Climate, 27,* 420–429. https://doi.org/10.1016/j.uclim.2018.12.014

Jamei, E., & Tapper, N. (2019). Chapter 19 – WSUD and urban heat island effect mitigation. In A. K. Sharma, T. Gardner, & D. Begbie (Eds.), *Approaches to water sensitive urban design* (pp. 381–407). Woodhead Publishing. https://doi.org/10.1016/B978-0-12-812843-5.00019-8

Jänicke, M. (2015). Horizontal and vertical reinforcement in global climate governance. *Energies, 8*(6), 5782–5799. https://doi.org/10.3390/en8065782

Järvinen, P. (2007). Action research is similar to design science. *Quality & Quantity, 41,* 37–54. https://doi.org/10.1007/s11135-005-5427-1

Jenerette, G. D., Harlan, S. L., Stefanov, W. L., & Martin, C. A. (2011). Ecosystem services and urban heat riskscape moderation: Water, green spaces, and social inequality in Phoenix, USA. *Ecological Applications, 21*(7), 2637–2651. https://doi.org/10.1890/10-1493.1

Jones, H. M. (2018). Climate change and increasing risk of extreme heat. In Y. Hosokawa (Ed.), *Human health and physical activity during heat exposure* (pp. 1–13). Springer. https://doi.org/10.1007/978-3-319-75889-3_1

Josseran, L., Caillère, N., Brun-Ney, D., Rottner, J., Filleul, L., Brucker, G., & Astagneau, P. (2009). Syndromic surveillance and heat wave morbidity: A pilot study based on emergency departments in France. *BMC Medical Informatics and Decision Making, 9,* 1–9. https://doi.org/10.1186/1472-6947-9-14

Kamruzzaman, M., Deilami, K., & Yigitcanlar, T. (2018). Investigating the urban heat island effect of transit oriented development in Brisbane. *Journal of Transport Geography, 66,* 116–124. https://doi.org/10.1016/j.jtrangeo.2017.11.016

Kamruzzaman, M., Hine, J., & Yigitcanlar, T. (2015). Investigating the link between carbon dioxide emissions and transport-related social exclusion in rural Northern Ireland. *International Journal of Environmental Science and Technology, 12,* 3463–3478. https://doi.org/10.1007/s13762-015-0771-8

Karlson, K. (2020). *Urban heat islands are not an accident.* Sierra Club. https://www.sierraclub.org/sierra/urban-heat-islands-are-not-accident

Kohler, M., Tannier, C., Blond, N., Aguejdad, R., & Clappier, A. (2017). Impacts of several urban-sprawl countermeasures on building (space heating) energy demands and urban heat island intensities. A case study. *Urban Climate, 19,* 92–121. https://doi.org/10.1016/j.uclim.2016.12.006

Kolbe, K. (2019). Mitigating urban heat island effect and carbon dioxide emissions through different mobility concepts: Comparison of conventional vehicles with electric vehicles, hydrogen vehicles and public transportation. *Transport Policy, 80,* 1–11. https://doi.org/10.1016/j.tranpol.2019.05.007

Koomen, E., & Diogo, V. (2017). Assessing potential future urban heat island patterns following climate scenarios, socio-economic developments and spatial planning strategies. *Mitigation and Adaptation Strategies for Global Change, 22*(2), 287–306. https://doi.org/10.1007/s11027-015-9646-z

Kotharkar, R., Ramesh, A., & Bagade, A. (2018). Urban heat island studies in South Asia: A critical review. *Urban Climate, 24,* 1011–1026. https://doi.org/10.1016/j.uclim.2017.12.006

Kuznetsov, S., & Tomitsch, M. (2018). A study of urban heat: Understanding the challenges and opportunities for addressing wicked problems in HCI. *Proceedings of the 2018 CHI Conference on Human Factors in Computing Systems,* Montreal, Quebec, Canada. https://doi.org/10.1145/3173574.3174137

Lapidus, F. (2019). *Scientists pinpoint urban heat islands.* VOA News. https://www.voanews.com/episode/scientists-pinpoint-urban-heat-islands-3962296

Leavitt, H. J. (1965). Applied organizational change in industry: Structural, technological, and humanistic approaches. In J. G. March (Ed.), *Handbook of organizations* (pp. 1144–1170). Rand McNally.

Lee, S., & French, S. (2009). Regional impervious surface estimation: An urban heat island application. *Journal of Environmental Planning & Management, 52*(4), 477–496. https://doi.org/10.1080/09640560902868207

Lehmann, S. (2014). Low carbon districts: Mitigating the urban heat island with green roof infrastructure. *City, Culture and Society, 5*(1), 1–8. https://doi.org/10.1016/j.ccs.2014.02.002

Lemonsu, A., Viguié, V., Daniel, M., & Masson, V. (2015). Vulnerability to heat waves: Impact of urban expansion scenarios on urban heat island and heat stress in Paris (France). *Urban Climate, 14,* 586–605. https://doi.org/10.1016/j.uclim.2015.10.007

Levermore, G., Parkinson, J., Lee, K., & Laycock, P. (2018). The increasing trend of the urban heat island intensity. *Urban Climate, 24,* 360–368. https://doi.org/10.1016/j.uclim.2017.02.004

Li, C.-f, Shen, D., Dong, J.-s, Yin, J.-y, Zhao, J.-j, & Xue, D. (2014). Monitoring of urban heat island in Shanghai, China, from 1981 to 2010 with satellite data. *Arabian Journal of Geosciences, 7*(10), 3961–3971. https://doi.org/10.1007/s12517-013-1053-8

Li, D., Liao, W., Rigden, A. J., Liu, X., Wang, D., Malyshev, S., & Shevliakova, E. (2019). Urban heat island: Aerodynamics or imperviousness? *Science Advances, 5*(4), 1–4. https://doi.org/10.1126/sciadv.aau4299

Li, X., Zhou, Y., Yu, S., Jia, G., Li, H., & Li, W. (2019). Urban heat island impacts on building energy consumption: A review of approaches and findings. *Energy, 174,* 407–419. https://doi.org/10.1016/j.energy.2019.02.183

Lindgren, R., Henfridsson, O., & Schultze, U. (2004). Design principles for competence management systems: A synthesis of an action research study. *MIS Quarterly, 28*(3), 435–472. https://doi.org/10.2307/25148646

Louiza, H., Zéroual, A., & Djamel, H. (2015). Impact of the transport on the urban heat island. *International Journal for Traffic and Transport Engineering, 5*(3), 252–262. https://doi.org/10.7708/ijtte.2015.5(3).03

Marando, F., Salvatori, E., Sebastiani, A., Fusaro, L., & Manes, F. (2019). Regulating ecosystem services and green infrastructure: Assessment of urban heat island effect mitigation in the municipality of Rome, Italy. *Ecological Modelling, 392*, 92–102. https://doi.org/10.1016/j.ecolmodel.2018.11.011

Mavrogianni, A., Davies, M., Batty, M., Belcher, S. E., Bohnenstengel, S. I., Carruthers, D., Chalabi, Z., Croxford, B., Demanuele, C., Evans, S., Giridharan, R., Hacker, J. N., Hamilton, I., Hogg, C., Hunt, J., Kolokotroni, M., Martin, C., Milner, J., Rajapaksha, I., & Ridley, I. (2011). The comfort, energy and health implications of London's urban heat island. *Building Services Engineering Research & Technology, 32*(1), 35–52. https://doi.org/10.1177/0143624410394530

Mbuh, M. J., Wheeler, R., & Cook, A. (2019). Spatiotemporal analysis of urban heat island intensification in the city of Minneapolis-St. Paul and Chicago metropolitan areas using Landsat data from 1984 to 2016. *Geocarto International*, 1–26. https://doi.org/10.1080/10106049.2019.1655802

McDonald, R., Kroeger, T., Boucher, T., Longzhu, W., & Salem, R. (2016). Planting healthy air: A global analysis of the role of urban trees in addressing particulate matter pollution and extreme heat. *The Nature Conservancy*. https://toolkit.climate.gov/reports/planting-healthy-air-global-analysis-role-urban-trees-addressing-particulate-matter

Mees, H. L. P., Driessen, P. P. J., & Runhaar, H. A. C. (2015). "Cool" governance of a "hot" climate issue: Public and private responsibilities for the protection of vulnerable citizens against extreme heat. *Regional Environmental Change, 15*(6), 1065–1079. https://doi.org/10.1007/s10113-014-0681-1

Mirzaei, P. A. (2015). Recent challenges in modeling of urban heat island. *Sustainable Cities and Society, 19*, 200–206. https://doi.org/10.1016/j.scs.2015.04.001

Mirzaei, P. A., & Haghighat, F. (2010). Approaches to study urban heat island – Abilities and limitations. *Building and Environment, 45*(10), 2192–2201. https://doi.org/10.1016/j.buildenv.2010.04.001

Mohajerani, A., Bakaric, J., & Jeffrey-Bailey, T. (2017). The urban heat island effect, its causes, and mitigation, with reference to the thermal properties of asphalt concrete. *Journal of Environmental Management, 197*, 522–538. https://doi.org/10.1016/j.jenvman.2017.03.095

Mortoja, M. G., & Yigitcanlar, T. (2020). Local drivers of anthropogenic climate change: Quantifying the impact through a remote sensing approach in Brisbane. *Remote Sensing, 12*(14), 1–24. https://doi.org/10.3390/rs12142270

NYC Government. (2017). *Cool neighborhoods NYC: A comprehensive approach to keep communities safe in extreme heat*. https://www1.nyc.gov/assets/orr/pdf/Cool_Neighborhoods_NYC_Report.pdf

Oikonomou, E., Davies, M., Mavrogianni, A., Biddulph, P., Wilkinson, P., & Kolokotroni, M. (2012). Modelling the relative importance of the urban heat island and the thermal quality of dwellings for overheating in London. *Building and Environment, 57*, 223–238. https://doi.org/10.1016/j.buildenv.2012.04.002

Okoli, C. (2015). A guide to conducting a standalone systematic literature review. *Communications of the Association for Information Systems, 37*, 879–910. https://doi.org/10.17705/1CAIS.03743

Olivo, Y., Hamidi, A., & Ramamurthy, P. (2017). Spatiotemporal variability in building energy use in New York City. *Energy, 141*, 1393–1401. https://doi.org/10.1016/j.energy.2017.11.066

Parsaee, M., Joybari, M. M., Mirzaei, P. A., & Haghighat, F. (2019). Urban heat island, urban climate maps and urban development policies and action plans. *Environmental Technology & Innovation, 14*, 1–16. https://doi.org/10.1016/j.eti.2019.100341

Patorniti, N. P., Stevens, N. J., & Salmon, P. M. (2017). A systems approach to city design: Exploring the compatibility of sociotechnical systems. *Habitat International, 66*, 42–48. https://doi.org/10.1016/j.habitatint.2017.05.008

Peterson, T. C. (2003). Assessment of urban versus rural in situ surface temperatures in the contiguous United States: No difference found. *Journal of Climate, 16*(18), 2941–2959. https://doi.org/10.1175/1520-0442(2003)016<2941:AOUVRI>2.0.CO;2

Pomerantz, M. (2018). Are cooler surfaces a cost-effect mitigation of urban heat islands? *Urban Climate, 24*, 393–397. https://doi.org/10.1016/j.uclim.2017.04.009

Price, A., Jones, E. C., & Jefferson, F. (2015). Vertical greenery systems as a strategy in urban heat island mitigation. *Water, Air, & Soil Pollution, 226*(8), 247. https://doi.org/10.1007/s11270-015-2464-9

Qin, Y. (2015). A review on the development of cool pavements to mitigate urban heat island effect. *Renewable and Sustainable Energy Reviews, 52*, 445–459. https://doi.org/10.1016/j.rser.2015.07.177

Ramakreshnan, L., Aghamohammadi, N., Fong, C. S., Ghaffarianhoseini, A., Ghaffarianhoseini, A., Wong, L. P., Hassan, N., & Sulaiman, N. M. (2018). A critical review of urban heat island phenomenon in the context of greater Kuala Lumpur, Malaysia. *Sustainable Cities and Society, 39*, 99–113. https://doi.org/10.1016/j.scs.2018.02.005

Räsänen, A., Heikkinen, K., Piila, N., & Juhola, S. (2019). Zoning and weighting in urban heat island vulnerability and risk mapping in Helsinki, Finland. *Regional Environmental Change, 19*(5), 1481–1493. https://doi.org/10.1007/s10113-019-01491-x

Richards, D. R., & Edwards, P. J. (2018). Using water management infrastructure to address both flood risk and the urban heat island. *International Journal of Water Resources Development, 34*(4), 490–498. https://doi.org/10.1080/07900627.2017.1357538

Romero Rodríguez, L., Sánchez Ramos, J., Sánchez de la Flor, F. J., & Álvarez Domínguez, S. (2020). Analyzing the urban heat island: Comprehensive methodology for data gathering and optimal design of mobile transects. *Sustainable Cities and Society, 55*, 1–18. https://doi.org/10.1016/j.scs.2020.102027

Rosenfeld, A. H., Akbari, H., Romm, J. J., & Pomerantz, M. (1998). Cool communities: Strategies for heat island mitigation and smog reduction. *Energy and Buildings, 28*(1), 51–62. https://doi.org/10.1016/S0378-7788(97)00063-7

Rossi, F., Cotana, F., Filipponi, M., Nicolini, A., Menon, S., & Rosenfeld, A. (2013). Cool roofs as a strategy to tackle global warming: Economical and technical opportunities. *Advances in Building Energy Research, 7*(2), 254–268. https://doi.org/10.1080/17512549.2013.865555

Roxon, J., Ulm, F. J., & Pellenq, R. J. M. (2020). Urban heat island impact on state residential energy cost and CO_2 emissions in the United States. *Urban Climate, 31*, 1–10. https://doi.org/10.1016/j.uclim.2019.100546

Roy, S. S., & Singh, R. B. (2015). Role of local level relative humidity on the development of urban heat island across the Delhi metropolitan region. In R. B.

Singh (Ed.), *Urban development challenges, risks and resilience in Asian mega cities* (pp. 99–118). Springer. https://doi.org/10.1007/978-4-431-55043-3_6

Ruefenacht, L. A., & Acero, J. A. (2017). *Strategies for cooling Singapore: A catalogue of 80+ measures to mitigate urban heat island and improve outdoor thermal comfort*. https://doi.org/10.3929/ethz-b-000258216

Sanchez-Guevara, C., Núñez Peiró, M., Taylor, J., Mavrogianni, A., & Neila González, J. (2019). Assessing population vulnerability towards summer energy poverty: Case studies of Madrid and London. *Energy & Buildings, 190,* 132–143. https://doi.org/10.1016/j.enbuild.2019.02.024

Sánchez-Guevara Sánchez, C., Núñez Peiró, M., & Neila González, F. J. (2017). Urban heat island and vulnerable population. The case of Madrid. In P. Mercader-Moyano (Ed.), *Sustainable development and renovation in architecture, urbanism and engineering* (pp. 3–13). Springer. https://doi.org/10.1007/978-3-319-51442-0_1

Santamouris, M. (2013). Using cool pavements as a mitigation strategy to fight urban heat island—A review of the actual developments. *Renewable and Sustainable Energy Reviews, 26,* 224–240. https://doi.org/10.1016/j.rser.2013.05.047

Santamouris, M. (2014). Cooling the cities – A review of reflective and green roof mitigation technologies to fight heat island and improve comfort in urban environments. *Solar Energy, 103,* 682–703. https://doi.org/10.1016/j.solener.2012.07.003

Santamouris, M., Ding, L., & Osmond, P. (2019). Urban heat island mitigation. In P. Newton, D. Prasad, A. Sproul, & S. White (Eds.), *Decarbonising the built environment* (pp. 337–355). Palgrave Macmillan. https://doi.org/10.1007/978-981-13-7940-6_18

Santamouris, M., Haddad, S., Saliari, M., Vasilakopoulou, K., Synnefa, A., Paolini, R., & Fiorito, F. (2018). On the energy impact of urban heat island in Sydney: Climate and energy potential of mitigation technologies. *Energy and Buildings, 166,* 154–164. https://doi.org/10.1016/j.enbuild.2018.02.007

Sein, M. K., Henfridsson, O., Purao, S., Rossi, M., & Lindgren, R. (2011). Action design research. *MIS Quarterly, 35*(1), 37–56. https://doi.org/10.2307/23043488

Sharma, R., Hooyberghs, H., Lauwaet, D., & De Ridder, K. (2019). Urban heat island and future climate change—Implications for Delhi's heat. *Journal of Urban Health, 96*(2), 235–251. https://doi.org/10.1007/s11524-018-0322-y

Shickman, K., & Rogers, M. (2019). Capturing the true value of trees, cool roofs, and other urban heat island mitigation strategies for utilities. *Energy Efficiency*. https://doi.org/10.1007/s12053-019-09789-9

Short, C. A., Lomas, K. J., & Woods, A. (2004). Design strategy for low-energy ventilation and cooling within an urban heat island. *Building Research & Information, 32*(3), 187–206. https://doi.org/10.1080/09613210410001679875

Si, Y., Zhou, Z., Su, M., & Chen, X. (2019). Climate change and quality of health care: Evidence from extreme heat. *The Lancet, 394*(S94), 94. https://doi.org/10.1016/S0140-6736(19)32430-4

Silva, H., & Fillpot, B. S. (2018). Modeling nexus of urban heat island mitigation strategies with electricity/power usage and consumer costs: A case study for Phoenix, Arizona, USA. *Theoretical and Applied Climatology, 131*(1), 661–669. https://doi.org/10.1007/s00704-016-1985-5

Silva, H. R., Phelan, P. E., & Golden, J. S. (2010). Modeling effects of urban heat island mitigation strategies on heat-related morbidity: A case study for Phoenix, Arizona, USA. *International Journal of Biometeorology, 54*(1), 13–22. https://doi.org/10.1007/s00484-009-0247-y

Stone, B., Hess, J. J., & Frumkin, H. (2010). Urban form and extreme heat events: Are sprawling cities more vulnerable to climate change than compact cities? *Environmental Health Perspectives, 118*(10), 1425–1428. https://doi.org/10.1289/ehp.0901879

Stone, B., & Rodgers, M. O. (2001). Urban form and thermal efficiency: How the design of cities influences the urban heat island effect. *Journal of the American Planning Association, 67*(2), 186–198. https://doi.org/10.1080/01944360108976228

Stone, B., Vargo, J., & Habeeb, D. (2012). Managing climate change in cities: Will climate action plans work? *Landscape and Urban Planning, 107*(3), 263–271. https://doi.org/10.1016/j.landurbplan.2012.05.014

Su, Y., Cheng, L., Cai, W., Lee, J. K. W., Zhong, S., Chen, S., & Huang, C. (2020). Evaluating the effectiveness of labor protection policy on occupational injuries caused by extreme heat in a large subtropical city of China. *Environmental Research, 186*, 1–9. https://doi.org/10.1016/j.envres.2020.109532

Sun, Y., Zhang, X., Zwiers, F. W., Song, L., Wan, H., Hu, T., Yin, H., & Ren, G. (2014). Rapid increase in the risk of extreme summer heat in Eastern China. *Nature Climate Change, 4*, 1082–1085. https://doi.org/10.1038/nclimate2410

Sung, C. Y. (2013). Mitigating surface urban heat island by a tree protection policy: A case study of the Woodland, Texas, USA. *Urban Forestry & Urban Greening, 12*(4), 474–480. https://doi.org/10.1016/j.ufug.2013.05.009

Susca, T., & Pomponi, F. (2020). Heat island effects in urban life cycle assessment: Novel insights to include the effects of the urban heat island and UHI-mitigation measures in LCA for effective policy making. *Journal of Industrial Ecology, 24*(2), 410–423. https://doi.org/10.1111/jiec.12980

Takebayashi, H., Kimura, Y., & Kyogoku, S. (2014). Study on the appropriate selection of urban heat island measure technologies to urban block properties. *Sustainable Cities and Society, 13*, 217–222. https://doi.org/10.1016/j.scs.2014.01.008

Takebayashi, H., & Moriyama, M. (2012). Relationships between the properties of an urban street canyon and its radiant environment: Introduction of appropriate urban heat island mitigation technologies. *Solar Energy, 86*(9), 2255–2262. https://doi.org/10.1016/j.solener.2012.04.019

Targino, A. C., Coraiola, G. C., & Krecl, P. (2019). Green or blue spaces? Assessment of the effectiveness and costs to mitigate the urban heat island in a Latin American city. *Theoretical and Applied Climatology, 136*(3), 971–984. https://doi.org/10.1007/s00704-018-2534-1

Teferi, E., & Abraha, H. (2017). Urban heat island effect of Addis Ababa City: Implications of urban green spaces for climate change adaptation. In W. Leal Filho, S. Belay, J. Kalangu, W. Menas, P. Munishi, & K. Musiyiwa (Eds.), *Climate change adaptation in Africa: Fostering resilience and capacity to adapt* (pp. 539–552). Springer. https://doi.org/10.1007/978-3-319-49520-0_33

Thompson, R., Hornigold, R., Page, L., & Waite, T. (2018). Associations between high ambient temperatures and heat waves with mental health outcomes: A systematic review. *Public Health, 161*, 171–191. https://doi.org/10.1016/j.puhe.2018.06.008

Tonekaboni, N. H., Ramaswamy, L., Mishra, D., Grundstein, A., Kulkarni, S., & Yin, Y. (2018). *Scouts: A smart community centric urban heat monitoring framework. Proceedings of the 1st ACM SIGSPATIAL Workshop on Advances on Resilient and Intelligent Cities*, Seattle, WA. https://doi.org/10.1145/3284566.3284570

Tuczek, M., Degirmenci, K., Desouza, K. C., Watson, R. T., Yigitcanlar, T., Garcia Hansen, V., Omrani, S., Bamdad, K., & Breitner, M. H. (2019). Toward a decision support system for mitigating urban heat. *Proceedings of the SIGGreen Pre-ICIS Workshop 2019*, Munich, Germany, 1–12.

Ulpiani, G. (2021). On the linkage between urban heat island and urban pollution island: Three-decade literature review towards a conceptual framework. *Science of the Total Environment, 751,* 1–31. https://doi.org/10.1016/j.scitotenv.2020.141727

Vargo, J., Stone, B., Habeeb, D., Liu, P., & Russell, A. (2016). The social and spatial distribution of temperature-related health impacts from urban heat island reduction policies. *Environmental Science & Policy, 66,* 366–374. https://doi.org/10.1016/j.envsci.2016.08.012

Vasilakopoulou, K., Kolokotsa, D., & Santamouris, M. (2014). Cities for smart environmental and energy futures: Urban heat island mitigation techniques for sustainable cities. In S. T. Rassia & P. M. Pardalos (Eds.), *Cities for smart environmental and energy futures: Impacts on architecture and technology* (pp. 215–233). Springer. https://doi.org/10.1007/978-3-642-37661-0_12

Veena, K., Parammasivam, K. M., & Venkatesh, T. N. (2020). Urban heat island studies: Current status in India and a comparison with the international studies. *Journal of Earth System Science, 129*(1), 15. https://doi.org/10.1007/s12040-020-1351-y

vom Brocke, J., Simons, A., Riemer, K., Niehaves, B., Plattfaut, R., & Cleven, A. (2015). Standing on the shoulders of giants: Challenges and recommendations of literature search in information systems research. *Communications of the Association for Information Systems, 37,* 205–224. https://doi.org/10.17705/1CAIS.03709

Wagner, M., & Viswanathan, V. (2016). Analyzing the impact of driving behaviour at traffic lights on urban heat. *Procedia Engineering, 169,* 303–307. https://doi.org/10.1016/j.proeng.2016.10.037

Wang, Y., & Akbari, H. (2016). Analysis of urban heat island phenomenon and mitigation solutions evaluation for Montreal. *Sustainable Cities and Society, 26,* 438–446. https://doi.org/10.1016/j.scs.2016.04.015

Wang, Y., Berardi, U., & Akbari, H. (2016). Comparing the effects of urban heat island mitigation strategies for Toronto, Canada. *Energy and Buildings, 114,* 2–19. https://doi.org/10.1016/j.enbuild.2015.06.046

Ward, K., Lauf, S., Kleinschmit, B., & Endlicher, W. (2016). Heat waves and urban heat islands in Europe: A review of relevant drivers. *Science of the Total Environment, 569-570,* 527–539. https://doi.org/10.1016/j.scitotenv.2016.06.119

Weber, S., Sadoff, N., Zell, E., & de Sherbinin, A. (2015). Policy-relevant indicators for mapping the vulnerability of urban populations to extreme heat events: A case study of Philadelphia. *Applied Geography, 63,* 231–243. https://doi.org/10.1016/j.apgeog.2015.07.006

Webster, J., & Watson, R. T. (2002). Analyzing the past to prepare for the future: Writing a literature review. *MIS Quarterly, 26*(2), xiii–xxiii.

Wilhelmi, O., & Hayden, M. (2016). Reducing vulnerability to extreme heat through interdisciplinary research and stakeholder engagement. In S. Steinberg & W. Sprigg (Eds.), *Extreme weather, health, and communities* (pp. 165–186). Springer. https://doi.org/10.1007/978-3-319-30626-1_8

Wondmagegn, B. Y., Xiang, J., Williams, S., Pisaniello, D., & Bi, P. (2019). What do we know about the healthcare costs of extreme heat exposure? A comprehensive literature review. *Science of the Total Environment, 657,* 608–618. https://doi.org/10.1016/j.scitotenv.2018.11.479

Wonorahardjo, S., Sutjahja, I. M., Mardiyati, Y., Andoni, H., Thomas, D., Achsani, R. A., & Steven, S. (2020). Characterising thermal behaviour of buildings and its effect on urban heat island in tropical areas. *International Journal of Energy and Environmental Engineering, 11*(1), 129–142. https://doi.org/10.1007/s40095-019-00317-0

Xiang, J., Hansen, A., Pisaniello, D., & Bi, P. (2016). Workers' perceptions of climate change related extreme heat exposure in South Australia: A cross-sectional survey. *BMC Public Health*, *16*(1), 1–12. https://doi.org/10.1186/s12889-016-3241-4

Xiao, X. D., Dong, L., Yan, H., Yang, N., & Xiong, Y. (2018). The influence of the spatial characteristics of urban green space on the urban heat island effect in Suzhou Industrial Park. *Sustainable Cities and Society*, *40*, 428–439. https://doi.org/10.1016/j.scs.2018.04.002

Xu, S.-X., Liu, T.-L., Jia, N., Wang, P., Liu, P., & Ma, S. (2020). The effects of transportation system improvements on urban performances with heterogeneous residents. *Journal of Management Science and Engineering*, 1–16. https://doi.org/10.1016/j.jmse.2020.09.002

Yamaguchi, K., & Ihara, T. (2020). Countermeasures to urban heat island considering urban energy usage. In N. Enteria, H. Awbi & M. Santamouris (Eds.), *Building in hot and humid regions* (pp. 15–57). Springer. https://doi.org/10.1007/978-981-13-7519-4_2

Yang, J., & Gakenheimer, R. (2007). Assessing the transportation consequences of land use transformation in urban China. *Habitat International*, *31*(3–4), 345–353. https://doi.org/10.1016/j.habitatint.2007.05.001

Yang, J., Wang, Z.-H., & Kaloush, K. E. (2015). Environmental impacts of reflective materials: Is high albedo a 'silver bullet' for mitigating urban heat island? *Renewable and Sustainable Energy Reviews*, *47*, 830–843. https://doi.org/10.1016/j.rser.2015.03.092

Yigitcanlar, T., Dodson, J., Gleeson, B., & Sipe, N. (2007). Travel self-containment in master planned estates: Analysis of recent Australian trends. *Urban Policy and Research*, *25*(1), 129–149. https://doi.org/10.1080/08111140701255823

Yigitcanlar, T., Kankanamge, N., Preston, A., Gill, P., Rezayee, M., Ostadnia, M., Xia, B., & Ioppolo, G. (2020). How can social media analytics assist authorities in pandemic-related policy decisions? Insights from Australian states and territories. *Health Information Science and Systems*, *8*(1), 1–21. https://doi.org/10.1007/s13755-020-00121-9

Yu, Q., Ji, W., Pu, R., Landry, S., Acheampong, M., O'Neil-Dunne, J., Ren, Z., & Tanim, S. H. (2020). A preliminary exploration of the cooling effect of tree shade in urban landscapes. *International Journal of Applied Earth Observation and Geoinformation*, *92*, 1–13. https://doi.org/10.1016/j.jag.2020.102161

Zhang, K., Chen, T.-H., & Begley, C. E. (2015). Impact of the 2011 heat wave on mortality and emergency department visits in Houston, Texas. *Environmental Health*, *14*(11), 1–7. https://doi.org/10.1186/1476-069X-14-11

Zhang, L., Fukuda, H., & Liu, Z. (2019). The value of cool roof as a strategy to mitigate urban heat island effect: A contingent valuation approach. *Journal of Cleaner Production*, *228*, 770–777. https://doi.org/10.1016/j.jclepro.2019.04.338

Zhu, Y., & Jia, Z. (2011). Soil water utilization characteristics of *Haloxylon ammodendron* plantation with different age during summer. *Acta Ecologica Sinica*, *31*, 341–346. https://doi.org/10.1016/j.chnaes.2011.09.004

Zinzi, M., Carnielo, E., & Federici, A. (2014). Preliminary studies of a cool roofs' energy-rating system in Italy. *Advances in Building Energy Research*, *8*(1), 84–96. https://doi.org/10.1080/17512549.2014.890539

Index

Note: Italics and **bold** page numbers refer to figures and tables, respectively.

N

O

P

Q

R

S

For Product Safety Concerns and Information please contact our EU representative GPSR@taylorandfrancis.com
Taylor & Francis Verlag GmbH, Kaufingerstraße 24, 80331 München, Germany

www.ingramcontent.com/pod-product-compliance
Lightning Source LLC
Chambersburg PA
CBHW070626230726
48635CB00003B/38
* 9 7 8 1 0 4 1 2 3 4 2 8 9 *